quantum
physics

mcgraw-hill book company

New York
St. Louis
San Francisco
Düsseldorf
Johannesburg
Kuala Lumpur
London
Mexico
Montreal
New Delhi
Panama
Rio de Janeiro
Singapore
Sydney
Toronto

quantum physics

berkeley physics course—volume **4**

The preparation of this course was supported by a grant from the National Science Foundation to Education Development Center, Incorporated.

Eyvind H. Wichmann

Professor of Physics
University of California, Berkeley

QUANTUM PHYSICS

04861

1234567890 MW MW 79876543210

Preface to the Berkeley Physics Course

This is a two-year introductory college physics course for students majoring in science and engineering. The intention of the writers has been to present physics in so far as possible as it appears to physicists working on the forefront of their field. We have sought to make a course which would vigorously emphasize the foundations of physics. Our specific objectives were to introduce students early to the ideas of special relativity, of quantum physics, and of statistical physics. At the same time we wanted our presentation to be of an elementary character, and it was never our intention to develop a course limited to honors students or to students in advanced standing. This course is intended for any student who has had physics in high school. A mathematics course including calculus should be taken concurrently.

The five volumes of the course include: °

    I. Mechanics (Kittel, Knight, Ruderman)
   II. Electricity and Magnetism (Purcell)
 III. Waves (Crawford)
 IV. Quantum Physics (Wichmann)
  V. Statistical Physics (Reif)

The initial work on this course led Alan M. Portis to devise a new elementary physics laboratory, now known as the Berkeley Physics Laboratory. Because the course emphasizes the principles of physics, some teachers may feel that it does not deal sufficiently with experimental physics. The laboratory is rich in important experiments, and is designed to balance the whole course.

In recent years several new college physics courses have been planned and developed in the United States. The idea of making a new course has come to many physicists, aware of developments in science and engineering, and of the increasing emphasis on science in elementary schools and in high schools. Our own course was conceived in a conversation between Philip Morrison, now of the

---

° Of these Volumes I, II, III, and V have appeared previously. The changes in our common preface for the present volume reflect some organizational changes under which this volume was prepared.

Massachusetts Institute of Technology, and C. Kittel late in 1961. We were encouraged by John Mays and his colleagues of the National Science Foundation and by Walter C. Michels, then the Chairman of the Commission on College Physics. An informal committee was formed under the Chairmanship of C. Kittel to guide the course through the initial stages. The committee met first in May 1962, in Berkeley, and at that time drew up a tentative outline, incorporating topics and attitudes which we believed should and could be taught to beginning college students of science and engineering. The membership of the committee has since undergone some changes, and it now consists of the undersigned. In the development of the course each author has been free to choose that style and method of presentation which seemed to him appropriate to his subject.

The financial support of the course development was provided by the National Science Foundation, with considerable indirect support by the University of California. The funds were administered by Education Development Center, Incorporated* (EDC), a nonprofit organization established to administer curriculum improvement programs. We are particularly indebted to Gilbert Oakley, James Aldrich, and William Jones, all of EDC, for their sympathetic and vigorous support. EDC established an office in Berkeley, most recently under the very competent direction of Mrs. Lila Lowell, to assist the development of the course and the laboratory.

The University of California has had no official connection with our program, but it has aided us in important ways. For this help we thank Professor Burton J. Moyer, Chairman of the Department of Physics, and the faculty and nonacademic staff of the Department. We are particularly grateful to those of our colleagues who have tried this course in the classroom and who, on the basis of their experience, have offered criticism and suggestions for improvements.

Your corrections and suggestions will always be welcome.

Frank S. Crawford, Jr.
Charles Kittel
Walter D. Knight
Alan M. Portis
Frederick Reif
Malvin A. Ruderman
Eyvind H. Wichmann
A. Carl Helmholz ⎫
Edward M. Purcell ⎭ *Chairmen*

---

* Formerly Educational Services Incorporated.

Preface to Volume IV

This volume of the Berkeley Physics Course is devoted to Quantum Physics. It is an introductory book, intended for the student whose background in physics corresponds to a fair fraction of the material in the preceding volumes of the series. My ideal reader-learner is thus a student of science or engineering in his sophomore year. To postpone the study of all aspects of quantum phenomena beyond the sophomore level does not seem reasonable and fair today in view of the developments in physics during the last fifty years. A well-balanced introductory course should certainly reflect some of these developments.

I do not believe that the study of quantum physics is *intrinsically* more difficult than the study of any other branch of physics. In every domain of physics we encounter phenomena which we feel are simple and transparent, as well as phenomena which are very difficult to account for in a quantitative fashion. It is true, of course, that all quantum phenomena were at one time regarded as highly mysterious and perplexing. In the period during which the field was first explored, physicists experienced very real psychological difficulties which derived partly from understandable prejudices in favor of the classical view of the world, and partly from the fragmentary nature of the experimental picture. There is, however, no reason why these same difficulties have to be re-created for the beginning student of today. It is now known with certainty that the classical description is only approximately correct, and there is now available an enormous body of experimental results which support and illuminate various aspects of present theoretical ideas. I firmly believe that among the known facts one can find topics for discussion which are clear-cut and simple enough to be dealt with in an elementary manner, but which at the same time illuminate important ideas and principles. I doubt very much that the student who is guided to think about a well-chosen sequence of simple but significant physical facts will feel that quantum phenomena are *more* mysterious than, say, the phenomenon of universal gravitation.

My purpose in this book is to present characteristic examples of quantum phenomena, to acquaint the reader with typical orders of magnitude of physical parameters in microphysics, and to introduce him to quantum-mechanical thinking. I have tried to include among my topics for discussion, phenomena and issues which are of particular importance for the understanding of physics. At the same time I have tried to keep the discussion as elementary as possible. I have selected my topics from the various domains of microphysics, but I have made no attempt to present a detailed systematic account of any of these domains. Such accounts should, in my opinion, be left for courses at the junior-senior level.

The requirements with respect to mathematical preparation are modest. I assume that the reader has had a course in calculus, including a first introduction to ordinary differential equations and some vector analysis. In order to prevent a shift of attention from the physical issues to technical mathematical questions, I have tried to avoid topics which at this stage might appear mathematically difficult. Topics which call for some knowledge of the properties of special functions, or of the method of separation of variables in the theory of partial differential equations, are not discussed at all. With regard to algebra I have concluded, with some regrets, that I should not assume familiarity with matrix theory, and I have therefore avoided topics for which matrix theory is the natural mathematical tool.

I do not by any means believe that the fulfillment of the general aims of this part of the course demands that *all* the material in this book be taught in class. On the contrary I wanted to leave the instructor considerable latitude in the selection of topics for discussion. To aid the instructor in the planning of his course I discuss the specific aims of the various chapters in the teaching notes following this preface, and I try to outline what might be regarded as a *minimum* program. I feel there is no harm in having more material available for reading than is actually taught in class, as there will always be some students who will want to read beyond the lectures.

Eyvind H. Wichmann

October, 1967
Berkeley, California

Acknowledgments

I am greatly indebted to the other members of the Berkeley Physics Course Committee for their continual help and encouragement during the past few years. I wish to thank in particular Professors C. Kittel, A. M. Portis and A. C. Helmholz for their many suggestions for improvements and for their constructive criticism.

Very many of my colleagues in the Department of Physics at Berkeley have helped me at one time or another, and I here want to express my gratitude to all of them. I am particularly grateful to Professors S. P. Davis, W. B. Fretter, W. D. Knight, L. B. Loeb, J. H. Reynolds, A. H. Rosenfeld, E. G. Segrè, and C. H. Townes, and to Dr. W. Hines for providing me with photographs, and for their comments on my manuscript.

This book has been developed from earlier drafts used in teaching this part of the course in Berkeley and elsewhere. The earliest version was used by myself when I taught a small group of students in Berkeley during the spring term of 1964. I wish to thank these students for their interest and for the very many helpful comments and suggestions which they made. Later drafts have been subsequently used as texts in the same course by Professors K. Dransfeld, F. S. Crawford, L. T. Kerth, and A. C. Helmholz. I thank these colleagues of mine for sharing with me their experiences in teaching this course.

My manuscripts have been typed by Mrs. Lila Lowell, and I wish to thank her for her infinite patience and her careful work. The manuscript has been critically read and checked for errors by Dr. J. D. Finley and Dr. L. J. Landau, and I am greatly indebted to them for their many useful comments. I also wish to thank Dr. J. Crichton, who similarly read the manuscript for the preliminary edition.

I am indebted to my wife, Marianne Wichmann, for illuminating my serious discussion of some topics with a few drawings of a not entirely serious character. All the other drawings were prepared in their final form by Mr. Felix Cooper, and it is a pleasure to thank him for his careful work.

Notes for Teaching and Study

The material is presented in nine chapters. Each chapter is divided into many short, consecutively numbered sections, each corresponding roughly to one idea or one step in the train of thought. Equations, figures, and tables within the text are numbered by the numbers of the sections within which they occur or to which they refer. Specific references to particular topics in the text are given in footnotes. General references are given at the end of the chapters. Tables of physical data are given in the Appendix, as well as on the insides of the covers of the book.° Problems for individual study are given at the end of the chapters. The serious student should do a very substantial fraction of these problems.

My references are to original papers, other textbooks, and elementary review articles of the kind which can be found in *Scientific American*. To my student-reader I want to say the following. You will get a distorted picture of physics if you confine your reading to textbooks alone. A textbook provides a framework for orderly and systematic studies, but it cannot possibly mirror the richness and variety of the intellectual effort in physics. This book, for instance, is very deficient in its description of experimental procedures. To encourage you to begin to become acquainted with the literature, I have included references to papers which report on original research. I certainly do not expect you to read more than a small fraction of these papers, but when you encounter a subject which you find particularly interesting I urge you to go to the library to look up the original sources. You will probably find other papers which also interest you, and soon you will become a habitual reader. Do not try to read papers for which you clearly do not possess the necessary background. There are many papers, especially on experiments, which you *can* read with your present preparation, and you should select among these. Your instructor can give you further advice on where to look. The elementary review articles

---

° (Note added in proof.) The completed manuscript was submitted to the publisher at the end of 1967, and the book accordingly does not contain references to more recent work. It may be stated, however, that nothing has happened in the meantime which would significantly affect the contents of this book.

in the *Scientific American*, which require very little preparation for reading, also can be very useful at this stage. There you can learn about current experiments and current topics of interest.

The question of units is not an issue in this book. The instructor can use the cgs system or the MKS system according to his preference. (The only place where it makes any difference is in the expression for the fine-structure constant.) Constants are presented in both systems of units. Experimental results are presented in the practical system. In the theoretical discussion I frequently write the equations in a dimensionless form in which the macroscopic units do not appear at all.

I wish to comment next on the contents of each chapter, to explain my intentions, and to indicate how cuts can be made. Some of the material is specifically identified in the text as "Advanced Topics." These topics are not necessarily more advanced than the other topics discussed, nor are they necessarily more difficult. They represent, however, a digression from the main line of presentation, and they can therefore safely be omitted without making the rest of the book incomprehensible.

Chapter 1 is a general introduction. The scope of quantum physics is discussed, and some aspects of the history of quantum physics are commented upon. The most important message is, perhaps, that quantum physics is relevant for *all* of physics, and not merely concerned with "microscopic" phenomena. In a minimum program most of Chapter 1 might well be left as a reading assignment, and the instructor might restrict his lecture discussion to the material contained in Sections 27–52, which are concerned with the entry of Planck's constant into the world of physics. The problems at the end of the chapter require no particular preparation and all can be assigned within the minimum program.

Chapter 2 is concerned with magnitudes of physical quantities in microphysics. The aim is to familiarize the student with these magnitudes, to discover "natural" combinations of physical constants, and to show the student how one makes simple estimates based on simple models. I regard these aims as very important, and the chapter, with the exercises at the end, is worth careful attention. In a minimum program Sections 47–57 can be omitted.

Chapter 3 is concerned with energy levels, but not with the theoretical explanation of the occurrence of levels. This explanation comes later, in Chapter 8. The reason for this somewhat peculiar order of presentation is that I wanted to place all topics requiring some knowledge of differential equations as late as possible in the book. Depending on the students' preparation this order might be changed. In Chapter 3 I wanted to give realistic examples of level systems and term schemes, and to show how simple conclusions can be drawn, based on the empirical fact that level systems occur in nature. Part of the chapter can well be left as a reading assignment. An important point which should be fully discussed is the connection between lifetime and level width (Sections 14–26).

Chapter 4 is concerned with the wave and particle properties of photons. Important experimental facts are presented, and the reader is guided to think quantum-mechanically about these facts. I feel that this chapter should not be cut.

Chapter 5 discusses the wave nature of all material particles. The student who has read Chapters 4 and 5 will thus have learned that all real particles found in nature have wave properties, and he will have some ideas about the immediate implications of this simple experimental fact. He will also have learned that the wave nature of particles does not directly contradict our experience with macroscopic physics, and why. Chapter 5 is thus to a large extent concerned with very fundamental issues. The derivation of the Klein-Gordon equation (Sections 36–46) should not be omitted. The discussion of the interpretation of the solutions of a wave equation as corresponding to vectors in a vector space (Sections 47–54) might be left as a reading assignment, or perhaps omitted entirely. The discussion of diffraction of waves by a periodic structure (Sections 16–22) might also be omitted in a minimum program, although it is a shame to omit a theory which has so many beautiful and clear-cut experimental applications.

In the first part of Chapter 6 the uncertainty relations are discussed (Sections 1–19). This material is of *crucial* importance and should not be cut. In the remainder of Chapter 6 an attempt is made to formulate and discuss some general rules of quantum-mechanical thinking. A theory of measurements is presented, and the notion of a statistical ensemble and of coherent and incoherent superpositions are discussed. I have tried to keep this discussion as physical and concrete as possible. It cannot be denied, however, that the discussion in this chapter goes much further than has been customary in introductory books, and many readers might feel that this material could wait until later. On the other hand I feel that some of the main ideas of this chapter are not particularly difficult if presented in an orderly fashion, and that it is worthwhile to try to present these ideas early.

An introduction to the Schrödinger theory is presented in Chapters 7 and 8. My purpose was to show in some detail how a wave-mechanical theory works out in practice. Sections 49–51 of Chapter 7 and Sections 49–58 of Chapter 8 can be omitted in a minimum program. The discussion of barrier penetration in alpha-decay (Sections 37–48, Chapter 7) should probably *not* be omitted, as the comparison between theory and experiment is bound to have a strong impact.

Chapter 9 is concerned with the problem of how to describe interactions between elementary particles. Sections 1–18 contain an elementary discussion of collision processes. Some known facts and some theoretical ideas about particles are discussed in Sections 19–31. This discussion is followed by a qualitative discussion of some basic ideas of quantum field theory. The tangible result of this discussion is the simplified derivation of the Yukawa potential in Sections 47–55. In a minimum program Chapter 9 might be omitted as a whole, but I do feel that some remarks about the interaction problem ought to be made somewhere in the course. Irrespective of whether the material in Chapter 9 is taught or not, I felt that it should be readily available to the interested student. The issues are, after all, at the center of attention in present-day physics.

The problems at the end of each chapter are intended to further illustrate the topics under discussion. They vary considerably in degree of difficulty. There are comparatively few problems of the kind which merely involve the substitution of numerical values into a formula which occurs somewhere in the book. A certain number of such problems do serve a useful purpose in giving the reader a feeling for relevant orders of magnitude. In my selection I wanted, however, to emphasize problems which really test the reader's understanding of the text, and I did not want to submerge these in a mass of very trivial problems. I have furthermore assumed that every instructor will want to make up a certain number of his own problems, appropriate for his particular course, and if the need arises some of these can well be of the simple substitution variety. If an instructor omits certain portions of the text, he will naturally omit the corresponding problems and perhaps replace these by others.

Besides these definite suggestions about what might be omitted, an instructor has the option of further omitting an occasional section here and there, and of shortening and simplifying the discussion, without thereby upsetting the aims of this book. In a minimum program the lectures might thus cover as little as from one-half to two-thirds of the material in the book. I estimate that this might correspond to about twenty hours of lectures, and this is thus the *minimum* amount of time which should be devoted to the Quantum Physics part of the course.

Contents

quantum
physics

Chapter 1

Introduction

Chapter 1   Introduction

The Scope of Quantum Physics

1   In this part of our course we shall study physics in the realm of atoms, nuclei and elementary particles. In so doing we will encounter new aspects of nature: new in the sense that we have not discussed them systematically in the preceding volumes. These aspects of nature are commonly referred to as *quantum phenomena,* and we therefore call the subject matter of this volume *quantum physics.* The currently accepted basic mathematical theory of quantum physics is known as *quantum mechanics.*

Now it should not be thought that "quantum physics" is something which does not concern the macroscopic world. Actually *all* of physics is quantum physics; the laws of quantum physics as we know them today are our most general laws of nature.

2   In the preceding volumes of the Berkeley Physics Course we have studied physical phenomena in the macroscopic world. The laws of nature which we have discovered are the laws of *classical physics.* Generally we can say that classical physics is concerned with those aspects of nature for which the question of the ultimate constitution of matter is not of *immediate* concern. In this volume, on the other hand, we will specifically study the elementary particles, and we must now try to discover the laws which govern the behavior of these particles. We will naturally focus our attention on physical situations in which these laws stand out as clearly as possible, and this means that we study situations involving the interactions of only a few particles at a time. Most of the physics studied in this volume could, therefore, be called *microphysics:* we study "small" systems consisting of a small number of elementary particles.

However, if we know the basic laws governing the elementary particles we can also, in principle, predict the behavior of macroscopic physical systems, consisting of a very large number of elementary particles. This means that the laws of classical physics follow from the laws of microphysics, and in this sense quantum mechanics is just as relevant in the macroscopic world as in the microscopic world.

3   When we apply the laws of classical physics to macroscopic systems we try to describe only certain gross features of the behavior of the system. We consider, for instance, the motion of a "rigid

An example of a quantum-mechanical system. The behavior of this electric motor (and the flashlight battery used as a power supply) is governed by the laws of quantum mechanics, although the author never suspected this when the motor was given to him about thirty years ago.

The design of an electric motor can, and should, be based on classical electromagnetic theory and classical mechanics, which are limiting forms of quantum mechanics. No engineer in his right mind would attempt to describe a macroscopic system such as this one in terms of the interactions between all the elementary particles which make up the system.

body" as a whole, but we do not try to discuss the motions of all the elementary constituents of the body. This is a characteristic feature of classical theories of physics as applied to macroscopic systems; the finer details of the behavior of the system are ignored and no attempt is made to consider all aspects of the situation. In this sense the laws of classical physics are approximate laws of nature. We should regard them as limiting forms of the more basic and comprehensive laws of quantum physics.

The classical theories are, in other words, *phenomenological theories*. A phenomenological theory attempts to describe and summarize experimental facts within some limited domain of physics. It is not intended to describe everything in physics, but if it is a good phenomenological theory it does describe everything within the limited domain very accurately. The philosophically minded reader may want to remark that ultimately *every* physical theory is "phenomenological," and that the difference between a basic theory and a phenomenological theory is only a question of degree. As physicists we recognize, however, a clear difference between the two kinds of theories. Our *basic* laws of nature are distinguished by their great generality; we are not aware of any exceptions to what they state. We regard them as true and exact and universally valid until there is clear experimental evidence to the contrary. In contrast to this, the laws contained in a phenomenological theory are recognized *not* to be of universal validity; we *know* that they are valid (i.e., sufficiently accurate) only in some limited domain of physics, and that outside this domain the phenomenological theory may be completely meaningless.

4 We should, of course, not be contemptuous of phenomenological theories. They serve the very useful purpose of summarizing our practical knowledge of the various domains of physics. There are many instances in physics in which we do believe that we have available a basic theory, but where the complexity of the phenomena prevents us from making accurate predictions based on "first principles." In such a case we try a simplified phenomenological theory which is partly based directly on the experimental facts, and partly based on some general features of the basic theory. We let, in other words, "the physical systems do some of our theoretical work." There are, furthermore, many instances in physics where the basic theory is missing. Any phenomenological theory which we can construct (based on some simple model) is then useful as a steppingstone in the search for a more comprehensive theory.

When we try to understand an unfamiliar physical phenomenon it is clearly rational to try the simplest thing first, i.e., to try a theory,

Albert Einstein. Born 1879 in Ulm, Germany; died 1955. Studied at the Institute of Technology (ETH) in Zürich, Switzerland. After receiving his diploma in 1900 he held a position as a patent examiner in the Swiss Patent Office in Bern. During this time he wrote three famous papers, all of which appeared in the 1905 issue of *Annalen der Physik*, dealing with the photo-electric effect, Brownian motion, and special relativity. Subsequently he held positions in Bern, Zürich, and Prague, and as director of the Kaiser Wilhelm Institute in Berlin. In 1933 he became a member of the Institute for Advanced Study in Princeton, N.J., settling permanently in the United States. He received the Nobel prize in 1921.

Einstein is generally regarded as the most outstanding physicist of this century, and as among the greatest scientists of all time. He possessed to an extraordinary degree the ability to grasp the essence of physical phenomena, and no short summary could do justice to his numerous, always profound contributions on the fundamental problems of physics. His theory of General Relativity stands out as one of the most remarkable intellectual creations of all time. *(Photograph by courtesy of Physics Today.)*

or model, which has worked successfully in a seemingly analogous situation. If our model turns out to be successful we have learned something, but if it turns out to be unsuccessful we have *also* learned something.

The important thing to keep in mind is that models are only models and that all of physics need not be described in terms of a single model.

5 People often talk about the "revolution" in physics brought about through the discovery of quantum mechanics. "Revolution" is a dramatic word (with a strange appeal, it seems) which suggests that something has been completely overturned. It should be noted, however, that the laws of classical physics, as applied to those situations which the classical theory was designed to describe, have not been overturned. The motion of a pendulum, for example, is described today in the same way as it was described in the nineteenth century.

It is furthermore the case that classical concepts often can be successfully employed to gain *some* understanding of phenomena in microphysics: they are of approximate validity. It is important that we understand the limits of applicability of classical ideas, and in this chapter we will try to give the reader a *rough* idea of these limits. As we learn more about quantum phenomena in later chapters, the reader will reach a more precise understanding of this important question.

That the classical theories of physics are not of universal validity has been convincingly established through many experiments performed during this century. In this volume we shall present some of the relevant experimental evidence to convince the reader of this fact of life.

6 When we think about the changes that have taken place in physics during this century we should keep in mind that no *comprehensive* classical theory of matter ever existed. The laws of classical physics are good phenomenological laws, but they do not tell us everything about macroscopic bodies. In terms of these laws we can describe the behavior (motion) of a mechanism consisting of springs, levers, flywheels, etc., if we are given some "material constants," such as the density, modulus of elasticity, etc., of the materials of which the mechanism is built. However, if we ask *why* the densities are what they are, *why* the elastic constants have the values they have, *why* a rod will break if the tension in the rod exceeds a certain limit, and so on, then classical physics is silent. Classical physics does not tell us why copper melts at 1083°C, why sodium vapor emits yellow light, why hydrogen has the chemical properties it has, why the sun shines, why the uranium nucleus disintegrates spontaneously, why silver conducts electricity, why sulfur is an insulator, nor why permanent magnets can be made of steel. We could go on and on listing everyday observational facts about which classical physics has very little or nothing to tell us.

7 The reader wants to know, do we *now* have a comprehensive theory of matter? The answer is no; we do not have a detailed theory for *everything* taking place in our world. However, our knowledge about nature has expanded enormously during the last sixty years. We have discovered aspects of nature never before dreamed of, and we have succeeded in solving many old problems. It is, for instance, fair to say that we now understand the facts of chemistry and the properties of matter in bulk quite well: in these domains of physics we can answer the questions that could not be discussed within classical theory.

Atoms and Elementary Particles

8 Let us talk about the idea of elementary particles. Some ancient Greek philosophers are credited with being the first to introduce the concept of atoms into the theory of matter. (This does not exclude the possibility that other people might have speculated along similar lines long before.) It should be stated immediately that the "atoms" of the ancients are most certainly not the same things as the atoms of today. It is in fact not an easy matter to understand precisely what the Greek philosophers really meant by the term, but the central problem with which they were concerned was the question of whether matter is, or is not, infinitely divisible. If matter is *not* infinitely divisible, then we must discover, on a sufficiently small scale, elementary constituents of matter, or "atoms." We take a chunk of matter, and we divide it again and again into smaller and smaller pieces. Eventually this splitting comes to an end; we find something which cannot be split further, and that is the "atom" (the word actually means "indivisible").

The Greek atomists believed that all matter is indeed built of "atoms," and presumably they felt that all the extremely varied aspects of matter are somehow explainable in terms of different configurations (and motions?) of "atoms." We believe something

It might well have occurred to some early natural philosopher that the strikingly regular and beautiful shapes of crystals reflect the way in which they are built from small particles, or atoms. Today this seems a very natural idea. It would appear, however, that this idea did *not* occur early. So far as the author knows there is no indication in the historical record that the Greek atomists speculated about crystals in this manner.

Crystallography as a science began to develop at the end of the eighteenth century. Among early workers we can mention Romé de Lisle and Haüy, who made precise measurements of the angles between cleavage planes. Before them both Robert Hooke and Christian Huygens had speculated about how crystals might be built of small (invisible) parts.

vaguely similar today, but there is certainly an enormous difference between our quantitative theories and the nebulous speculations of the ancients.

9 We are not going to discuss the early history of the atomic theory of matter in this book, but the reader is urged to ponder the remarkable understanding of natural phenomena which was achieved during the nineteenth century on the hypothesis that matter is made of atoms. On this assumption we can understand the basic fact of chemistry, namely that a given chemical compound always consists of certain basic chemical elements in fixed, definite proportions, characteristic of the compound. Consider, in particular, the striking fact that we can represent chemical compounds by such simple formulas as H_2O, H_2SO_4, Na_2SO_4, and NaOH. What is striking about these formulas is the occurrence of *small* integers which tell us that *two units* of hydrogen combine with *one unit* of oxygen to form *one unit* of water, and so on. If we assume that matter is made of atoms we can immediately understand these empirical facts: Chemical compounds consist of molecules, which in turn are composite systems of a small number of atoms. Two hydrogen atoms combine with one oxygen atom to form one water molecule. Clear and simple.

As further evidence in favor of the atomic hypothesis we point to the successes of the *kinetic theory of gases*, developed in particular by J. C. Maxwell and L. Boltzmann. This theory could explain many properties of gases on the hypothesis that a gas in a container is a swarm of molecules moving randomly inside the container, incessantly colliding with each other and with the walls of the container. The kinetic theory could furthermore be used to estimate Avogadro's number, $N_0 = 6.02 \times 10^{23}$, which is the number of molecules in a mole of any gas. (By a *mole* of any chemical compound is understood a quantity of the substance which has a mass in grams equal to the molecular weight of the compound.) The first crude estimate of N_0 was given by Loschmidt in 1865.

In view of such evidence for the existence of atoms it is hard to understand a certain school of thought, which persisted until the turn of the century, and which rejected the atomic hypothesis on the grounds there was no *direct* (!) evidence that matter is made of atoms.

10 The "atoms" of the Greek philosophers do not correspond to our atoms of today, because our atoms are not indivisible: they are made of protons, neutrons and electrons. It is rather the protons, neutrons, electrons, and a host of other elementary particles which

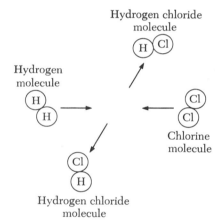

Fig. 9A A *very* schematic representation of the chemical reaction $H_2 + Cl_2 \rightarrow 2HCl$, in which a hydrogen molecule reacts with a chlorine molecule to form two molecules of hydrogen chloride. The figure symbolizes the idea that a chemical reaction consists in the redistribution of the "elementary" constituents.

The details of the processes which actually take place when hydrogen gas burns in an atmosphere of chlorine are very complex. Energy, in the form of light and of kinetic energy of the reaction products, is liberated in the process. The resulting heating of the gases leads to a partial dissociation of the hydrogen and chlorine molecules into atoms, which can then combine into hydrogen chloride molecules. Other processes, in which the atoms and molecules are excited internally through collisions or by light, also play an important role.

play the role of the Greek "atoms." What do we mean by an "elementary particle"? The *precise* definition of this term is somewhat controversial today, but for our purposes we can give a simple and practical answer to the question: A particle is to be regarded as elementary if it cannot be described as a composite system of other more elementary entities. An elementary particle has no "parts," it is not "built" of anything simpler. Our mental splitting attempts have come to an end. With this definition, the proton, the neutron and the electron are all elementary, but the hydrogen atom or the uranium nucleus is not.

We can say that the essence of the idea that matter is not infinitely divisible is this: We cannot go on forever analyzing things in terms of the parts of which the things are built. Finally this process loses its meaning; we encounter irreducible entities, and these are our elementary particles.

11 How can we assert that the electron is *really* elementary? Might not what is regarded as elementary today be found to be composite tomorrow? After all, the atoms of today were the elementary particles of the nineteenth century: could not history repeat itself?

There are many experimental facts which strongly suggest that history will *not* repeat itself, and that particles such as the electron, the proton or the neutron will never be found to be composite in the same sense as the hydrogen atom was found to be composite. Let us try to describe the nature of this evidence.

If two marbles collide with a sufficiently high relative velocity they will break into smaller fragments. In the same way two hydrogen molecules colliding with a high relative velocity will break into fragments. Unless the velocity is *very* high we will find among the fragments such things as hydrogen atoms, or protons, or electrons; in other words, the components of which hydrogen molecules are built. In both these cases it is fair to describe what has happened as follows: the violence of the collision overcame the cohesive forces which keep the parts together in a marble, or in a hydrogen molecule, and the objects therefore broke apart. A similar interpretation can be given to many nuclear reactions. Nuclei are made of protons and neutrons, and if an energetic proton collides with a nucleus it may happen that a few protons and neutrons are knocked out of the nucleus.

12 However, if we study a violent collision of two elementary particles, such as two protons, we discover phenomena which are *qualitatively different* from the phenomena considered above. For

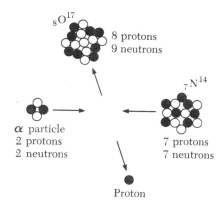

Fig. 11A Schematic representation of a nuclear reaction, in which an alpha particle (helium nucleus) collides with a nitrogen nucleus to produce an oxygen nucleus and a proton. This particular reaction, discovered by Rutherford in 1919, was the first observation of the transmutation of stable nuclei. [E. Rutherford, *Philosophical Magazine* **37**, 581 (1919).] In Rutherford's experiment nitrogen was bombarded by alpha-particles from a radioactive source, and the occurrence of the reaction was established through observation of the emitted protons.

The figure, quite analogous to Fig. 9A, symbolizes the ideas that nuclei are made of protons and neutrons, and that (low-energy) nuclear reactions consist in the rearrangement of these particles among nuclei. It should, of course, not be taken literally: in no sense of the word do nuclei "look" like this.

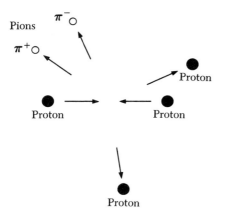

Fig. 12A Schematic representation of the creation of two pi mesons in a high-energy collision of two protons. One pion carries the charge $+e$, and the other the charge $-e$, where e is the magnitude of the electronic charge. The total charge is thus conserved in this event.

Since the two protons remain after the collision and two new particles appear it is strikingly obvious that naive models of the kind shown in Figs. 9A and 11A cannot apply here: the event cannot be thought of as a "rearrangement of the elementary constituents (?) of the two protons."

instance, if a proton of very high energy collides with another proton it may happen that the two protons remain after the collision and that we find in addition one, or several, new elementary particles such as pi mesons, among the reaction products. We say that the pi mesons, also called pions, are *created* in the reaction. This is not the only thing which can happen in a proton-proton collision: the protons may disappear and a number of entirely new particles, known as *K*-mesons and hyperons, may appear instead.

Similarly it can happen, in a violent collision of two electrons, that the final reaction products consist of *three* electrons and one positron. (The positron is an elementary particle similar to the electron, except that it carries the opposite charge.) On the other hand, if an electron and a positron collide with each other it can happen that these two particles disappear (we say that they are *annihilated*) and we are left with only electromagnetic radiation in the form of gamma rays.

13 An interesting example of a creation process is the creation of an electron-positron pair when a gamma ray passes through the electric field in an atom. Material particles can thus be created from electromagnetic radiation. Figure 13A, which is a cloud-chamber photograph of so-called cascade showers, "shows" many instances of this phenomenon. The explanation (see also Figs. 13B and C) for what is seen is as follows. If an energetic charged particle, say an electron or a positron, passes through one of the horizontal lead plates seen in the photograph it may be very slightly deflected in the field of one of the atoms in the plate. Such a deflection constitutes accelerated motion, and consequently electromagnetic radiation in the form of an energetic gamma ray is emitted. (The particle may, of course, be deflected by several atoms in a single plate, in which case several gamma quanta will be emitted.) The gamma rays arising in this manner then create electron-positron pairs in the fields of the atoms which they encounter when *they* traverse the plates. These charged particles in turn give rise to more gamma rays as they are deflected in the plates, and the new gamma rays give rise to new pairs, and so on. A single energetic charged particle, or a single gamma ray, can thus give rise to a cascade of gamma rays, electrons and positrons. The charged particles leave visible tracks in the cloud chamber; these are the tracks we see in Fig. 13A. The gamma rays are not visible in the figure.

The cascade shower in the right part of the photograph appears to have been initiated by a gamma ray, incident from above. The energy of this gamma ray was probably about 20 BeV. The shower

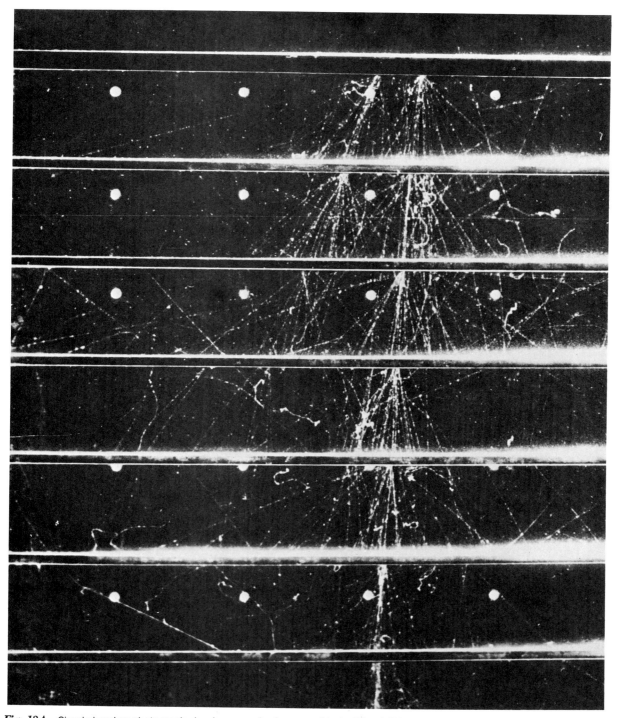

Fig. 13A Cloud chamber photograph showing cascade showers. Most of the visible tracks are due to electrons and positrons, which generally move toward bottom of picture. The particle entering at the top right and penetrating three plates before stopping in the fourth may be a pion. See text for further comments. *(Courtesy of Professor W. B. Fretter, Berkeley.)*

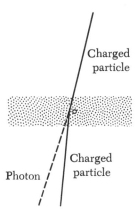

Fig. 13B An energetic charged particle (say a positron or an electron) is deflected by the electric field within an atom, and as a result of this accelerated motion a gamma ray (i.e., an energetic photon) is emitted. This is the physical phenomenon of *bremsstrahlung*. The shaded part of the figure represents matter in bulk, say a portion of a lead plate in a cloud chamber. (The size of the atom is slightly exaggerated for clarity.)

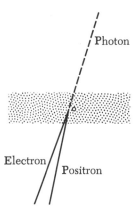

Fig. 13C An energetic gamma ray colliding with the electric field inside an atom gives rise to an electron-positron pair: this is the physical phenomenon of *pair production*. The two basic processes shown in the above two figures are responsible for the development of the cascade shower shown in Fig. 13A.

at the left appears to have been initiated by a charged particle, of somewhat lower energy. Both showers probably originated from some event which took place in the wall of the chamber, outside the field of view. Most of the particles seen in the showers moved in the downward direction. It is a characteristic feature of these processes that the most energetic particles tend to be emitted in the direction of the incident particle, whereas less energetic particles can be emitted in other directions. If we look closely at the photograph we notice that the secondary showers due to particles emitted in directions other than those of the principal showers soon "die out." A cascade shower naturally stops when the original energy has been distributed among so many charged particles and photons that none of them has enough energy for the creation of additional pairs. The low-energy particles are then absorbed by the lead plates.

The energy of the particle initiating a shower can be estimated from the number of charged secondary particles which are produced.

14 The creation and annihilation processes which we have mentioned are important aspects of nature. It is obvious that these phenomena are in no way analogous either to the shattering of marbles or to chemical reactions. We can describe a chemical reaction by saying that new molecules are formed from the elementary constituents of other molecules, and for the purpose of such a description the *atoms* are the elementary constituents of molecules. Consider, in contrast to this, a collision event in which the two particles originally present remain after the collision, along with a number of new particles created in the event. Clearly we cannot describe such an event in terms of a rearrangement of the elementary constituents of the original particles into new composite systems. Nor can this description be applied to events in which some of the original particles disappear. A striking example of the latter phenomenon is the annihilation of an electron-positron pair, in which the material particles originally present disappear completely and we are left with gamma rays.

15 To decide experimentally whether a particle is elementary or composite we try to shatter it by letting it collide with another particle and observe the reaction products. In this way we can shatter molecules into atoms, and atoms into electrons and nuclei, and it is fair to say that molecules are built of atoms, which in turn are built of electrons and nuclei. The nineteenth century physicists were really mistaken when they thought that atoms are indestructible and indivisible: atoms can in fact be shattered readily. In the same way nuclei can be shattered, and it is fair to say that nuclei are built of protons and neutrons.   To shatter a nucleus, however, requires much more energy than to shatter an atom, and in this sense nuclei are much less "destructible" than atoms.

With modern particle accelerators we can produce beams of very energetic particles, and we thus have the means for shattering such particles as protons if they could indeed be shattered.   But protons do not shatter like atoms and nuclei: something quite different happens.   We must conclude that when we study electrons, protons, neutrons, etc., we have reached a limit: it does not appear to be sensible and useful to regard these particles as made up of other more elementary particles.

16   Nobody would attempt today to create a comprehensive theory of matter based on the proposition that matter is infinitely divisible: such an undertaking would be futile.   Let us, however, speculate a bit on what features such a theory *might* have.   If we take a chunk of copper, and divide it into smaller and smaller pieces, we never get anything but small chunks of copper.   No matter how small the pieces are, they are still recognizable as chunks of copper. What does this mean?   It means that the physical laws governing the behavior of *small* chunks of copper are the same as the laws governing the behavior of *large* chunks of copper; physical systems can be "scaled down" indefinitely.   Now it must be admitted that our theory need not *necessarily* have this feature, but it would be a very natural feature of a theory describing matter which is infinitely divisible.   We note that in many respects our classical theories of physics do have this feature.   The laws of physics which we use to describe some machine weighing a ton are not qualitatively different from the laws we use to describe a wrist watch. Macroscopic physical systems can be scaled over a considerable range.

This "preservation of the form of physical laws" which might appear natural if matter were infinitely divisible is certainly totally implausible if matter is made of elementary particles.   An *atom* of copper is in no way like a macroscopic chunk of copper; it is something entirely different.   We have absolutely no a priori reasons to believe that the laws of physics which describe macroscopic systems sufficiently accurately would also be adequate to describe the structure of atoms and elementary particles.

17   To admit, as a matter of abstract principle, that classical ideas might not be appropriate to atoms, and that the electron is really an elementary particle is one thing, but to live up to such principles fully in one's thinking is quite a different thing. Experience shows that our thinking tends to be prejudiced, and that we do not easily give up ideas which we have once absorbed. Since our first conscious observations of physical phenomena concern macroscopic systems, we acquire a set of "classical prejudices" which have to be overcome when we wish to study quantum physics.[†] Let us try to illustrate the meaning of these remarks by considering two closely related problems which have been the objects of much speculation in this century.

18 Let us ask the following: What are the forces which keep an electron together? What fraction of the mass of an electron is of an *intrinsic* nature and what fraction is due to the energy of the electrostatic field of the electron? To try to deal with these questions we assume a not unreasonable model according to which the electron is a small uniformly charged sphere of radius r. The different parts of this sphere repel each other electrostatically, and there must therefore be some other kind of force which keeps the sphere together. What is the nature of this force?

In Volume II of this series[‡] we have learned how to compute the total energy "residing" in an electrostatic field: we integrate $(1/8\pi)\mathbf{E}^2$ over all space, where $\mathbf{E}$ is the local electric field. For our model we obtain the expression $W = \frac{3}{5}(e^2/r)$ for the electrostatic energy,[§] where e is the electronic charge. (The coefficient in front of the expression e^2/r depends on the details of the model: for a uniformly charged sphere it happens to have the value $\frac{3}{5}$. What is important here is not the value of this coefficient but the proportionality of W to the expression e^2/r. That W depends in this way on e and r is immediately obvious on dimensional grounds.) We can now write the mass of the electron in the form $m = m_e + m_i$, where $m_e = W/c^2$ is the electromagnetic contribution and m_i is the "intrinsic" part. The problem is: How large is m_e? Could it perhaps be that $m = m_e$, in which case the entire mass would be of electromagnetic origin? If we make this assumption we can

---

[†] It is not only the beginning physics student who has such prejudices, the senior physicist has them too. Since rigidity of mind appears to increase with age it is plausible that the senior physicist actually suffers more from his "classical prejudices" than the beginning student does.

[‡] Berkeley Physics Course, Vol. II, *Electricity and Magnetism*, Chap. 2, p. 51.

[§] This holds for the cgs-system of units. In the MKS system we have

$$W = \frac{3}{5}\left(\frac{e^2}{4\pi\varepsilon_0 r}\right).$$

compute the radius r and we find $r = 1.7 \times 10^{-13}$ cm. There are many experimental facts which suggest that the electron must be very "small," and it is therefore comforting that we did obtain something small. Note that we cannot make r much smaller, unless we wish to contemplate the possibility that m_i is negative.

Since the electron is supposed to be elementary it might appear particularly tempting to try a model with $r = 0$, in which case the electron would be a "point-particle" with no extent and no structure. This, however, would lead to an infinite electromagnetic self-energy W, and to a negative infinite intrinsic mass m_i, which hardly makes sense. (This circumstance, which raises an insurmountable obstacle in the way of the *mathematically* simple and attractive model of a point-electron, is referred to in the literature as "the difficulty of the infinite self-energy of the electron.")

19   Let us now think critically about the above speculations: do they really make any sense?   In asking our questions we have clearly made many assumptions which reflect our prejudices. We have assumed that the electron is a small charged sphere, and we have assumed that Coulomb's law can be applied to the "parts" of this sphere. How do we know that Coulomb's law applies to this situation? And what about the idea that a force has to hold together the "parts" of the electron against the electrostatic forces of repulsion? We have said earlier that the electron has no "parts"; *it is an elementary particle*. To ask what holds the electron together means we contemplate the possibility that it could break into parts, but this is a very questionable idea. Note that the electrostatic self-energy of the particle is the work which we could obtain by letting the "parts" of the particle disperse completely; this is how we originally derived the result that the electrostatic energy of any system of charges equals the integral of the square of the electric field strength over all space. If the particle *cannot* be dispersed, then the electrostatic self-energy is a doubtful concept. This is particularly true of the nonsensical infinite self-energy of the "point-electron."

Most physicists have realized by now that attempting to create some kind of classical model for the electron is meaningless. The electron does not behave like a charged sphere, and all discussions about what would keep it together if it were like a charged sphere or what its classical self-energy might be, are irrelevant in physics. Our classical prejudices led us to ask questions to which no sensible answers can be expected.

We should mention, however, the amusing circumstance that the ghost of the infinite self-energy has not yet been completely exorcised from physics; remnants of this confusion still persist in quantum mechanics.

The Limits of Applicability of Classical Theory

20 In the theory of special relativity the velocity of light plays a fundamental role. This velocity, $c = 3 \times 10^{10}$ cm/sec, is the upper limit on the velocity of any material particle and the upper limit on the velocity by which energy or information can be transmitted in physical space. The existence of this velocity provides us with a simple and natural criterion in terms of which we can decide when a physical phenomenon may be discussed "non-relativistically" and when it must be discussed "relativistically." Roughly speaking a non-relativistic treatment is adequate, i.e., sufficiently accurate, whenever all relevant velocities are small compared to the velocity of light.

We may ask whether an analogous criterion exists which tells us when we must apply quantum mechanics and when the theory of classical physics is adequate. Is there a constant of nature, "analogous" to the constant c, in terms of which the desired criterion may be formulated?

Such a constant does exist, and it is known as *Planck's constant.* It is denoted by h, and it has the value

$$h = 6.626 \times 10^{-27} \text{ erg sec}$$
$$= 6.626 \times 10^{-34} \text{ joule sec}$$

The physical dimension of this constant is thus [time] $\times$ [energy] = [length] $\times$ [momentum] = [angular momentum]. Such a physical quantity is known as *action,* and accordingly Planck's constant is also called the (fundamental) *quantum of action.*

The criterion is roughly the following. If for a physical system any "natural" dynamical variable † which has the dimension of action assumes a numerical value comparable to Planck's constant h, then the behavior of the system must be described within the framework of quantum mechanics. If, on the other hand, every variable having the dimension of action is very large when measured against h, then the laws of classical physics are valid to sufficient accuracy.

---

† A *dynamical variable* is any variable which characterizes the state of the system; for example, a position coordinate, a component of momentum, a component of angular momentum, a component of velocity, the total energy, etc.

We emphasize that this is a *rough* criterion which merely tells us when we have to be careful. The fact that an action variable is small in any particular case does not necessarily mean that classical theory is *totally* inapplicable. In many cases classical theory will give us at least *some* insight into the behavior of the system, especially if tempered by some quantum-mechanical notions.

21 We note immediately that Planck's constant is "small," which means that h is numerically small when measured in units which are appropriate for a description of macroscopic phenomena, i.e., MKS or cgs units. Differently stated, a quantity of action in the macroscopic world is *enormous* when measured in units of h.

We may, for instance, think about a pendulum in a pendulum clock. To find a quantity which has the dimension of action we form the product of the period of the pendulum with its total energy when it is swinging. The period is of the order of a second, and the energy is certainly much larger than an erg, which means that the product of these two quantities is much larger than 10^{26} times h. According to our criterion a classical description of the swinging pendulum should be, and indeed is, completely adequate.

Consider similarly a rotating body. Let the moment of inertia be 1 gm cm^2, and let the angular velocity be 1 rad/sec. The angular momentum is then 1 gm cm^2/sec = 1 erg sec $> 10^{26} h$, and the angular momentum is thus enormous compared with h, Even if the body were just a small grain of sand with a rotational period of an hour, its angular momentum would still be extremely large when measured in units of h.

Consider finally a small, but macroscopic, harmonic oscillator. Let the mass be 1 gm, the maximum velocity be 1 cm/sec, and the maximum amplitude be $x = 1$ cm. The maximum momentum is then $p = 1$ gm cm sec^{-1}. The quantity $xp = 1$ erg sec is an action variable which is again $> 10^{26} h$.

This discussion illustrates how our test, when applied to such macroscopic systems, will always tell us what we already knew, namely that the systems can be described classically.

22 Let us now try to obtain a deeper understanding of what our criterion really means.

In classical physics we assume that every dynamical variable of a system can be specified and measured to arbitrary precision. This does not mean that we can do it in actual practice; it rather means that we do not admit that there is any limit to the precision *in principle*. The set of dynamical variables in classical physics includes such variables as position coordinates, components of

Max Karl Ernst Ludwig Planck. Born 1858 in Kiel, Germany; died 1947. After studies in Munich and Berlin, Planck obtained his doctoral degree in 1879. His thesis dealt with the second law of thermodynamics. After holding a position at the University of Kiel, Planck was appointed professor of Theoretical Physics at the University of Berlin in 1899. He retired at the age of seventy in 1928. He received the Nobel prize in 1919.

In the beginning of his career Planck devoted himself to the study of thermodynamics: a subject in which he remained interested throughout his life. In Berlin he became acquainted with the experimental work on thermal radiation carried out by Lummer, Pringsheim, Rubens and Kurlbaum, and he set himself the task of deriving a theoretical black-body-radiation law. His successful effort marks the beginning of quantum physics, and what is now called Planck's constant first appeared in his paper of 1900. After this monumental discovery Planck continued to play an active role in the development of quantum physics. *(Photograph by courtesy of Physics Today.)*

momenta, components of angular momenta, etc., for a system of particles or a single particle, as well as such variables as the components of the electric and magnetic field vectors at a given point in space at a given time.

A careful analysis of the actual behavior of microphysical systems shows, however, that there is a *fundamental limit* to the accuracy to which variables of this kind can be specified and measured. The very penetrating and beautiful analysis by which this limit was established was carried out by W. Heisenberg in 1927. We refer to the existence of such limits as the *uncertainty principle;* a specific quantitative expression of this principle in any particular case is known as an *uncertainty relation.*

A particular uncertainty relation concerns the pair of variables (q, p), where q is the position coordinate of a particle and p is the momentum of the particle. This relation reads

$$\Delta q \, \Delta p \geqq \frac{h}{4\pi} \tag{22a}$$

Here Δq is the root-mean-square error in q and Δp is the root-mean-square error in p, and the inequality thus asserts that the two variables q and p cannot be known more accurately than that the product of the "uncertainties" of the two variables is of the order of Planck's constant.

We note immediately that because of the smallness of Planck's constant, h, the uncertainty relation is of no importance in macrophysics; other sources of errors in q and p always mask the fundamental uncertainty expressed in the inequality (22a). The relation (22a) therefore does not in any sense contradict our *empirical knowledge* of macrophysics even though it does contradict our classical *theories* about macroscopic systems.

23 The uncertainty principle is often "explained" as follows. Dynamical variables such as position, momentum, angular momentum, etc., must be defined *operationally*, i.e., in terms of the experimental procedures through which they are measured. If we now analyze realistic procedures of measurement in microphysics, it turns out that a measurement will always *perturb* the system; there is a characteristic unavoidable interaction between the system and the measuring apparatus. If we try to measure the position of a particle very accurately, we will perturb it in such a way that its momentum after the measurement is very uncertain. If we try to measure its momentum very accurately we perturb it in such a way that its position will be very uncertain. If we try to measure both the position and the momentum of the particle *simultaneously,* then these two measurements will necessarily interfere with each other in such a way that the accuracy of the final result will be subject to the inequality (22a). The discussion then proceeds to show how these perturbations arise in particular cases.

This kind of description of the meaning of the uncertainty principle is very common in texts on quantum mechanics. The author does not want to maintain that it is completely wrong, but he does feel that it is misleading and that it can give rise to grave misunderstanding. It misses the essential point, which is this: *The uncertainty relations state the limits beyond which the concepts of classical physics cannot be employed.* The "classical physical system," described through classical dynamical variables which are definite functions of the time, and which can be known in principle to arbitrary precision, is a figment of imagination; it does not exist in the real world. Experiments have been performed which tell us that this is so. If we describe an actual system as a "classical system" then we make an approximation, and the uncertainty relations tell us how far we can go.

24 To elucidate these ideas further, let us consider the one-dimensional motion of a particle. According to classical dynamics we describe the instantaneous position of the particle through the position variable $q = q(t)$. If the mass of the particle is m and if it moves sufficiently slowly, then its momentum p is given by $p = p(t) = m \, dq(t)/dt$. Now we might think that the uncertainty relation merely expresses an unfortunate property of our measuring instruments which prevents us from determining $q(0)$ and $p(0)$ to arbitrary accuracy, although we can very well think about precise values of these variables and about the precise subsequent motion of the particle. We might, in other words, think that we can go on using a classical description according to which every particle follows a definite trajectory but with the refinement that we introduce an uncertainty about *which one* of the trajectories the particle follows by imposing uncertainty relations on the initial conditions which determine the trajectories.

This is not so. Experiments tell us that we have to modify our ideas in a much more profound manner. *The motion of a classical trajectory must be rejected;* to ask for, or think about, the simultaneous values of $q(t)$ and $p(t)$ is as meaningless as to ask for the hair color of the King of the United States.

25 Now it may seem that our discussion is logically contradictory. First we state an uncertainty relation, and then we declare that the variables q and p which occur in this relation make no sense. If they make no sense, how then can the relation make any sense? The answer is as follows. In the quantum-mechanical description of the behavior of a particle we can introduce certain mathematical objects q and p which in many respects *correspond* to the classical position and momentum variables. These objects, however, are *not* identical with the classical variables. The relation (22a) tells us that if we try to interpret the quantum-mechanical objects q and p as "position" and "momentum," and thus interpret the motion in classical terms, then there is a fundamental limitation on the accuracy with which "position" and "momentum" can be known. In other words, the relation tells us that if we *try* to introduce classical variables, and *try* to interpret the motion classically, then the precision with which these variables can be specified is limited.

26 It should be clearly understood that no analysis of the measuring process in purely classical terms could ever lead to an uncertainty relation. The uncertainty relations reflect experimentally discovered facts about nature. The particles occurring in nature do not behave like classical point-particles, nor do they behave like small billiard balls: †they behave quite differently, and that is why certain kinds of measurements cannot be made or even imagined.

In subsequent chapters we will learn about the properties of the particles in the real world, and we will then see how naturally the seemingly strange uncertainty relations fit into the scheme of things.

The Discovery of Planck's Constant

27 Let us now consider the early history of Planck's constant, and let us see how it was discovered, and how it made its way into physics. We shall go back in time to the beginning of this century, and consider some of the outstanding problems in physics of that period, namely the following:

(i) The problem of the black-body radiation law
(ii) The problem of the photoelectric effect
(iii) The problem of the stability and size of atoms

These three problems were not, of course, the only ones which

---

†For some reason the *billiard ball* has come to play the role of the prototype of a classical particle in textbooks on quantum mechanics. The author, of course, conforms to this tradition. It may amuse the reader to know that the author has never played billiards and has never held a billiard ball in his hand. His knowledge of the alleged properties of billiard balls is, therefore, book knowledge, derived from texts on quantum mechanics.

occupied the physicists of the time, but we single them out because they illustrate the dilemma of classical physics in a particularly clear-cut way.

The reader should realize that our discussion is extremely deficient as a historical account: we could not possibly hope to do justice to the very interesting development of quantum physics in a few pages. We are looking at the situation at the beginning of this century in retrospect, and it is then easy to see that these three problems were key problems. However, if we examine the publications for the year 1900 in *Annalen der Physik* (which was one of the leading journals in physics at the time) we find that the majority of physicists were concerned with very different things. The ability to distinguish the truly significant from the insignificant is a rare ability indeed (at any time), and we have every reason to admire the remarkable insight and imagination of the early pioneers of quantum physics.

28 To dramatize the situation we shall regard these three problems as three different aspects of the fundamental "Mystery of the Missing Constant." This is certainly not the way in which a physicist in the year 1900 would have formulated the difficulties facing him, but it is instructive to consider the matter in retrospect from this point of view.

The missing constant is of course Planck's constant h. In a purely classical theory of matter this constant would not occur. Let us therefore consider some of the other basic constants of physics which *would* play a role in a classical description.

(i) The velocity of light, $c = 3.00 \times 10^{10}$ cm/sec. This constant was known with fair accuracy in the year 1900.

(ii) Avogadro's number, $N_0 = 6.02 \times 10^{23}$, which is the number of molecules in a mole of any gas. A *crude* value of this constant, based on the kinetic theory of gases, was available in 1900.

(iii) The mass of the hydrogen atom, $M_H = 1.67 \times 10^{-24}$ gm. To an accuracy of 1 part in 2000 this is also the mass of the proton, M_p. Since a mole of hydrogen has a mass which is very close to 2 gm, we have

$$N_0 M_H \cong N_0 M_p \cong 1 \text{ gm} \tag{28a}$$

and if we know Avogadro's number we can thus find M_H.

(iv) The elementary charge, $e = 1.6 \times 10^{-19}$ coul $= 4.8 \times 10^{-10}$ esu. The charge of the electron is $-e$, and the charge of the proton is $+e$. The charge carried by a mole of singly charged ions (i.e., each ion carries the charge e) is known as Faraday's constant F. We thus have

Joseph John Thomson. Born 1856 near Manchester, England; died 1940. For many years Thomson (affectionately nicknamed "J.J.") held the position of Cavendish Professor of Physics at Cambridge University, and also that of Professor of Physics at the Royal Institution, London. His many and varied contributions to physics included researches on the conduction of electricity through gases, the charge and mass of the electron, and the properties of positive rays. Thomson discovered the electron in 1897. His work on the positive rays led to his discovery of the isotopes of neon. He received the Nobel prize in 1906. *(Photograph by courtesy of Professor L. B. Loeb, Berkeley.)*

$$F = N_0 e = 96{,}500 \text{ coul} \qquad (28\text{b})$$

The Faraday constant F is readily measurable in an electrolysis experiment. F is, for instance, the quantity of charge which has to pass through an electrolysis cell in order to deposit one gram-equivalent of silver (i.e., 107.88 gm of silver since the atomic weight of silver is 107.88).

(v) The ratio of charge to mass for the electron, $e/m = 1.76 \times 10^8$ coul/gm, and the ratio of charge to mass for the proton, $e/M_p = 9.6 \times 10^4$ coul/gm. These constants can be determined in deflection experiments with beams of electrons or protons in electric and magnetic fields. In this way e/m had been determined by J. J. Thomson in 1897.[†] We should note that

$$\frac{e}{M_p} = \frac{F}{N_0 M_p} \qquad (28\text{c})$$

and this constant is therefore not independent of the constants mentioned earlier.

We should also note that given precise values of e/m and e/M_p we can find a precise value for

$$\frac{M_p}{m} = \frac{e/m}{e/M_p} \qquad (28\text{d})$$

even if the charge e is not known with corresponding precision. This assumes, of course, that the charge of the proton is equal in magnitude to the charge of the electron.

(vi) The mass of the electron, $m = 9.11 \times 10^{-28}$ gm. This constant can be inferred from e and e/m.

29 Avogadro's number N_0 is the link which connects microphysics with macrophysics. The enormous size of this number tells us how small atoms and molecules really are, and why the granular structure of matter is not more apparent in the world of macrophysics. As we have said N_0 was not very well known at the end of the last century. The constants F, e/m, m/M_p were, however, much better known, and an independent good measurement of either N_0 or e would thus lead to a better knowledge of the basic constants e, m, and M_p. One important aspect of Planck's theory of black-body radiation was, as we shall see, that it permitted an independent, and better, evaluation of N_0.

About a decade later R. A. Millikan, in his famous oil-drop experiment, measured e directly by observing the motion of small

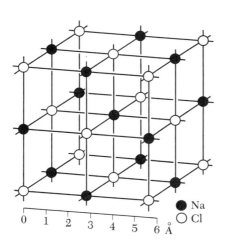

Fig. 30A The crystal structure of sodium chloride. The lattice is cubical, with chlorine atoms and sodium atoms alternating at the vertices. The centers of the small spheres in the figure indicate the mean positions of the sodium and chlorine nuclei. The size of these spheres is not intended to convey any information about the sizes of either the nuclei or the atoms.

● Na
○ Cl

0 1 2 3 4 5 6 Å

[†]J. J. Thomson, "Cathode Rays," *Philosophical Magazine* **44**, 293 (1897).

charged drops of oil floating in air under the combined influence of gravity and an electric field.[†] Although this kind of experiment can hardly be expected to yield a high-precision value of e, it was of great importance as an independent, and *conceptually simple*, measurement of this constant.

30 We want to go ahead of our story and mention that Avogadro's number N_0 can also be measured very directly, by counting the number of atoms in a crystal. The atoms in the crystal are arranged in a regular lattice, say a cubical lattice, and if we can determine the spacing between neighboring atoms in the crystal, the *lattice constant*, we can clearly find N_0. This spacing can be determined through X-ray diffraction experiments, provided we also determine the wavelength of the X-rays used, for instance through measurements with a mechanically ruled "macroscopic" grating. N_0 was eventually determined in this manner.

The clever idea that Nature provides us with ready-made diffraction gratings in the form of crystals first occurred to M. von Laue, and on his suggestion diffraction experiments with X-rays in crystals were carried out in 1912 by W. Friedrich and P. Knipping[‡] This was in fact the first conclusive experimental proof that X-rays are indeed waves of short wavelength.

31 In order to be able to understand the issues connected with black-body radiation we have to digress to discuss heat and temperature.[§] These concepts are relevant to the description of the behavior of matter in bulk under conditions of thermal equilibrium. This topic has nothing to do with the structure, or behavior, of *isolated* atoms, molecules, or nuclei, but it is of importance in many manifestations of quantum phenomena. The reason for this is, of course, that we do not, in general, carry out our measurements on isolated atoms, molecules, or nuclei; we observe these particles "embedded" in bulk matter.

Thermal energy is energy associated with the disorganized motion of the constituents of a macroscopic body. Heat is thermal energy in transfer (from one body to another). What is temperature?

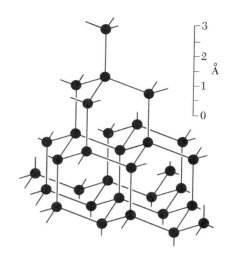

Fig. 30B The crystal structure of diamond. Each carbon atom has four nearest neighbors located at the vertices of a tetrahedron (nearest neighbors are joined by solid lines).

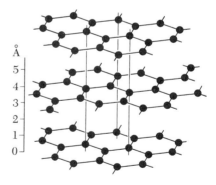

Fig. 30C The crystal structure of graphite. Diamond and graphite both consist only of carbon. The marked differences in the physical properties of these two materials derive from the different forms of their crystal lattices. The graphite lattice consists of equally spaced parallel planes in which the carbon atoms are arranged in a hexagonal pattern. Compare the lattice shown above with the diamond lattice in Fig. 30B.

† R. A. Millikan, "The Isolation of an Ion, a Precision Measurement of Its Charge, and the Correction of Stokes's Law," *The Physical Review* **32**, 349 (1911).

‡ W. Friedrich, P. Knipping, and M. Laue, "Interferenzerscheinungen bei Röntgenstrahlen," *Annalen der Physik* **41**, 971 (1913).

§ These topics are discussed more fully in Berkeley Physics Course, Vol. V, *Statistical Physics*. See also Physical Science Study Committee, *Physics*, Chaps. 9 and 26, 2nd ed. (D. C. Heath and Company, Boston, 1965).

32 It is not quite so easy to give, in a single sentence, a *precise* definition of the concept of temperature. In a sense we all "know," what temperature is, and we know how to measure temperature with a thermometer. A thermometer is any body, or system, for which a change in temperature produces a readily observable change, such as a change in length; a change in volume; a change in electrical resistance; etc. Let us consider, as an example, a mercury thermometer. Temperature is read by observing the level of a column of mercury in a capillary of uniform cross section. To establish a temperature scale we might assign the temperature 0° to the melting point of ice, and the temperature 100° to the boiling point of water, and then define the intervening "degrees" by dividing the interval between the two reference points on the capillary into a hundred equal parts. In this manner we can indeed define a measure of temperature, but our procedure has the grave defect (from the point of view of basic physical theory) that our temperature scale depends on the special properties of an arbitrarily selected substance, in this case mercury. If we follow the same procedure using another substance, say alcohol, we might find that 30° on the alcohol scale is not the same thing as 30° on the mercury scale.

For scientific purposes it is clearly desirable to have a measure of temperature which is independent of the special properties of any particular substance. In the next volume of this series, which is devoted to thermal physics, we shall discuss in detail how such a measure can be defined. The resulting scale of temperature is the *absolute scale,* on which temperature is measured in degrees Kelvin, denoted °K. On the absolute scale 0°K is the lowest possible temperature: this temperature corresponds approximately to -273°C. For convenience the magnitude of the Kelvin degree is so chosen that a temperature *difference* has the same numerical value on the absolute scale as on the centigrade scale, and we thus have, by definition,

$$\text{(temperature in °K)} = \text{(temperature in °C)} + 273.15$$

33 Let us try to obtain a qualitative idea of what temperature "means" from the standpoint of microphysics. The basic idea is as follows. As the temperature increases the average energy associated with the random motions of the elementary constituents of a macroscopic body increases. At the temperature 0°K all *random* motions cease, and this is the physical significance of the lowest possible temperature. (The emphasis is on the word "random.")

In statistical mechanics we often idealize the properties of a real gas in terms of a model: we assume that the gas consists of a large number of small identical particles (molecules), moving at random, and with *negligible* mutual interactions. This model provides us with a good description of a *dilute* real gas. If the particles in the (model) gas are monatomic molecules, we speak of an ideal monatomic gas. It is possible to prove that for one mole of an ideal gas,

$$PV = \tfrac{2}{3}N_0 E_{\text{kin}} \tag{33a}$$

where P is the pressure, V the volume of the container, and E_{kin} the average kinetic energy per (monatomic) molecule.

The absolute temperature is so defined that, within this model, it simply expresses the average kinetic energy by $E_{\text{kin}} = \tfrac{3}{2}kT$, where the constant of proportionality k is known as Boltzmann's constant. We can therefore write (33a) in the form

$$PV = N_0 kT = RT \tag{33b}$$

where the constant $R = N_0 k$ is the *universal gas constant*. It is an *experimental fact* that this law holds accurately for all sufficiently rare gases: i.e., for any real gas the law holds better the less dense the gas is. We can exploit this fact to construct a gas thermometer calibrated to show absolute temperature.

34 The universal gas constant has the value

$$R = N_0 k = 8.3 \times 10^7 \text{ erg } (^\circ\text{K})^{-1} \text{ (mol)}^{-1}$$
$$= 1.99 \text{ cal } (^\circ\text{K})^{-1} \text{ (mol)}^{-1} \tag{34a}$$

It is a macroscopic constant which can be readily measured on the basis of the relation (33b).

Boltzmann's constant, $k = R/N_0$, is the gas constant per molecule. It can be found provided N_0 is known, and it has the value

$$k = 1.38 \times 10^{-16} \text{ erg } (^\circ\text{K})^{-1} \tag{34b}$$

It is in effect a conversion factor from temperature to energy. That temperature and energy are related in this way should not, however, lead anybody to believe that energy and temperature are "the same thing."

35 After this survey of the basic constants of classical physics we now consider the problem of the black-body radiation law. The empirical facts are as follows. The surface of a material body held at a high temperature emits light of all frequencies, or wavelengths. If we plot the amount of radiant energy emitted per unit time per unit area per unit wavelength interval versus wavelength we obtain

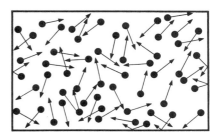

Fig. 33A The relation $PV = \tfrac{2}{3}N_0 E_{\text{kin}}$ can be easily understood. Consider a container of volume V in which there are N_0 molecules. Let us first assume that all the molecules move to the right, with velocity v. The number of molecules colliding with a unit area of the wall per unit time is $v(N_0/V)$. Each molecule transfers an amount of momentum $2mv$ to the wall. The pressure P' is equal to the total momentum transfer per unit area per unit time, and hence we have $P' = 2mv^2(N_0/V) = 4\,E_{\text{kin}}(N_0/V)$.

In reality the direction of motion is random, and the true pressure P is related to the pressure P' computed above by $P = \tfrac{1}{6}P'$, which leads to Eq. (33a). (We can understand the factor $\tfrac{1}{6}$ if we imagine that the molecules move in six standard directions: in both directions along three perpendicular axes. Only one-sixth of the molecules will then contribute to the pressure on the right wall.)

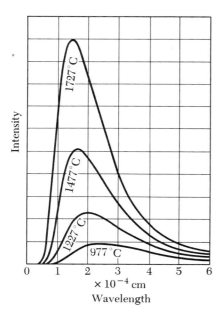

Fig. 35A Graphs showing power emitted by a blackbody radiator, per unit area per unit wavelength interval, for four different temperatures. The *total* power emitted is proportional to the areas under the curves; it is proportional to the fourth power of the absolute temperature. Note how the location of the maximum depends on the temperature; the precise relationship is expressed by Wien's law.

a curve which goes to zero for very long, and for very short wavelengths; in general the curve will have a single maximum at some wavelength λ_{max}, which depends on the temperature. The location of this maximum, and the total amount of radiation emitted, is *very roughly* the same for all material surfaces. Instead of studying the radiation from a *material surface* one may observe the radiation emerging from a *small hole* in the wall of a closed material surface kept at a fixed temperature. In this kind of measurement we thus have an enclosure, or "oven," made of any suitable refractory material, with a small hole in the side (i.e., a hole small compared with the linear dimensions of the cavity). We direct our instruments at the hole and thereby measure the radiant energy emerging from the *interior* of the enclosure. It was found in such measurements that

(i) A graph (see Fig. 35A), of the intensity of radiation from the hole versus wavelength is a smooth curve falling to zero for long, as well as for short wavelengths, and with a maximum at a wavelength λ_{max} which depends on the temperature T of the walls in a very simple manner, namely

$$\lambda_{max}T = C_0 = 0.2898 \text{ cm } °\text{K} \qquad (35a)$$

(ii) The spectral distribution of the emitted radiation, i.e., the shape of the curve mentioned in (i), is independent of the shape of the cavity, and independent of the material of which the walls are made. The constant C_0 in equation (35a) which expresses *Wien's displacement law*, is thus a universal constant which describes a remarkable property of cavities *in general*.

(iii) The intensity of the radiation emerging from the hole is always larger, at every wavelength, than the corresponding intensity of emission from a material surface kept at the same temperature as the walls of the cavity; the order of magnitude of the intensity, however, is the same.

36 A surface which absorbs all radiation incident upon it is called a *black-body surface*. To an *outside* observer a small hole in the wall of a cavity appears more or less like a black-body surface, especially if the interior walls of the cavity are rough and blackened. The reason is simply that any radiation (light) incident from the outside upon the hole will be almost completely absorbed in multiple reflections inside the cavity, even if the interior walls are not perfectly absorbing.

Because of these circumstances we refer to the radiation emerging from a hole in the wall of a cavity as *black-body radiation*. It was shown by G. R. Kirchhoff, from very general thermodynamic considerations, that, for any wavelength, the ratio of the rate of emission from an arbitrary material surface to the rate of emission from a black-body surface equals the absorption coefficient for the material at the wavelength in question. The black-body surface is, therefore, a suitable standard emitter, and we shall limit our considerations to black-body radiation; i.e., radiation from a hole in the wall of a cavity.

37 By the end of the nineteenth century careful measurements had been carried out on the black-body-radiation law, and in particular the relation (35a) had been established. The outstanding theoretical problem was to derive the radiation law from first principles. That the hole should emit radiation is not in itself surprising; we know that the constituents of matter are charged, and the thermal vibrations of the constituents in the walls naturally lead to the emission of radiant energy into the cavity. This radiation may also be absorbed by the walls, and if the walls are kept at a fixed temperature we will have some kind of equilibrium between the radiant energy in the cavity and the walls, i.e., the emission and absorption rates will be equal. The problem is thus to derive an expression for the density of radiant energy in the cavity as a function of wavelength and temperature.

We here focus our attention on just one detail of this problem, namely the relation (35a). To see what is involved we rewrite this equation in the form

$$\frac{\lambda_{\max}}{c} \times kT = X_1 = \frac{C_0 k}{c} \tag{37a}$$

where c is the velocity of light, k is Boltzmann's constant, and X_1 is a new constant. Since the left-hand side of (37a) has the physical dimension [time] $\times$ [energy] = [action] the constant X_1 is a quantity of action. How can we derive a theoretical expression for X_1? How are we to produce a quantity having the physical dimension of action from the constants of nature available? This is certainly a dilemma because it is hard to see how the constants m, M_H and e could possibly enter into the expression for X_1. The physical situation seems very clear-cut; the radiant energy inside the cavity is in thermal equilibrium with the walls. The radiation emitted from the cavity is, however, *independent of the size and shape of the cavity and also independent of the material in the walls;* how then could constants like m and e, which refer to properties of the walls, be relevant? Our suspicion that X_1 cannot be derived from the remaining constants seems quite justified and, as a matter of fact, the relation (37a) cannot be understood on the

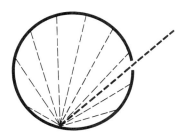

Fig. 36A To an outside observer a small hole in the wall of a cavity with a (partially) absorbing interior surface is a black-body surface: it absorbs incident radiation almost completely. A ray of light entering through the hole is partly absorbed and partly diffusely reflected when it hits the interior surface. The reflected rays are again partly absorbed, and partly diffusely reflected, and only a very small fraction of the incident light will find its way out again through the hole.

Differently stated: a *photon* entering the cavity has a very small probability of escaping through the hole.

The reader can easily test for himself that a black-body surface can be realized in this manner. Paint the interior of a small cardboard box black, and cut a small hole in one side. Viewed from outside the hole appears considerably "blacker" than any "black" *material* surface.

basis of classical physics. The situation was actually quite desperate in the year 1900, before Planck's discovery. The application of classical statistical mechanics had led to the absurd black-body-radiation law stating that the intensity radiated increases monotonically with the frequency in such a way that the total radiated intensity is infinite, which means that radiation cannot be in thermal equilibrium with matter at any temperature!

38 On December 14, 1900 Max Planck presented a derivation of the black-body-radiation law at a meeting of the German Physical Society in Berlin, and this date can be regarded as the birthday of the theory of quantum physics.[†] In his derivation of the theoretical expression for the intensity of radiation as a function of the wavelength and the temperature, Planck departed from classical physics by making a radical ad hoc assumption the essence of which can be stated as follows. An oscillator of natural frequency ν can take up or give off energy only in parcels of magnitude $E = h\nu$, where h is a new fundamental constant of nature. Planck was then able to derive, among other things, an expression for our constant X_1, namely

$$\frac{\lambda_{\max}}{c} \times kT = \frac{C_0 k}{c} = X_1 = 0.2014 \times h \tag{38a}$$

This then is how Planck's constant made its first appearance.

Planck himself was very reluctant to accept this departure from classical physics, and after his great discovery he tried very hard for many years to understand the phenomenon of black-body radiation on a purely classical basis. Of these fruitless efforts he later said that he did not regard them as useless labor; only by his repeated failures did he reach the final conviction that an explanation could not be found within classical physics.

39 In its full glory Planck's radiation law reads as follows:

$$E(\lambda,T) = \left(\frac{8\pi hc}{\lambda^5}\right) \times \frac{1}{\exp{(hc/\lambda kT)} - 1} \tag{39a}$$

where $E(\lambda,T)$ is the density of radiant energy in the cavity per unit wavelength interval, at the wavelength λ, and at the temperature T. The constant k is Boltzmann's constant, and c is the velocity of light.

[†]M. Planck, "Über das Gesetz der Energieverteilung in Normalspektrum," *Annalen der Physik* **4**, 553 (1901).

The intensity of the radiation emitted from a small hole in the wall of the cavity is proportional to the energy density inside the cavity, and the expression (39a) is thus the mathematical expression for the relationships illustrated in Fig. 35A.

To find the location of the maximum of $E(\lambda,T)$, as a function of λ, with T constant, we set the derivative of $E(\lambda,T)$ with respect to λ equal to zero, and solve for λ_{max}. This is how we obtain the relation (38a), or the equivalent relation

$$\lambda_{max}T = C_0 = 0.2014 \times \frac{hc}{k} \qquad (39b)$$

Since λ_{max} and T can be readily measured, and since c is known, we can thus determine h/k experimentally on the basis of the relation (39b). Furthermore, by a detailed comparison of the experimentally measured $E(\lambda,T)$ with the theoretical expression (39a) we can also determine the constant h. This permits a computation of Boltzmann's constant k, and finally, by virtue of the relation $N_0 = R/k$, a computation of N_0. The value for k which Planck obtained in this way is about 2.5 per cent smaller than the best modern value.

40 The detailed history of Planck's radiation law is fascinating. Before Planck succeeded in deriving the expression (39a) from a "microscopic" point of view he had actually guessed the correct dependence of $E(\lambda,T)$ on λ and T. This guess was partly based on some careful measurements by H. Rubens and F. Kurlbaum, and partly on some general theoretical considerations. [The relationship expressed by (39a) is obviously too complicated to arrive at on purely empirical grounds.] Planck presented his preliminary result to the German Physical Society on October 19, 1900. In this version the formula contained two constants without any physical interpretation, say the constants which we would now write as $(8\pi hc)$ and (hc/k). This formula was again checked against the experimental results by Rubens, and by O. Lummer and E. Pringsheim, and was found to agree with the facts with remarkable precision.† Planck was thus faced with the problem of finding some kind of *basic* theoretical justification for a formula which appeared to be the correct one. This he succeeded in doing, after about eight weeks of strenuous labor.

---

† For *later* tests of Planck's law, see H. Rubens and G. Michel, "Prüfung der Planckschen Strahlungsformel," *Physikalische Zeitschrift* **22**, 569 (1921).

The Photoelectric Effect

41 Around the turn of the century it was known experimentally that when light (in the visible or ultraviolet region) is incident on a metal surface, electrons will be ejected from the surface.† This phenomenon is in itself not surprising since we know that light is electromagnetic radiation, and we may thus expect that the electric field of the light wave could exert a force on the electrons in the metal surface and cause some of these to be ejected. What is surprising, however, is to find that the kinetic energy of the ejected electrons is independent of the *intensity* of the light, but does depend on the *frequency* in a very simple way: it increases linearly with the frequency. If we increase the intensity of the light we merely increase the number of electrons emitted per unit time, but not their energy. This is *very* hard to understand on a classical basis as we would expect that when the intensity of the light wave, and hence the amplitude of the electric field in the wave, increases, then the electrons would be accelerated to higher velocities.

These facts had been established before 1905 by P. Lenard and others. *Precise* measurements of the relationship between the frequency of the light and the energy of the ejected electrons were not carried out until 1916, when the subject was studied very carefully by R. A. Millikan.

42 In 1905 Albert Einstein suggested an explanation of these phenomena.‡ According to this explanation the energy in a beam of monochromatic light comes in parcels of magnitude $h\nu$, where ν is the frequency; this *quantum* of energy can be transferred completely to an electron. The electron acquires, in other words, the energy $E = h\nu$ while it is still inside the metal. If we now assume that a certain amount of work, W, has to be performed to remove the electron from the metal, then the electron will emerge from the metal with the kinetic energy $E_{\text{kin}} = E - W$, or

$$E_{\text{kin}} = h\nu - W \tag{42a}$$

The quantity W, known as the *work function* of the material, is assumed to be a constant characteristic of the metal, independent of the frequency ν.

The equation (42a) is Einstein's celebrated photoelectric equation. The energy of the emitted electrons increases linearly with the frequency, but is independent of the intensity of the light.

---

† See PSSC, *Physics,* Chap. 33.

‡ A. Einstein, "Über einen die Erzeugung und Verwandlung des Lichtes betreffenden heuristischen Gesichtspunkt," *Annalen der Physik* **17,** 132 (1905).

The number of electrons emitted is naturally proportional to the number of the incident quanta, and hence proportional to the intensity of the incident light. In this manner Einstein was able to account for the qualitative aspects of the photoelectric effect, such as they were known to him at the time.

43   Einstein arrived at this idea by noticing that certain aspects of Planck's strange black-body-radiation law could be understood by assigning corpuscular properties to the electromagnetic radiation in the cavity, i.e., by assuming that the radiant energy consists of quanta of magnitude $h\nu$. We should note here that the real meaning of Planck's assumption was at that time shrouded in obscurity, and Einstein's new way of looking at the phenomenon of black-body radiation was therefore an important step forward. The most important aspect of this matter was, however, that Einstein could apply his insight into the phenomenon of black-body radiation to a *new* physical situation, namely the photoelectric effect.

44   The equation (42a) was a precise theoretical prediction, and as such susceptible to quantitative experimental tests. It furthermore offered the opportunity for a new measurement of Planck's constant, assuming that Einstein's ideas were correct. As we have mentioned before these extremely important questions were studied by R. A. Millikan in a series of very careful and beautiful measurements,† in which he found complete agreement with Einstein's equation (42a).

Millikan's method is illustrated schematically in Fig. 44A. Monochromatic light is incident on a metal surface, usually an alkali metal, and causes the ejection of photoelectrons. A collecting electrode, which can be kept at an arbitrary potential $-V$ with respect to the photocathode, is placed near the photosensitive surface, and the current of photoelectrons is measured. If we now assume that the electrons are all emitted with the same kinetic energy E_{kin}, as given by equation (42a), then it is clear that none of the electrons can reach the collecting electrode if $eV > E_{\text{kin}}$. We can therefore observe the current as a function of the retarding potential V, and if V_0 is the potential at which the current just becomes zero we have

$$V_0 = \left(\frac{h}{e}\right)\nu - \frac{W}{e} \tag{44a}$$

A plot of the cutoff retarding potential V_0 versus the frequency

† R. A. Millikan, "A Direct Photoelectric Determination of Planck's 'h'," *The Physical Review* **7**, 355 (1916).

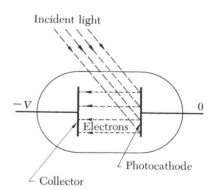

Fig. 44A   Very schematic figure to illustrate the principle of Millikan's experiment. Electrons are ejected with an energy $E_{\text{kin}} = h\nu - W$, where W is the work function characteristic of the cathode material, when light of frequency ν is incident on the photocathode. The electron current to the collector will cease when the retarding potential $V > (h\nu - W)/e$. Observation of the critical retarding potential $V_0 = (h\nu - W)/e$ versus ν yields the constant h/e. (See Fig. 44B.)

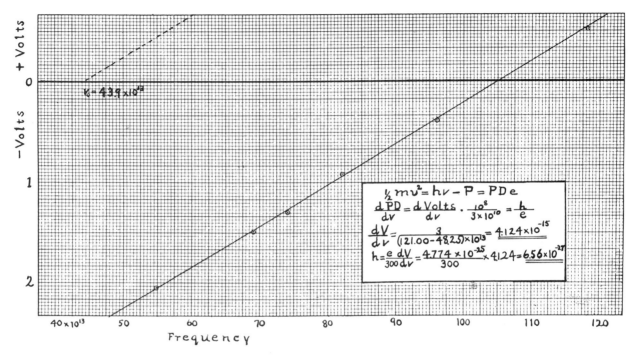

Fig. 44B A graph taken from Millikan's paper [R. A. Millikan, *Physical Review* **7**, 355 (1916)] showing the linear relation between the critical retarding potential V_0 and the frequency of the light, for a photosensitive surface of sodium. As we see, Millikan presented his computation of Planck's constant on the basis of his curve on the graph. *(Courtesy of The Physical Review.)*

ν will thus be a straight line, as shown in Fig. 44B taken from Millikan's paper. From the slope of this line we can find the constant h/e, and from its intercept with the V_0-axis we can find the material constant W/e.

This is a conceptually simple and clear-cut experiment, but to obtain accurate and reproducible results considerable care is required.

45 Let us consider the relation (44a) numerically. With $h = 6.63 \times 10^{-27}$ erg sec $= 6.63 \times 10^{-34}$ joule sec, and $e = 1.60 \times 10^{-19}$ coul, we obtain $h/e = 4.14 \times 10^{-15}$ volt sec. For visible light the wavelength lies in the range 4000 Å to 7000 Å, where 1 Å(ngström) $= 10^{-8}$ cm.† This corresponds to the frequency range $(4.3$ to $7.5) \times 10^{14}$ sec^{-1}. Blue light corresponds to a frequency of about 7×10^{14} sec^{-1}, and for this case we obtain $(h/e)\nu \sim 2.8$ volts. For light in the visible or near-ultraviolet region the retarding potential will thus be of the order of magnitude of one volt. It is an experimental fact that the material constant W/e is also typically of this

---

† In honor of the memory of the Swedish physicist A. J. Ångström, English-speaking people might make an effort to pronounce his name correctly. The first vowel "Å" is pronounced like *o* in *long*, and the second vowel "ö" like the same letter in German or like the vowel *eu* in the French word *deux*.

order. It is particularly small for alkali metals, which is why photo-cells intended for visible light have photocathodes made of these materials. A photocell will obviously not respond to radiation for which $W > h\nu$.

46 The qualitative features of photoemission which had been discovered before 1905 were certainly remarkable, although it required the insight of Einstein to fully appreciate the significance of these phenomena. Had Millikan's quantitative results been available at that time they would certainly have been recognized more generally as a major challenge to classical ideas.

The essence of the matter is clearly the strange relation

$$\frac{E}{\nu} = X_2 \qquad (46a)$$

where E is the energy which can be transferred to an electron from a beam of monochromatic light of frequency ν and where X_2 is a constant independent of the intensity of the light, independent of the frequency of the light, and independent of the material in which the electrons are embedded. (That the electrons emerge with a kinetic energy *smaller* than E would not have been regarded as a mystery in 1905 any more than today: the work function W simply represents the binding energy of the electrons in the material.) To understand a relation such as (46a) on a classical basis, and to express this understanding in a formula giving the mysterious constant X_2 in terms of the basic constants of classical physics, appears to be a truly hopeless task. The constant X_2 has the phys-ical dimension of action, and it is true that such a quantity can indeed be formed from the basic constants, namely $(e^2/c) \cong (h/860)$. We now know that $X_2 = h$, and the quantity (e^2/c) is thus of the wrong order of magnitude, about 1000 times too small, which is not encouraging. Playing with dimensional arguments in this manner does not, however, really lead us anywhere unless we can propose a *classical* mechanism which leads to the equation (46a). Nobody has been able to do that, and the facts concerning the photoelectric effect very strongly support Einstein's idea that radiant energy is *quantized*.†

As we will learn later the relation (46a) expresses a very basic principle of quantum physics, namely that energy and frequency are *universally* related by $E = h\nu$. Such a relationship is entirely foreign to classical physics, and the mysterious constant $X_2 \, (= h)$ in (46a) is a manifestation of secrets of nature unsuspected at the time.

---

† It may be mentioned here that Einstein did not use the term *photon* for the electro-magnetic quanta in his paper; that term was introduced much later.

The Problem of the Stability and Size of Atoms

47 Let us now turn to the third of our problems, namely the question of the stability and size of atoms, and let us in particular consider the latter question. We can define the "size" of an atom as the typical distance between neighboring atoms in a crystal or liquid. Experimentally this size is known to be of the order of $1 \text{ Å} = 10^{-8}$ cm. The order of magnitude of this distance is related to the order of magnitude of Avogadro's number N_0 as follows. One cubic centimeter of any liquid or solid has a mass of, very roughly, one gram. One gram of any substance contains, very roughly, N_0 atoms, and therefore the separation between neighboring atoms in a solid or liquid must be of the order of $(1/N_0)^{1/3}$ cm ~ 1 Å. A *precise* measurement of the interatomic spacing in a crystal leads, as we have said before, to a value for Avogadro's number.

The question is now whether we can find an explanation for the size of an atom within the framework of classical physics: whether we can compute the "radius" of an atom given the basic constants of classical physics.

48 After Rutherford's famous analysis† around 1910 of the alpha-particle scattering experiments of H. Geiger and E. Marsden, a certain picture of the atom had emerged, according to which the atom consists of a very small central nucleus surrounded by one or more electrons. There were good reasons to believe that both the nucleus and the electrons are very small compared with the size of the atom, smaller than, say, 10^{-11} cm at least. Furthermore, most of the mass of the atom seemed to be concentrated in the nucleus.

Under these circumstances it was very natural to try to create some kind of solar-system model of the atom, with the nucleus playing the role of the sun, and the electrons playing the role of the planets. The particles would move under their mutual electrostatic interactions, and most of the atom would consist of "empty space." The size of an atom would be related to the radius of the orbit of the outermost electron.

Let us accept this model temporarily for the sake of argument and let us also assume, at first, that the velocities of the particles will be small enough to permit a discussion within nonrelativistic mechanics. We must then answer the question: What determines the

Ernest Rutherford.   Born 1871 near Nelson, New Zealand; died 1937.   After holding a professorship at McGill University in Montreal, Canada, Rutherford accepted a position at Manchester University in 1907. In 1919 he succeeded J. J. Thomson in the Cavendish chair at Cambridge University.   He received the Nobel prize in 1908 (in chemistry).

Rutherford did pioneering work of outstanding importance in radioactivity and nuclear physics. His experimental work is characterized by exceptional skill and ingenuity, and his analysis of experimental facts reveals a deep physical insight. *(Photograph by courtesy of Professor L. B. Loeb, Berkeley.)*

---

†E. Rutherford, "The Scattering of α and β Particles by Matter and the Structure of the Atom," *Philosophical Magazine* **21**, 669 (1911). See also Berkeley Physics Course, Vol. I, *Mechanics*, Chap. 15, and PSSC, *Physics*, Chap. 32.

size of the orbit of the outermost electron? We note that in this model there is no place for the velocity of light. But then we cannot form a quantity having the physical dimension length from our remaining fundamental classical constants e, m, and M_H, and we may suspect that our problem is insoluble within classical mechanics. To see this very clearly we may argue as follows:

49 Consider an atom, consisting of Z electrons, each one of charge $-e$, and a nucleus of charge $+Ze$. Without loss of generality we can assume that these particles move in such a way that the center of mass of the system remains at rest. Each particle then moves along some orbit specified by the function $r_k(t)$ which gives the position vector of the kth particle as a function of the time t. (We select the center of mass of the system as the origin.)

The functions $r_k(t)$, $(k = 1, 2, \ldots, Z + 1)$, taken together thus constitute *one* solution of the equations of motion for the system. From this one solution we can construct a whole family of *new* solutions by a simple *scaling* as follows. If q is any non-zero constant, then the functions $r_k'(t)$ defined by

$$r_k'(t) = q^2 r_k(t/q^3) \tag{49a}$$

also satisfy the equations of motion. In other words: the function $r_k'(t)$ describes the trajectory of the kth particle in the new state of motion of the system. We can see this very easily as follows. The force $\mathbf{F}_{ij}$ which the jth particle exerts on the ith particle is given by

$$\mathbf{F}_{ij} = Q_i Q_j \frac{(\mathbf{r}_i - \mathbf{r}_j)}{|\mathbf{r}_i - \mathbf{r}_j|^3} \tag{49b}$$

where Q_i is the charge of the ith particle and Q_j is the charge of the jth particle. The new solution is obtained from the old solution by multiplying all distances by the factor q^2, which means that the forces in the new state of motion are obtained from the forces in the old state of motion by multiplication by the factor q^{-4}. This means that all accelerations must be scaled by the same factor q^{-4}. Since linear distances are scaled by the factor q^2 we conclude that all velocities must be scaled by the factor q^{-1}, and that all times must be scaled by the factor q^3. This is just what the equation (49a) expresses, and this equation thus defines a new solution, as asserted.

Let us furthermore note that all angular momenta are scaled by the factor q, and that all potential and kinetic energies, and hence also the total energy, are scaled by the factor q^{-2}.

The fact that we can obtain a new solution from a given solution through a scaling of the kind described is really an extension of Kepler's third law. Applied to the particular case of a single electron moving around a fixed nucleus, our argument tells us that for two elliptical orbits of the same eccentricity the ratio of the squares of the periods is proportional to the ratio of the cubes of the semi-major axes.

Since we can give any value we please to q we actually have a whole family of solutions, and there is no reason why we should prefer any particular one of these; there is, in other words, no principle which tells us why a particular "size" for the atom should be preferred. One might, of course, argue that the actual size of an atom is determined by "accident" but such an argument would hardly stand up. How is it possible that this "accident" always leads to the same size of orbit for a given species of atoms? Why don't we have a continuous distribution in size for, say, the hydrogen atom?

50 In view of this dilemma we may wonder whether we were justified in trying to discuss the problem non-relativistically. We note that it is indeed possible to form an expression having the dimension of length from the classical constants, if we include the velocity of light, namely

$$\frac{e^2}{mc^2} = 2.8 \times 10^{-13} \text{ cm} \qquad (50a)$$

This is essentially the "classical radius of the electron," which we discussed in Sec. 18. We would thus expect that if relativity really plays an essential role, i.e., if the electrons move with a velocity comparable with the velocity of light, then the size of an atom would be some multiple of the length e^2/mc^2. This length is, however, more than 10^4 times too small, and it does not seem likely that this line of approach will lead anywhere. It is true that our simple scaling argument in the preceding section does not apply as such in a relativistic model, but there is still no principle which would tell us why only certain orbits can occur, corresponding to the observed sizes of atoms.

51 We may regard our dilemma as a "mystery of the missing constant." Suppose now that we are bold and assume that the present mystery is related to our earlier "missing-constant mysteries," and that Planck's constant should play a role in the description of atomic structure. This constant has the dimension of angular momentum, and we might try the ad hoc assumption that only those solutions of the equations of motion for which the total angular momentum of the atom is a definite multiple of h are realized in nature. If we accept this principle we must give up our scaling argument since under the transformations described by Eq. (49a) the angular momentum will be scaled by the factor q, which now is not allowed. This means that there will be preferred solutions, and hence we now have a principle available for the determination of the size of an atom.

In 1913 Niels Bohr presented a theory of the hydrogen atom along these lines.[†] In the simplest version of this theory a single electron moves in a circular orbit of radius a_0 around a proton. The orbit is determined by the equation of motion

$$m\left(\frac{v^2}{a_0}\right) = \frac{e^2}{a_0{}^2} \tag{51a}$$

together with the *quantum conditions of Bohr;*

$$J = mva_0 = \frac{h}{2\pi} \tag{51b}$$

where v is the velocity of the electron and J is the angular momentum. The quantum condition thus states that the angular momentum equals $h/2\pi$. If we eliminate v from the above equations we obtain

$$a_0 = \frac{h^2}{(2\pi)^2 me^2} = 0.53 \times 10^{-8} \text{ cm} \tag{51c}$$

which is of the desired order of magnitude. Furthermore, we should note that the problem of atomic size is intimately related to the problem of atomic binding energies; once we know the size of an atom we can also estimate the work required to break it up into its elementary constituents.

52 As the reader undoubtedly knows, Bohr was able to go much further; he was actually able to explain the spectrum of the hydrogen atom quantitatively, which was a spectacular success for the new ideas. His quantum condition was certainly foreign to classical physics, and in addition Bohr had to assume that the motion of the electron in the ground state of the hydrogen atom does not lead to emission of electromagnetic radiation; otherwise, according to classical electromagnetic theory, the electron would spiral in toward the nucleus in a very short time (of the order of 10^{-9} sec).

[†] N. Bohr, "On the Constitution of Atoms and Molecules," *Philosophical Magazine* **26**, 1 (1913).

Niels Henrick David Bohr. Born 1885 in Copenhagen, Denmark; died 1962. After studies at the University of Copenhagen Bohr went to Cambridge, and a few months later to Manchester to work with Rutherford. In 1913 he published his celebrated paper on atomic structure. In 1916 Bohr became professor of theoretical physics at the University of Copenhagen. His Institute for Theoretical Physics (established in 1921) became a world center at which most of the world's outstanding physicists spent some time as visitors. Bohr was awarded the Nobel prize in 1922.

After his pioneering work on the hydrogen atom Bohr made many important and outstanding contributions to the development of atomic physics, and later also to nuclear physics. Through his publications, and through his personal contacts with other physicists, his influence as a champion of new ideas has been very considerable. (*Photograph by courtesy of Professor L. B. Loeb, Berkeley.*)

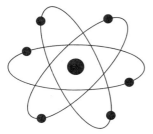

Fig. 52A A symbol of the Atomic Age, which has absolutely nothing whatever to do with the structure of atoms. Figures of this general appearance occur widely as emblems of companies, government agencies and other organizations which have something to do with "Atoms." Very fanciful versions can sometimes be seen in advertisements, in which the tremendous speed of the electrons is indicated by what look like vapor trails (presumably vapor trails in the ether?).

No harm is done so long as it is understood that this is only a symbol, but there is always the danger that some people might be misled into believing that atoms really look this way.

This planetary theory of atoms should not be taken seriously; it is really quite wrong. That it works so well in the particular case of the hydrogen atom is a fortunate (or unfortunate) accident. Fortunate, because it encouraged Bohr and others to try to create a quantum theory of atoms. Unfortunate, if it leads anybody to believe that atoms in any way resemble planetary systems. Bohr himself was not fooled: he regarded his own theory merely as an intermediate step in the search for a more coherent theory, which now exists.

53 The three problems which we have considered may be regarded as three aspects of the discovery of Planck's constant. If we think in particular about the last problem we see that the addition of this constant to our list of fundamental constants of nature will have far-reaching consequences. We can now hope to be able to understand not only the sizes and binding energies of atoms, but also of molecules, and the road seems open to a *quantitative* atomic theory of matter in bulk.

It should be stressed that the *essential* aspect of the three problems is, of course, that the resolution of the difficulties required a departure from the classical macroscopic laws of physics. *The consideration of these problems therefore led to something much more than the discovery of a new constant; namely, to the discovery of new laws of physics.*

The development of physics was very rapid after these initial discoveries, and it became clear that the key had been found to the explanation of many phenomena in microphysics. The theoretical work culminated with the publication of two consistent mathematical theories of quantum physics; namely the *matrix mechanics* created by Werner Heisenberg in 1925 and the *wave mechanics* created by Erwin Schrödinger in 1926. These two theories were subsequently shown to be completely equivalent, and merely two different forms of what we today call *quantum mechanics*, which is the currently accepted fundamental theory on which the study of microphysics is based.

54 The reader may want to ask some questions. Are we sure that quantum mechanics is the ultimate truth? What is there left to discover in physics?

The author is happy to be able to reassure the reader with regard to these questions. We can never know whether any theory is "the ultimate truth." Nor can we ever know "what there is left to discover"; most likely very much, because, as we have said before, we most certainly do not now have a comprehensive theory of all phenomena occurring in nature. We have learned a lot, but there is more to be learned. *That* is one of the reasons why physics is interesting. The reader need have no fear that he was born too late to make discoveries in physics.

Let us try to answer these questions a bit more precisely. The *general* principles of quantum mechanics are "true" in the sense that there is no experimental evidence against them, but an enormous amount of evidence that they can be used for successful predictions.

The favorable evidence is particularly convincing in the field of quantum electrodynamics, which is the basic theory of atoms, molecules, electromagnetic radiation, and matter in bulk such as we know it on our earth. As we have said before, there never was any *fundamental* classical theory of physics in this realm; now we have a theory, and a highly successful one. This means that we believe that we now know the basic facts in terms of which we should be able to explain, say, such phenomena as superconductivity and superfluidity. To date, however, nobody has *really* been able to explain these two phenomena in a quantitative manner from first principles. Knowing the basic principles is one thing, but explaining a complex phenomenon involving many particles is quite another thing. Because we can explain the behavior of simple systems, consisting of a comparatively small number of particles (such as individual atoms and simple molecules), we believe in our basic principles. We are, however, quite unsophisticated mathematically, and as the complexity of the situation increases we find it harder and harder to make quantitative predictions, although we may have a general qualitative understanding of the phenomena. It is safe to predict that there will always be problems in physics which are hard in this sense, and that there will always be a place for clever ideas to overcome such difficulties. It is possible that quantum electrodynamics is an almost "closed" subject from a fundamental point of view, but it is certainly not closed in the sense that all possible consequences have already been extracted from the theory.

55 From the standpoint of physics around the turn of the century the "stable and indivisible" atoms were the elementary particles of the world. Today the atoms have lost this distinguished position; they have been explained in terms of more elementary things through the theory of quantum electrodynamics. The same is, in a sense, the case with the nuclei as well. In this latter case we cannot explain the properties of nuclei completely from first principles, but we nevertheless firmly believe that it is proper to regard the nuclei as composite systems essentially made of protons and neutrons.

After the composite nature of atoms and nuclei became clear the number of recognized elementary particles took a sharp dip. This number has subsequently steadily increased, and it is now about the same as it was around the turn of the century. Electrons, muons, neutrinos, protons, neutrons, hyperons, pions, kaons and many other particles have taken the place of the atoms. We have already explained in what sense these particles are elementary.

At present there does not exist a basic theory of the elementary particles. What form the possible future theories may assume is anybody's guess, and the field is wide open for brilliant ideas.

References for Further Study

1) It will be assumed in this book that the reader has *some* familiarity with the most basic facts of quantum physics, at the level at which these topics are discussed in a high school physics text such as *Physics,* by the Physical Science Study Committee (D. C. Heath and Company, Boston, 1960). (Part IV, in particular.)

In the cases in which the above assumption does not correspond to the facts some supplementary reading is called for. Any library will have a variety of semi-popular accounts of "atomic physics," some of which are bad and some of which are good. Taken with a grain of salt such a book might serve the purpose. Articles in magazines such as *Scientific American* can be very useful, and are *strongly* recommended. The reading of such articles will probably whet the reader's appetite and lead to further self-study and reading. *Whenever the reader's background permits it he should try to read original papers,* but highly technical, or mathematically complicated accounts are best avoided at this stage.

2) The reader may be interested in reading selected portions of some text on quantum physics in which a more complete account of experiments is given than in this book. Among the many such texts we mention the following:

a) E. Grimsehl and R. Tomaschek: *A Textbook of Physics,* vol. V, *Physics of the Atom* (Blackie and Son Limited, London, 1945).

b) G. P. Harnwell and J. J. Livingood: *Experimental Atomic Physics* (McGraw-Hill Book Company, New York, 1933).

3) The following books are historical surveys of the development of modern physics:

a) M. Jammer: *The Conceptual Development of Quantum Mechanics* (McGraw-Hill Book Company, New York, 1966). A magnificent piece of work, which, however, requires a substantial knowledge of quantum mechanics for a full appreciation. The beginning of the book, which deals with the early history, can be read with a modest background. The many carefully compiled references to original papers is a valuable feature.

b) E. Whittaker: *A History of the Theories of Aether and Electricity,* vols. I and II (Harper Torchbooks, Harper and Brothers, New York 1960). The second volume discusses the development of quantum mechanics. These books (as well as Jammer's) discuss the interesting false leads as well: the theories once taken seriously but now forgotten.

4) a) An extremely interesting and penetrating analysis of the development of quantum mechanics as well as of relativity is given by Albert Einstein (in German with an English translation) in the form of an autobiography published in: *Albert Einstein, Philosopher-Scientist*, vol. I, edited by P. A. Schilpp (Harper Torchbooks, Harper and Brothers, New York, 1959).

b) Planck's own account of the development of his ideas is presented in: M. Planck: *A Survey of Physical Theory* (Dover Publications, New York, 1960).

5) In the text of this chapter references to several of the important original papers have been given. The reader is urged very strongly to at least look at these: several of the papers can be read without undue difficulty. Collections of these, and other papers, have been published subsequently. We mention two sources of this kind:

a) *Great Experiments in Physics*, edited by M. H. Shamos (Holt, Rinehart and Winston, New York, 1962). (Translated and abbreviated versions, with editorial comments.)

b) *The World of the Atom*, edited by H. A. Boorse and L. Motz, vols. I and II (Basic Books, Inc., New York, 1966). This is a very complete collection, with editorial comments giving historical background and biographical information. Selective reading of these books is strongly recommended.

6) Many of the experimental discoveries and theoretical ideas discussed in this volume have in due time been rewarded by the Nobel Prize. Every recipient of this prize is required to give a semipopular lecture in Stockholm about his work. Extracts from these lectures, with short descriptions of the prize-winning work, can be found in N. H. de V. Heathcote: *Nobel Prize Winners in Physics 1901–1950* (Henry Schuman, New York, 1953).

Problems

1 (*a*) First consider, and then describe very briefly, the kind of reasoning and the kinds of measurements that have led to definite assignments of atomic and molecular weights.

(*b*) In 1815 William Prout suggested that all elements might be combinations of hydrogen, which would thus be the primordial material from which everything else is made. What might have led him to such a hypothesis, and why was his suggestion rejected during the nineteenth century?

2 Many atoms (or rather *nuclei*) disintegrate spontaneously, usually through the emission of either an electron or an alpha particle, which is nothing else than a helium nucleus. This is the phenomenon of *radioactivity*, first discovered by Henri Becquerel in 1896. [H. Becquerel, "Sur les radiations invisibles émises par les corps phosphorescents," *Comptes Rendus* **122**, 501 (1896).] The rate of disintegration is governed by a statistical law, which predicts that out of N_i atoms originally present $N(t) = N_i \exp(-\lambda t)$ will have survived at the time t. The constant λ, which describes the rate of disintegration, is characteristic of the atom (nucleus). The time T it takes for half of the atoms originally present to decay is known as the half-life. We obviously have $T = (1/\lambda) \ln 2$.

(*a*) Show that the above law of disintegration results if we assume that each atom disintegrates independently of the other atoms, and also assume that the probability that an atom, which has survived until the time *t*, will disintegrate during the time interval (*t, t* + Δ*t*) is independent of *t*.

(*b*) In the decay of radium atoms an alpha particle is emitted. If this alpha particle strikes a zinc sulfide screen, a flash of light (called a *scintillation*) will mark the point of impact. It is thus possible to count directly the number of alpha particles emitted per second from 1 gram of radium, and this number was determined by Hess and Lawson as 3.72×10^{10}. The atomic weight of radium is 226. Use these data to find the half-life of radium. (Measurements with radioactive substances have been used to give independent estimates of Avogadro's number. In the above problem the procedure is reversed, and we instead determine the half-life of radium.)

3 The moving parts of a wrist watch are pretty "small." Making reasonable estimates of the magnitudes of the physical parameters which characterize a "typical" wrist watch, show, however, on the basis of the general criterion given in Sec. 20, that quantum mechanics is totally irrelevant to the art of watchmaking.

4 In the spirit of the preceding problem, consider a simple electrical lumped-constant circuit, consisting of a capacitor of capacity 100 pF (100 $\mu\mu$F) and an inductance of 0.1 mH. Suppose that the circuit oscillates so that the maximum voltage across the capacitor is 1 mV. Try to find a "natural" physical quantity with the physical dimension action, and compute this quantity in units of Planck's constant *h*.

5 A broadcasting antenna emits radiation (radio waves) at the frequency 1 Mc/sec at the rate of 1 kW. What is the corresponding number of photons emitted per second? The size of this number explains why the quantum nature of electromagnetic radiation is not immediately apparent when we study radiation from antennas.

This example, and those in Probs. 3 and 4, are ridiculous in the sense that the relevant numbers are ridiculous. We shall not try to apply quantum mechanics to obviously macroscopic problems in the rest of this book. It may, however, be edifying to have done problems of this nature *once*—if only to learn the lesson that they *are* ridiculous.

6 To see that the statement that electromagnetic radiation comes in packets of energy $E = h\nu$, where ν is the frequency, does not violate common sense (that is, does not violate your experience with macroscopic phenomena), compute the number of photons emitted per second by a light source of strength 1 *candela*. We assume for simplicity that the light emitted is of yellow color, the wavelength being 5600 Å (Ångström units). A source of strength 1 candela emits light energy at the rate 0.01 watt.

Suppose that an observer looks at an isotropic light source of strength 1 candela at a distance of 100 meters. Compute the number of photons entering one of his eyes per second; assume that the entrance pupil of the eye has the diameter 4 mm. Because the number of photons is as large as it is, we do not observe any "flickering" even though the luminous flux received by the eye is small by macroscopic standards.

7 We all know that stars "twinkle." To see whether this may be a manifestation of the quantum nature of light, estimate the number of photons entering the eye of an observer when he looks at a star of first apparent visual magnitude. Such a star produces a flux at the surface of the earth of about 10^{-6} lumen/meter^2. One lumen at the wavelength of maximum visibility, which is about 5560 Å, corresponds to 0.0016 watt. A star of the first apparent visual magnitude is a fairly bright star, easily visible to the naked eye, although not among the very brightest stars. The star Aldebaran is an example.

Suppose that N photons per second enter the eye of the observer. How large is the mean fluctuation in this number? After you have determined N, decide what is the likely explanation of the twinkling. Why do planets appear to twinkle much less, or not at all?

8 (a) Consider Wien's displacement law, and suppose that we have a black-body radiator kept at a temperature of 2500°K. Compute the wavelength in Ångström units, as given by Wien's law, for which the emission is a maximum. Does this wavelength lie in the visible region?

(b) Derive Wien's displacement law from Planck's formula (39a).

(c) On the basis of Planck's radiation law (39a), show that the total rate at which radiation is emitted by a black-body radiator (i.e., all frequencies included) is proportional to the fourth power of the temperature T.

9 In our historical discussion of the black-body-radiation law we mentioned that Planck, in his derivation, made the assumption that a harmonic oscillator, of frequency ν, can take up energy only in packets of magnitude $h\nu$. (The reader should note that we did not attempt explanations in our historical survey, and at this time the reader is therefore under no obligation to understand how Planck arrived at his final result.) It is of interest to see what connection there may be between Planck's assumption, and the assumption which Bohr made in his derivation of the characteristics of the hydrogen atom. Let us therefore consider the following: A harmonic oscillator, of mass m and spring constant K, behaves in accordance with Planck's assumption. This means that the energy of the oscillator can change only by integral multiples of $h\nu$, where ν is the frequency of the oscillator. Let us introduce the action variable $J = \pi\, q_0 p_0$, where q_0 is the maximum displacement of the oscillating mass point, and where p_0 is the maximum momentum.

Chapter 2

Magnitudes of Physical Quantities

in Quantum Physics

Chapter 2   Magnitudes of Physical Quantities
in Quantum Physics

Units and Physical Constants

1 One of the aims of this chapter is to give the reader a feeling for the orders of magnitude of various physical quantities in the realm of quantum physics. Many of the important physical quantities, such as the electronic charge, the electronic mass, Planck's constant, etc., have numerical magnitudes when expressed in our familiar macroscopic units which are inconvenient and untransparent because they are so very small. It is difficult to grasp directly what it *means* that Planck's constant has the value $h = 6.6 \times 10^{-27}$ erg sec. It is therefore important that we study in detail how these various constants occur in physics, and what their numerical values actually mean.†

Every domain of physics has what we may call natural units for the physical quantities involved, which means that when we express any physical quantity in terms of these natural units then the numerical values are reasonable in the sense that their significance can be readily grasped. The numerical values may vary over the range 10^{-6} to 10^6, but we will not encounter numbers such as 10^{-27}. Our familiar macroscopic units (in the MKS system) are particularly suited for our everyday experiences with physical phenomena, and they are based on readily available macroscopic standards. We note that they are really "human units"; units such as the meter and the kilogram, and the second unmistakably refer to human characteristics. The so-called "scientific," or cgs system of units is more appropriate for small animals, such as cockroaches. We shall try to free our discussion from the arbitrary standards of the human system, or the cockroach system, and try to identify the natural units in the various domains of quantum physics.

2 We begin by listing some physical constants in a table. These constants are commonly referred to as "the fundamental constants of microphysics," but there is really nothing fundamental at all about the individual numbers in Table 2A because the macroscopic

---

† In so doing we will sometimes anticipate the later, more detailed, discussion. In case the reader encounters some passages which now seem mystifying he need not be overly concerned; he should return to the present chapter several times as we go along in our course. Hopefully most readers will already have *some* familiarity with the topics of our discussion.

standards are arbitrary and "accidental." This does not mean that the list is unimportant. Once we have selected our macroscopic standards we naturally want to relate the basic parameters of quantum physics to these, and that is the purpose of the list.

We have also quoted the estimated errors for the constants to give the reader a feeling for how accurately they are known at present. At the level of this book the reader will almost never have occasion to carry out any computations to greater accuracy than can be obtained with the slide rule, which accuracy is about 0.2 per cent per multiplication or division. The reader should also learn to make simple estimates which may vary in accuracy from 10 per cent to merely an estimate of an order of magnitude. On the inside of the front cover of this book the reader will find a table of very rough values of the most important constants, and this table should properly be committed to memory. Other, more detailed tables of physical data can be found in the Appendix.

3 The definition of Avogadro's number requires discussion. When chemists tabulated atomic weights in the past they employed a scale in which naturally occurring oxygen was assigned *by definition* the atomic weight 16 *exactly*. The atomic weight of hydrogen, for instance, was thus defined as

(atomic weight of hydrogen)

$$= 16 \times \frac{\text{(mass of hydrogen "atom")}}{\text{(mass of oxygen "atom")}} \quad (3a)$$

The word "atom" is embellished with quotation marks because the weight of the "atom" always refers to the element as it occurs in nature. The atomic weights, as defined in Eq. (3a), are determined by the chemist through careful *weighing operations;* he will, for instance, determine the amount, in grams, of naturally occurring hydrogen which will combine with 16 grams of naturally occurring oxygen to form water with nothing left over. The resulting number divided by two is the atomic weight of hydrogen.

The atomic weights determined by the chemists in this manner are called *atomic weights on the chemical scale.* Many of the elements have atomic weights which are close to integers, but there are also notable exceptions; the atomic weight of chlorine, for instance, is 35.5.

4 As the reader knows, the mass of an atom is mainly concentrated in the nucleus. Nuclei are built of protons and neutrons. The number of protons plus the number of neutrons is known as the *mass number* of the nucleus. This integer is commonly denoted

TABLE 2A   *Some Physical Constants*

Planck's constant:
$$h = 2\pi \hbar = (6.62559 \pm 0.00015) \times 10^{-27} \text{ erg sec}$$
$$\hbar = h/2\pi = (1.05449 \pm 0.00003) \times 10^{-27} \text{ erg sec}$$

Velocity of light:
$$c = (2.997925 \pm 0.000001) \times 10^{10} \text{ cm sec}^{-1}$$

Electronic charge:
$$e = (4.80298 \pm 0.00006) \times 10^{-10} \text{ esu}$$
$$= (1.60210 \pm 0.00002) \times 10^{-19} \text{ coul}$$

Electron mass:
$$m = (9.10908 \pm 0.00013) \times 10^{-28} \text{ gm}$$

Proton mass:
$$M_p = (1.67252 \pm 0.00003) \times 10^{-24} \text{ gm}$$

Avogadro's number:
$$N_0 = (6.02252 \pm 0.00009) \times 10^{23} \text{ (mol)}^{-1}$$

Boltzmann's constant:
$$k = (1.38054 \pm 0.00006) \times 10^{-16} \text{ erg } (^\circ\text{K})^{-1}$$

TABLE 3A   *Atomic Weights (Atomic Masses) of the Lightest Elements* †

Element	Z	Atomic weight
H	1	1.00797
He	2	4.0026
Li	3	6.939
Be	4	9.0122
B	5	10.811
C	6	12.01115
N	7	14.0067
O	8	15.9994
F	9	18.9984
Ne	10	20.183
Na	11	22.9898
Mg	12	24.312
Al	13	26.9815
Si	14	28.086
P	15	30.9738
S	16	32.064
Cl	17	35.453
A	18	39.948

† See Table C in Appendix for a complete listing.

TABLE 4A Naturally Occurring Isotopes of Selected Light Elements

Ele-ment	Z	Iso-tope A	Atomic mass	Natural abundance percent
H	1	1	1.007825	99.985
		2	2.01410	0.015
He	2	3	3.01603	0.00013
		4	4.00260	100
Li	3	6	6.01513	7.42
		7	7.01601	92.58
Be	4	9	9.01219	100
B	5	10	10.01294	19.6
		11	11.00931	80.4
C	6	12	12.000000	98.89
		13	13.00335	1.11
N	7	14	14.00307	99.63
		15	15.00011	0.37
O	8	16	15.99491	99.759
		17	16.99914	0.037
		18	17.99916	0.204
F	9	19	18.99840	100
...	...	...	...	...
S	16	32	31.97207	95.0
		33	32.97146	0.76
		34	33.96786	4.22
		36	35.96709	0.014
Cl	17	35	34.96885	75.53
		37	36.96590	24.47
...	...	...	...	...

by A. The number of protons is called the *atomic number* of the nucleus. It is denoted by Z, and the nuclear charge is thus eZ, where e is the elementary charge. The chemical properties of an atom are determined almost exclusively by the nuclear charge, and Z is thus a characteristic of the *chemical element*. It is found that there are many instances of families of nuclei with the *same* charge, but with different mass numbers, and these different nuclei are referred to as different *isotopes* of the element. Isotopes differ in the number of neutrons. The mass of the proton is almost equal to the mass of the neutron, and the masses of all nuclei are very closely proportional to the integral mass number A. The explanation for the occurrence of markedly non-integral atomic weights is that many naturally occurring chemical elements are mixtures of two or more different isotopes, in which case the "atomic weight" of the element as measured by the chemists is an average of the more fundamental atomic weights of the different isotopes.† It is an experimental fact that the relative abundances of the different isotopes occurring in an isotopic mixture of an element are very nearly the same all over the surface of the earth. Furthermore the different isotopes have, for all practical purposes, identical chemical properties and are therefore almost impossible to separate from each other by "chemical" means. If this were not the case, the chemists' tables of atomic weights would be worthless.

5 When the chemist writes an equation for a chemical reaction he employs symbols such as H (hydrogen), Li (lithium), Fe (iron), etc., to designate the naturally occurring chemical element which may or may not be a mixture of isotopes. From the standpoint of the nuclear physicist, however, the oxygen isotopes of mass number 16 and mass number 18 are entirely different objects, and he has to distinguish between these objects when he writes *his* formula for a nuclear reaction. This is done by subscripts and superscripts, and an isotope is commonly denoted by

$$_Z(\text{chemical symbol})^A \quad \text{or} \quad (\text{chemical symbol})^A$$

Naturally, occurring oxygen is a mixture of three stable isotopes, namely O^{16}, O^{17}, and O^{18}, of which the isotope O^{16}, which occurs with relative abundance 99.759 per cent, is the dominant component.

---

† That a chemical element can consist of different isotopes was firmly established by J. J. Thomson. [J. J. Thomson, "Rays of Positive Electricity," *Proceedings of the Royal Society* (London, Series A) **89**, 1 (1913).]

6 Physicists and chemists have recently agreed upon a new standard for atomic weights based on the mass of the carbon isotope C^{12}. The *atom* (not the nucleus) of this carbon isotope is thus assigned a mass of precisely 12 *atomic mass units*, abbreviated 12 amu. This convention, to which we shall adhere, gives rise to the *new scale* of atomic masses. Hence:

$$1 \text{ amu} = \tfrac{1}{12} \text{ (mass of one } C^{12} \text{ atom)}$$
$$= (1.66043 \pm 0.00002) \times 10^{-24} \text{ gm} \qquad (6a)$$

Avogadro's number N_0 is defined as the number of atoms in 12 grams of isotopically pure C^{12}, and this is the number quoted in Table 2A.

On the new scale the atomic weight of naturally occurring oxygen is 15.9994. This is very close to the number 16, which is the atomic weight of oxygen on the old chemical scale. For most practical purposes the difference between the new scale and the old chemical scale can therefore be ignored.

7 Avogadro's number N_0 is the link which connects microphysics with macrophysics. Let us mention some important quantities involving N_0 which express this connection.

(i) The mass of the proton is 1.0073 amu, and the mass of the neutral hydrogen atom (the isotope H^1) is 1.0078 amu. The product of Avogadro's number N_0 and the proton mass M_p is therefore

$$N_0 M_p = 1.0073 \text{ gm} \qquad (7a)$$

which is quite close to 1 gm. In rough computations we thus have

$$\text{(mass of proton)} \cong \text{(mass of hydrogen atom)} \cong \frac{1}{N_0} \text{ gm} \qquad (7b)$$

(ii) The product of N_0 and Boltzmann's constant k gives us the *universal gas constant R*,

$$N_0 k = R = 8.314 \times 10^7 \text{ erg } (°K)^{-1} \text{ (mol)}^{-1}$$
$$= 1.986 \text{ cal } (°K)^{-1} \text{ (mol)}^{-1} \qquad (7c)$$

since Boltzmann's constant is the gas constant per *molecule*.

(iii) The product of N_0 and the electronic charge e gives us the *Faraday constant F*,

$$N_0 e = F = 96,487 \text{ coulomb (mol)}^{-1} \qquad (7d)$$

This constant expresses the total charge carried by one mole of singly charged ions.

8 Let us next discuss Planck's constant, which occurs in two versions, denoted by h and $\hbar$, as shown in Table 2A. (The symbol $\hbar$ is read: "h-bar.") Both constants are called "Planck's constant," and both are commonly employed, although one may loosely say that $\hbar$ is favored; it is a "better" constant. The reason why both constants are employed is that it is easier to write an $\hbar$ with a bar than to write out factors 2π which would otherwise appear in many of our formulas; i.e., the same reason why "frequencies" occur in two versions.

In this book we denote a frequency, which is the number of repetitions of a periodic phenomenon per unit time interval, or the number of *cycles/unit time*, by the letter ν. We denote *angular velocities* by the letter ω, and we measure angular velocities in radians/unit time, or simply in 1/unit time. With every frequency ν we associate an angular velocity ω according to

$$\omega = 2\pi\nu \tag{8a}$$

It follows that

$$\hbar\omega = h\nu \tag{8b}$$

and both expressions thus give the energy of a photon of frequency ν. It should be noted that the quantity ω is also commonly called "frequency," or *angular frequency*, the relation (8a) being understood.

A corresponding notation is also employed for wavelengths. The true *wavelength*, which is the period of a periodic phenomenon in a spatial dimension, is denoted by the letter λ. With every wavelength λ we associate a quantity $\lambdabar$ by

$$\lambdabar = \frac{\lambda}{2\pi} \tag{8c}$$

For a monochromatic wave propagating with the phase velocity c we have

$$\lambda\nu = \lambdabar\omega = c \tag{8d}$$

The reader should carefully learn these generally accepted conventions.

9 The wavelength of a wave is often expressed in terms of its inverse, $\tilde{\nu} = 1/\lambda$, called the *wave number*. This mode of expression is employed extensively in optical spectroscopy, and the wave number is expressed in units of cm^{-1}. For a light wave in vacuum we have

$$\tilde{\nu} = \frac{1}{\lambda} = \frac{\nu}{c} \qquad (9a)$$

where ν is the frequency.   The wave number is proportional to the frequency, but should not be confused with the frequency.   It should be mentioned that in the optical region the wavelengths and wave numbers can be measured very accurately; much more accurately than the velocity of light has been measured.   Therefore, in the optical region the wave numbers are better known than the corresponding frequencies.   On the other hand the frequency can be measured very accurately in the microwave region, and in this region the frequencies are much better known than the corresponding wave numbers or wavelengths.

10   In Chap. 1 we mentioned some methods whereby the fundamental constants can be measured: historically these were the first methods.   The best numerical values for the basic constants are, however, not obtained nowadays through such conceptually transparent and simple measurements, and we mentioned the direct methods only to make it immediately clear that these constants are not beyond our reach.   The best values come from a number of measurements of *derived* quantities, i.e., expressions involving these constants (and other constants) in various combinations which we believe we understand well theoretically.   From the derived quantities we can compute the basic constants.   Since the number of all the derived quantities measured is actually larger than the number of basic constants, the equations are overdetermined, and this circumstance provides us with valuable checks on the internal consistency of all the measured quantities which have been taken into consideration in our determination of the constants.

Energy

11   Let us now consider the units which are used to describe energy in microphysics.   One of the most useful energy units is the *electron volt*, abbreviated eV.   It is defined as the energy acquired by an elementary charge of magnitude e when it passes through a potential drop of one volt.   Taking into account the magnitude of e as given in Table 2A we can express the electron volt in ergs:

$$1 \text{ eV} = (1.60210 \pm 0.00002) \times 10^{-12} \text{ erg} \qquad (11a)$$

In addition to the electron volt we also employ the derived units

$$1 \text{ keV} = 1000 \text{ eV}, \qquad 1 \text{ MeV} = 10^6 \text{ eV},$$
$$1 \text{ BeV} = 10^3 \text{ MeV} = 10^9 \text{ eV} \tag{11b}$$

where keV is an abbreviation for *kilo-electron volt,* MeV is an abbreviation for *million electron volts* and BeV is an abbreviation for *billion electron volts.*† The unit electron volt is particularly convenient in atomic physics since atomic binding energies are of the order of an eV, whereas the unit MeV is useful in nuclear physics since nuclear binding energies are of the order of an MeV. The unit BeV is used in discussions of very-high-energy interactions of the fundamental particles.

12 In Chap. 1 we discussed the fundamental role played by the constants c and h. These constants are so fundamental in relativistic quantum physics that one frequently employs a system of units in this field in which $\hbar = 1$ and $c = 1$; the constants $\hbar$ and c are dimensionless and equal to one. The reader may feel that such a definition violates our notion of physical dimension. However, it must be understood that our assignment of physical dimensions to various physical quantities is arbitrary, and purely a matter of convention. Strictly speaking only *directly comparable physical quantities* have the same "physical dimension"; i.e., quantities which can be directly measured off against each other. All other assignments of dimension are based on some relations between physical quantities which we believe are particularly fundamental. Because of the fundamental nature of the velocity of light we may certainly relate distance, x, and time, t, through the relation $x = ct$ if we like, and hence measure distance and time in the same units. This is in effect what the astronomers do when they measure distances in light years.

To set $\hbar = c = 1$ leads to clear and simple and esthetically appealing formulas, and we will sometimes take advantage of this possibility. The author was strongly tempted to set $\hbar = c = 1$ consistently throughout the book; it would really be the right thing to do. To take this step might, on the other hand, cause the reader needless difficulties in reading other introductory books on quantum physics, since almost all such books are based on the conventional MKS or cgs systems. For this reason we will mostly adhere to the conventions of the cgs system of units.

13 Let us now explore some relations between various physical

---

† Outside the United States this unit is also written GeV.

quantities which arise because of the existence of the distinguished constants c and $\hbar$. We consider a mass, m, and associate with this mass a number of other physical quantities constructed from m, $\hbar$, and c, and we give the conventional physical dimensions of these quantities.

$$m = [\text{mass}] \qquad \frac{mc^2}{\hbar} = [\text{time}]^{-1}$$

$$mc = [\text{momentum}] \qquad \frac{\hbar}{mc^2} = [\text{time}] \qquad (13a)$$

$$mc^2 = [\text{energy}] \qquad \frac{\hbar}{mc} = [\text{length}]$$

The reader should check that the physical dimensions (with the conventions of the cgs system) are correctly stated. All these quantities "hang together" through the constants h and c. Based on the above relations an *energy* can be associated with a mass or a frequency or an inverse length, and the magnitude of the energy can be expressed in terms of the magnitudes of the associated quantities.

14   With an energy E we thus associate the frequency E/h, the wave number $E/(hc)$, and the mass E/c^2. We find the following conversion factors:

$$\frac{(\text{energy})}{(\text{mass})} = (9.31478 \pm 0.00005) \times 10^8 \, (\text{eV})/(\text{amu}) \qquad (14a)$$

$$\frac{(\text{frequency})}{(\text{energy})} = (2.41804 \pm 0.00002)$$
$$\times 10^{14} \, (\text{cycles/sec})/(\text{eV}) \qquad (14b)$$

$$\frac{(\text{wave number})}{(\text{energy})} = (8.06573 \pm 0.00008) \times 10^3 \, (\text{cm}^{-1})/(\text{eV}) \qquad (14c)$$

The table on the inside of the back cover of this book is partly based on the above conversion factors. Each horizontal row shows a set of corresponding quantities associated with the quantity in the first column. The second and the third columns give the energy E in eV and ergs. The seventh column gives the corresponding mass E/c^2, in atomic mass units, amu; the eighth column the corresponding frequency E/h, in cycles/sec; and the ninth column gives the wave number $E/(hc)$, in cm^{-1}. The tenth column gives the associated wavelength $(hc)/E$, in Ångström-units, and this is the only quantity in the table which is not directly proportional to E.

15 In chemistry the units *calorie,* abbreviated *cal,* and kilocalorie, abbreviated *Cal,* or *kcal,* are commonly used to express an energy. (The calorie is often called "small calorie," and the kilocalorie called "large calorie.") These units are defined by

$$1 \text{ kcal} = 1000 \text{ cal}, \quad 1 \text{ cal} = 4.186 \text{ joule} = 4.186 \times 10^7 \text{ erg} \quad (15a)$$

It is of interest to relate an energy E which refers to a single atom, or molecule, to the corresponding bulk energy E_{bulk}, of N_0 of these particles: i.e., the energy associated with one gram-atom, or one mole. We then have

$$\frac{E_{\text{bulk}}}{E} = N_0 = 23{,}050 \text{ (cal)/(eV)}$$
$$= 9.6487 \times 10^{11} \text{ (erg)/(eV)} \quad (15b)$$

In the table on the inside of the back cover the fourth and fifth columns give bulk energies in erg/mol and cal/mol.

16 In Secs. 31–34, Chap. 1, we briefly discussed the concepts of heat and temperature. We noted that Boltzmann's constant k is in effect a conversion factor from temperature to energy. It is in fact common practice to express a temperature in terms of the corresponding energy, and vice versa, where the correspondence is *defined, arbitrarily,* by

$$\text{(equivalent energy)} = k \times \text{(temperature)} \quad (16a)$$

For the purpose of such a conversion it is convenient to express Boltzmann's constant in the form

$$k = 8.617 \times 10^{-5} \text{ (eV)/}^{\circ}\text{K}, \qquad \frac{1}{k} = 11{,}605\,^{\circ}\text{K/(eV)} \quad (16b)$$

In this correspondence *"room temperature"* $(= 20^{\circ}\text{C} = 293^{\circ}\text{K})$ is equivalent to the energy

$$k \times 293^{\circ}\text{K} \simeq (1/40) \text{ eV} \quad (16c)$$

The sixth column in the table on the inside of the back cover gives the equivalent temperature, in Kelvin degrees.

17 That energy and temperature can be expressed in the same units should not lead anybody to believe that energy and temperature are the "same thing." It is, for instance, *not true* that the thermal energy of an arbitrary macroscopic body at a temperature T equals the number of atoms in the body times kT. The internal energy of a macroscopic body depends not only on the temperature, but also on other (macroscopic) parameters, and furthermore the precise relation between energy and temperature depends on the nature of the system. This is an important point, and the formula (16a) must not be misinterpreted.

However, we can make a highly useful statement. It is often the case (but not always), that if a macroscopic body is kept at a temperature T, then the average "disorganized" energy per atom (or molecule), of the body is of the *order* of kT.

This statement enables us to *estimate* the mean energy of an atom, or molecule, in disorganized thermal motion once we know the temperature. For many special systems we can make *precise* statements. An important example concerns a gas of molecules at the temperature T. The average kinetic energy E_{tr} associated with the *translational* motion of a molecule is given by

$$E_{tr} = \frac{3}{2} kT \tag{17a}$$

and this relation holds irrespective of whether the molecules are monatomic or not. The derivation of this formula is a problem of statistical mechanics, and we postpone it to the next volume. We shall make occasional use of this result, even though we have not derived it yet.

18   As we have said before, the concepts of heat and temperature are not relevant when we consider *isolated* nuclei, atoms or molecules: these concepts apply to matter in bulk. However, we can in general not carry out our measurements on isolated particles: we have to observe these particles embedded in macroscopic quantities of matter. The disorganized thermal motion is therefore often an important factor to take into consideration when we wish to understand the behavior of quantum-mechanical systems, and in particular when we study macroscopic manifestations of quantum phenomena.

The important feature of the thermal motion in a system is that it is, from our standpoint, a *random* motion. It introduces an apparent element of chance in the behavior of the system, as observed by us. We can say that the thermal random motion is "noise in the symphony of pure quantum mechanics." And we can add that often the noise is so loud that the music cannot be heard. In principle the thermal motion could be suppressed by keeping the system under study, and its surroundings, at a temperature in the immediate neighborhood of $0°K$, because the thermal motion ceases at absolute zero. In practice this cannot be done; the thermal motion is an essential feature of the world we live in.

The Energy Spectrum of Physical Phenomena

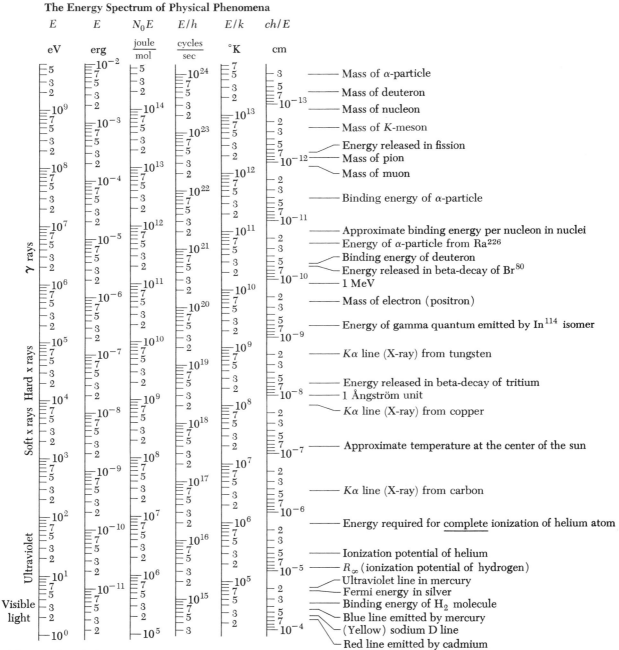

Characteristic energies of physical phenomena. The selected data on this page and the next are presented to give the reader a general idea of typical energies for various phenomena. The energies are expressed in several commonly used units: see Secs. 14–16 for an explanation.

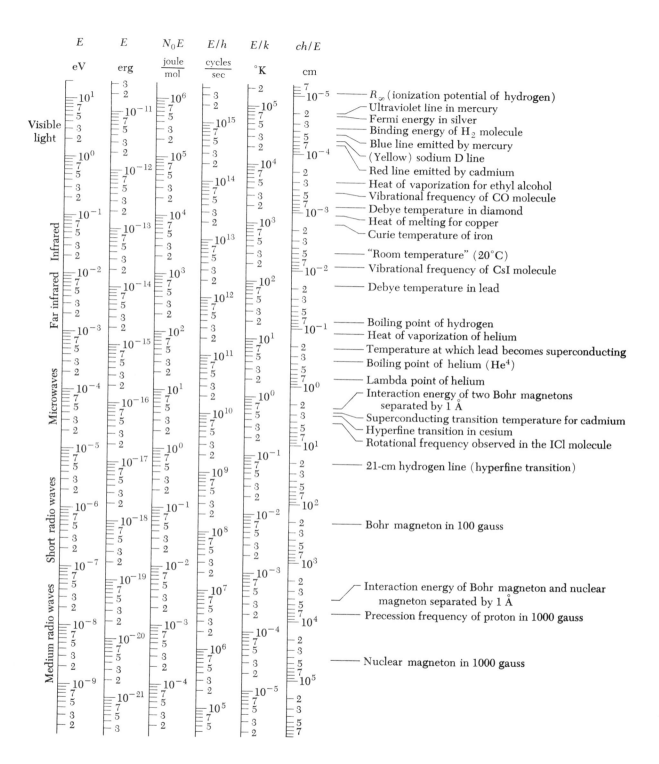

E	E	$N_0 E$	E/h	E/k	ch/E
eV	erg	$\dfrac{\text{joule}}{\text{mol}}$	$\dfrac{\text{cycles}}{\text{sec}}$	°K	cm

Visible light

Infrared

Far infrared

Microwaves

Short radio waves

Medium radio waves

R_∞ (ionization potential of hydrogen)
Ultraviolet line in mercury
Fermi energy in silver
Binding energy of H_2 molecule
Blue line emitted by mercury
(Yellow) sodium D line
Red line emitted by cadmium
Heat of vaporization for ethyl alcohol
Vibrational frequency of CO molecule
Debye temperature in diamond
Heat of melting for copper
Curie temperature of iron
"Room temperature" (20°C)
Vibrational frequency of CsI molecule
Debye temperature in lead
Boiling point of hydrogen
Heat of vaporization of helium
Temperature at which lead becomes superconducting
Boiling point of helium (He^4)
Lambda point of helium
Interaction energy of two Bohr magnetons separated by 1 Å
Superconducting transition temperature for cadmium
Hyperfine transition in cesium
Rotational frequency observed in the ICl molecule
21-cm hydrogen line (hyperfine transition)
Bohr magneton in 100 gauss
Interaction energy of Bohr magneton and nuclear magneton separated by 1 Å
Precession frequency of proton in 1000 gauss
Nuclear magneton in 1000 gauss

Magnitudes Characteristic of Atomic and Molecular Physics

19 Consider the atom as a dynamical system consisting of a nucleus, very small in size, surrounded by a cloud of electrons. The electrons are attracted by the nucleus, and they interact mutually, through electromagnetic forces. The belief that the electromagnetic forces are the only forces of importance in determining the structure of atoms and molecules is based on the comparison between theory and experiment which has been carried out to date.

The quantum theory of the interaction of charged particles with the electromagnetic field is known as *quantum electrodynamics*. This theory, which incorporates the Principle of Special Relativity, is presently our most successful theory of fundamental processes involving elementary particles. It is the theory within which we are to describe the structure of atoms and molecules and the emission and absorption of electromagnetic radiation by these objects.

20 Experimentally the order of magnitude of the size of a nucleus is about 10^{-13} cm, whereas the size of an atom is about 10^{-8} cm. The nucleus is thus very small compared to an atom.

The mass of a nucleus is large compared to the mass of an electron, which is 0.0005486 amu, the ratio of the electron mass to the proton mass being

$$\frac{m}{M_p} = \frac{1}{1836} \tag{20a}$$

It is therefore reasonable to expect that, at least in a first approximation, the *motion* of the nucleus does not play an essential role, and in this approximation we may regard the nucleus as being "infinitely" heavy, and hence fixed in space. Since the nucleus is furthermore very small we make the further approximation that it is a "point": it plays no other role than to provide an electrostatic field described by the potential

$$V(r) = \frac{eZ}{r} \tag{20b}$$

where e is the elementary charge and Z is the atomic number.

The problem of atomic theory, in the first approximation, is thus to study the motion of the electrons in this electrostatic field, taking into account the mutual electrostatic repulsions between the electrons as well. The reader should be reminded that when we speak of "motion" we mean motion in the sense of quantum mechanics. We will explain later precisely what that means.

21 Quantum electrodynamics in a restricted sense is concerned with the interactions of electrons with the electromagnetic field. Let us consider the relevant physical quantities in this theory, namely the mass of the electron, m; the charge of the electron, $-e$; the velocity of light c; and Planck's constant $\hbar$. From the constants m, c, and $\hbar$ we may form *natural units of quantum electrodynamics*, as explained in Sec. 13: thus m is the unit of mass, mc^2 the unit of energy, $\hbar/mc$ the unit of distance, and $\hbar/mc^2$ the unit of time. Furthermore $\hbar$ is the unit of angular momentum, and c the natural unit of velocity.

So far we have not considered the elementary charge e. This constant plays the role of a *coupling constant;* it tells us how strongly the electrons are coupled to the electromagnetic field.†
We try to form a dimensionless quantity which measures the strength of this coupling, and we consider the *electrostatic energy of repulsion, in the natural units introduced above, of two electrons separated by a distance of one natural unit.* This quantity is denoted by α and we obtain

$$\alpha = \frac{e^2/(\hbar/mc)}{(mc^2)} = \frac{e^2}{\hbar c}$$

$$= (7.29720 \pm 0.00003) \times 10^{-3} \approx 1/137 \qquad (21a)‡$$

This constant α plays a fundamental role in atomic physics and it is known as the *fine structure constant.* It may be regarded as the square of the elementary charge in natural units, and it describes the magnitude of this charge in a manner independent of any arbitrary macroscopic physical standards. That α is numerically quite small reflects a basic "weakness" of the electromagnetic interactions; the electrostatic energy of two electrons placed at a unit distance apart is small compared to the rest energy of an electron. The *fine structure constant is one of the truly fundamental constants of nature,* and at present it is a purely empirical constant in the sense that we have no theoretical understanding of its magnitude. It might just as well have "turned out" to be large, in which case the world would look very different indeed. In fact, *unimaginably* different.

---

† This is a common mode of expression. It would, however, be more profound to say that the coupling constant tells us how strongly elementary charges interact with *each other.* The electromagnetic field is, after all, a mental construct introduced for the purpose of discussing interactions between *charges.*

‡ This expression for α applies in the cgs system of units. In the MKS system we have: $\alpha = e^2/(4\pi\varepsilon_0\hbar c)$.

The reader should note, by inspection of Eq. (21a), that the *mass of the electron does not enter into the expression for* α. Consequently α is the coupling constant which describes the coupling of *any* elementary particle carrying the elementary charge e to the electromagnetic field.

In Table 21A we list some important quantities which we can form from m, $\hbar$, c, and e, and we give the names under which these quantities are known.

22 In Sec. 51, Chap. 1, we discussed one aspect of Bohr's semiclassical theory of the hydrogen atom, namely the size of this atom, and we stated that the constant a_0, defined in Eq. (51c) in that chapter, is a typical atomic size. The reader will note that this constant a_0, known as the *first Bohr radius (in the hydrogen atom)*, is the same as the one given in Table 21A. In the discussion in Chap. 1 a_0 stood for the radius of a circular orbit of the electron in the planetary model of the atom, and hence the name. In the quantum-mechanical discussion of the hydrogen atom this constant has a different interpretation: $1/a_0$ is the average of $1/r$ in the ground state of the atom, where r is the distance between the electron and the proton. In either case a_0 can be regarded as the "typical" distance between the electron and the proton.

23 Let us continue with the semiclassical discussion of Chap. 1 and try to estimate the binding energy of the electron in the hydrogen atom. For an electron moving with velocity v (and hence momentum $p = mv$) at a distance r from the proton, the total energy E is given by

$$E = \frac{p^2}{2m} - \frac{e^2}{r} = \frac{1}{2}mv^2 - \frac{e^2}{r} \tag{23a}$$

For a circular orbit, of radius $r = a_0$, the condition for dynamical equilibrium reads

$$\frac{mv^2}{a_0} = \frac{e^2}{a_0{}^2} \tag{23b}$$

and combining this equation with equation (23a) we obtain

$$E = \frac{1}{2}\left(\frac{e^2}{a_0}\right) - \frac{e^2}{a_0} = -\frac{e^2}{2a_0} = -\frac{1}{2}\alpha^2 mc^2 = -R_\infty \tag{23c}$$

The energy of the electron in this orbit is thus $-R_\infty$, or about -13.6 eV. This energy should be compared with the total energy when the electron is infinitely far away from the proton and at rest; by inspection of equation (23a) we see that this energy equals zero.

TABLE 21A   *More Physical Constants*

Rest energy of the electron:

$$mc^2 = (0.511006 \pm 0.000002)\ \text{MeV}$$

Compton wavelength of the electron:

$$\lambdabar_e = \frac{\hbar}{mc} = (3.86144 \pm 0.00003) \times 10^{-11}\ \text{cm}$$

First Bohr radius:

$$a_0 = \frac{\hbar^2}{me^2} = \alpha^{-1}\lambdabar_e$$
$$= (5.29167 \pm 0.00002) \times 10^{-9}\ \text{cm}$$

Nonrelativistic ionization potential of hydrogen with infinite proton mass:

$$R_\infty = \tfrac{1}{2}\alpha^2 mc^2 = (13.6053 \pm 0.0002)\ \text{eV}$$

Rydberg constant for infinite proton mass:

$$\tilde{R}_\infty = \frac{\alpha}{4\pi a_0} = R_\infty/hc$$
$$= (109737.31 \pm 0.01)\ \text{cm}^{-1}$$

Therefore, to remove the electron completely from the circular orbit considered we have to supply an energy R_∞ to the atom. This energy is called the *ionization energy*. The ionization energy, expressed as the equivalent wave number, is known as the Rydberg constant. We denote it by $\tilde{R}_\infty$.[†]

It so happens, and in principle we should regard this as an "accident," that this simple estimate based on the otherwise not very convincing planetary model gives *precisely* the same ionization energy R_∞ as the rigorous quantum-mechanical theory, and R_∞ is therefore the ionization energy of hydrogen, or, differently stated, $-R_\infty$ is the *ground state energy* of the hydrogen atom.

It is furthermore the case that the ionization energies of *all* atoms (i.e., the work required to remove *one* electron from the atom) is roughly of the order of 10 eV; we will return to this question later.

24 Let us see how the weakness of the electromagnetic forces, i.e., the smallness of the coupling constant α, manifests itself in the structure of the hydrogen atom. If the coupling constant were of order unity, then we would expect the size of the atom to be of the order of the natural unit of length in quantum electrodynamics, i.e., the Compton wavelength $\lambdabar_e = \hbar/mc$. The coupling constant is, however, "small" ($\alpha \approx 1/137$), and the Coulomb field of the nucleus is therefore unable to keep the electron confined within a Compton wavelength. The orbit of the electron is *large* in the natural quantum-electrodynamical unit, namely of radius $a_0 = \lambdabar_e/\alpha$.

We find the velocity of the electron in the orbit by solving Eq. (23b) for v:

$$v = \sqrt{\frac{e^2}{ma_0}} = \alpha c \qquad (24a)$$

The velocity is thus 137 times smaller than the natural unit, which is the velocity of light c. This is the a posteriori justification for our nonrelativistic discussion of this problem.

The kinetic energy, E_{kin}, and the potential energy, E_{pot}, are given by

$$E_{\text{kin}} = \tfrac{1}{2}mv^2 = \tfrac{1}{2}m(\alpha c)^2 = R_\infty \qquad (24b)$$

$$E_{\text{pot}} = -\frac{e^2}{a_0} = E - E_{\text{kin}} = -2R_\infty = -2E_{\text{kin}} \qquad (24c)$$

On the basis of these considerations we can say that the hydro-

Arnold Sommerfeld. Born 1868, in Königsberg, Germany (now Kaliningrad, USSR); died 1951. For many years professor of physics at the University of Munich.

Sommerfeld made important contributions in the development of quantum physics, and in particular in early atomic theory. He refined Bohr's theory in two ways: to include elliptical orbits, and to take special relativity into account. His relativistic theory of the hydrogen atom introduced the fine structure constant into physics. (*Photograph by courtesy of Professor L. B. Loeb, Berkeley.*)

[†] The subscript ∞ in R_∞ and $\tilde{R}_\infty$ refers to the model in which the proton is infinitely heavy and stays fixed. The actual ionization energy is slightly smaller.

gen atom is a loosely bound extended structure. The reader should think very carefully about this and about the role which the fine structure constant α plays in the theory of the atom.

25 Since the velocity of the electron turned out to be small in our semiclassical description it is reasonable to expect that it is possible to describe the atom within a nonrelativistic version of quantum mechanics. In such a theory the velocity of light should play no role if we regard the constants m, $\hbar$ and e as the basic constants. In particular it should be possible to express the Bohr radius a_0 and the ionization energy R_∞ in terms of these constants only. This is in fact the case, and we find

$$a_0 = \frac{\lambda_e}{\alpha} = \frac{\hbar^2}{me^2} \tag{25a}$$

and

$$R_\infty = \frac{1}{2}\alpha^2 mc^2 = \frac{e^2}{2a_0} = \frac{e^4 m}{2\hbar^2} \tag{25b}$$

The velocity of light does not appear in the extreme right-hand side of these expressions. Furthermore the length a_0 is the only length, and the energy R_∞ the only energy, which we can form from the constants m, $\hbar$, and e. We can therefore argue that since these constants are the ingredients of the nonrelativistic quantum-mechanical theory (which is so far unknown to the reader), every length computed within the theory must be a numerical multiple of a_0, and similarly every energy must be a numerical multiple of R_∞. (Numerical multiple here means a number independent of the three constants. We expect that in a "reasonable" theory such numbers will be "of the order of unity.")

26 The reader probably feels that these "derivations" are quite outrageous. What value can there be in an argument based on the Bohr model, which we have earlier declared to be quite wrong? And how seriously are we to take the "dimensional argument" in the preceding section? Could it not happen that the constant "of order unity," which gives the correct energy in terms of R_∞, in fact turns out to be something like 4711, or maybe $(2\pi)^{-4}$? Such constants would obviously make quite a bit of difference in our estimate.

The answer is that it could very well happen, but the experienced author knows that it actually does not happen; the constant is equal to unity. About "simple derivations" of this kind, which occur frequently in physics texts, the cynic can always remark that the arguments seem to work out particularly well in all the cases in which either the experimental results or the results of a more complete theory are known.

In defense of what has been done we say the following: (i) we wish to form a picture of the orders of magnitude in atomic and molecular physics. Instead of just telling the reader that the ionization energy of hydrogen is 13.6 eV we should try to relate this 13.6 eV to expressions formed from the basic constants. It is nice to know that 13.6 eV equals $\alpha^2 mc^2/2$, and it is nice to know that 0.53 Å equals $(1/\alpha)(\hbar/mc)$. Our discussion of quantum electrodynamics and its relevance to the hydrogen atom gives us at least *some* understanding of how it all might hang together. The author would certainly not have presented these ideas if they did not have their counterparts in the precise theory. Our "derivations" are therefore at least useful as mnemotechnic devices.

(ii) The Bohr theory is admittedly wrong. On the other hand the reader undoubtedly knows that it was successful in some instances, although it failed badly in other cases. In a vague sense the theory therefore has *some* elements of truth in it. It introduces Planck's constant into physics and thereby introduces a relation between position and momentum which does not occur at all in a purely classical theory: something like $rp \sim \hbar$. We can take the point of view that our derivation based on the Bohr theory was really in essence an experimentation with a relation of this kind: $rp \sim \hbar$. Later we will play with this relation in a different way, and we will discuss a method of estimating the size and ionization energy of the hydrogen atom on the basis of the uncertainty relation. At the same time we will gain a much better understanding of why the hydrogen atom does not collapse.

(iii) The dimensional argument in Sec. 25 would be much more convincing in connection with a serious study of a definite equation for the quantum mechanical description of the hydrogen atom, such as the so-called Schrödinger equation. Without actually solving this equation we could quite easily conclude that numbers such as 4711 or $(2\pi)^{-4}$ will not occur. To draw a conclusion of this kind we must, of course, have some experience with the nature of solutions of differential equations. (The Schrödinger equation is a differential equation.) Dimensional arguments work best when they are tempered by a good understanding of the general features of a theory.

Our simple dimensional argument is an introduction to arguments of this kind. The reader has been told that a "good" theory exists; what can this theory be expected to give us? That is the question we have asked and answered.

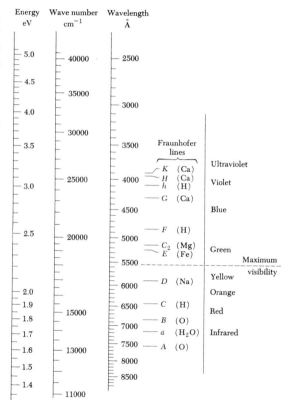

The spectral neighborhood of visible light. The Fraunhofer lines are prominent absorption lines (dark lines) in the solar spectrum. The left column gives the old letter designations of these lines, and the right column the chemical symbol identifying the atom or molecule causing the absorption.

The assignment of colors to the various spectral regions is, of course, only approximate. Note the point of maximum visibility at about 5500 Å.

27   Let us now continue with our discussion of atomic physics, and let us try to obtain a rough understanding of the structure of heavy atoms, i.e., for the case when the atomic number Z is large. The reader has undoubtedly heard that the electron cloud around such a nucleus has a shell structure in some sense, and we will try to base our discussion on this idea.   Let us imagine that we build up the atom by starting from the bare nucleus and then adding one electron at a time.   How tightly will the first electron be bound?

The expression for the energy of the system now takes the form

$$E = \frac{p^2}{2m} - \frac{e^2 Z}{r} \qquad (27a)$$

and a moment's reflection shows us that the discussion for the case of the hydrogen atom still applies, provided we replace the fine structure constant α by αZ.   In other words, the *first* electron will be bound at an energy

$$e_1 = -Z^2 R_\infty = -Z^2 \,(13.6 \text{ eV}) \qquad (27b)$$

and at a "distance" from the nucleus of

$$r_1 = \frac{a_0}{Z} \qquad (27c)$$

For large Z, this distance is small compared to the Bohr radius in hydrogen, a_0.   The next electron added will also be bound at a small distance, and the binding energy will be large compared to the ionization energy of hydrogen; the electrostatic force of repulsion between the two electrons is clearly Z times smaller than the force of attraction to the nucleus.   Let us consider the appearance of the *ion* after we have added a few electrons.   These electrons will all be bound within a small distance of the nucleus, and if there are n of these electrons the ion will present the appearance of a "nucleus" of charge $(Z - n)e$ beyond the distance at which the electrons are bound.   The next electron will therefore necessarily also be tightly bound *unless* $(Z - n)$ is small, but it will be *less* tightly bound than the *first* electron.   We may thus imagine that the successive electrons will be less and less tightly bound, and after we have added $(Z - 1)$ electrons the ion appears like a cloud of charge of magnitude e, and of a size comparable to the Bohr radius a_0.   The binding energy of the *last* electron added will therefore be of the order of R_∞, and thus of the order of ten electron volts.   The final size of the atom will be of the order of the Bohr radius a_0.

28   This picture is certainly very rough.   Note that we have neither proved, nor even made plausible, the idea that the electron

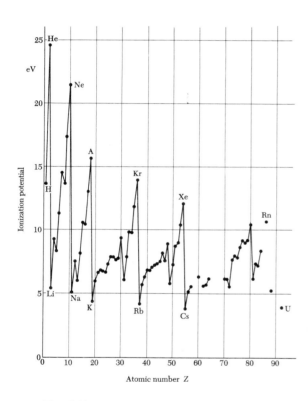

Fig. 27A Graph showing ionization potentials of atoms versus atomic number. The ionization potential is the energy necessary to remove *one* electron from the neutral atom. We see that this energy is *roughly* of the same order of magnitude for all atoms, i.e. of the order of 10 eV.

The reader who has some familiarity with chemistry will at once notice that there is a marked correlation between the magnitude of the ionization potential and the chemical properties of the element. The ionization potential is particularly large for the noble gases, and particularly small for the alkali metals.

cloud has a "shell structure." Our discussion was, however, based on that idea; we "built up" the atom in a particular way.

Now it is a fact that to really understand the structure of atoms we must observe a fundamental principle of physics which we have not mentioned so far and which is entirely foreign to classical physics. This principle is known as the *exclusion principle.* It says that *no two electrons can ever be in the same state of motion* in an atom. The electrons "avoid" each other. (This "avoidance" is something quite different from the Coulomb repulsion of two similarly charged particles. To really understand the meaning of the exclusion principle requires a knowledge of quantum mechanics.) The exclusion principle is the key to the explanation of atomic structure. It has profound consequences, and the world would be unimaginably different if it were not observed by nature. That this is so is most certainly not obvious at this stage.

The exclusion principle was discovered by Wolfgang Pauli in 1924 from a consideration of the empirical facts of atomic physics as they were known at that time.†

29 Our discussion is therefore *very* defective, but it nevertheless gives us some kind of picture of the nature of heavy atoms. It follows from this picture that transitions in the state of motion of the outermost or *optical electrons* will involve energies of the order of an electron volt, which roughly corresponds to wavelengths of the emitted photons in the *optical region*, i.e., in the energy range 1.8–3.0 eV, or the wavelength range 7000–4000 Å. Transitions involving the *innermost electrons* on the other hand correspond to comparatively much larger energies, ranging up to 70 keV (= 70,000 eV), which correspond to wavelengths down to 0.2 Å. These photons lie in the far ultraviolet or X-ray region. The dependence of these transition energies on the atomic number Z is shown by formula (27b).

We note that the atom, which has a typical size of 1 Å, is *small* compared to the wavelength of optical photons. This circumstance is a consequence of the smallness of the coupling constant α, and to see this we apply the following considerations:

The binding energy of an optical electron is of the order of $\alpha^2 mc^2$. The transition energies characteristic of the optical electrons are of the same order of magnitude; they could certainly not be larger. The transition of one of the outermost electrons between two quasi-stationary states is associated with the emission

Wolfgang Pauli. Born 1900 in Vienna, Austria; died 1958. After finishing his PhD dissertation in 1921 Pauli spent some time at the University of Göttingen, and at Bohr's Institute in Copenhagen. In 1928 he accepted the chair in theoretical physics at the Institute of Technology (ETH) in Zürich, Switzerland. He received the Nobel prize in 1945.

Pauli was one of the most outstanding theoretical physicists of this century. He made important contributions in many areas, ranging from atomic structure to the theory of quantum fields and elementary particles. Pauli's work was characterized by a deep physical insight, as well as by great mathematical skill, and he was known (and feared) as a stern critic of obscure thinking. His discovery of the exclusion principle, and of the connection between spin and statistics, are perhaps his most famous achievements. (*Photograph by courtesy of Physics Today.*)

---

† W. Pauli, "Über den Zusammenhang des Abschlusses der Elektronengruppen im Atom mit der Komplexstruktur der Spektren," *Zeitschrift für Physik* **31**, 765 (1925).

or absorption of a photon of an energy equal to the energy difference between the two levels, and the corresponding wavelength of this photon will therefore be of the order of

$$\lambda_{\text{opt}} \sim \frac{2\pi\hbar c}{\alpha^2 mc^2} = \frac{2\pi a_0}{\alpha} \simeq 1000 \, a_0 \qquad (29a)$$

which explains the order of magnitude of the ratio of wavelength to size of the atom.

30   We have now gained a considerable understanding of the relevant orders of magnitude in the realm of *atomic physics*. Let us try to say something about molecules. The crucial problem here is to understand the molecular binding; why do atoms sometimes form stable molecules and other times do not? To *really* understand these questions requires much more sophisticated methods than the ones we have used for atoms. Nevertheless we may try to answer a small part of the general question, and we may ask: Granted that atoms do form stable molecules in some cases, what is the characteristic binding energy and what is the characteristic separation of two atoms in a molecule?

Let us consider the simplest case, namely a hydrogen molecule, which is a bound state of two protons and two electrons. We try to estimate the binding energy and the internuclear distance through a dimensional argument; our argument thus concerns those favorable cases in which binding does occur, as in the hydrogen molecule.

Since the proton is so much heavier than the electron, the *motion* of the protons will again not play an essential role in the determination of the ground state of the hydrogen molecule. In a first approximation one may in fact regard the two protons as stationary, at a fixed separation d, and these two protons are then surrounded by the "cloud" of the two electrons. We imagine that we find the ground state energy of the two electrons, as a function of the interproton separation d. For a certain value of d this energy will assume its smallest value, and at this energy we have a stable molecule. Our problem is nonrelativistic, and since the protons are regarded as infinitely heavy we only have the constants m, $\hbar$ and e available. The only "natural" energy is then R_∞, and the only "natural" distance is the Bohr radius a_0. These quantities should thus be characteristic of the molecule. A more careful study confirms this expectation, which is also in agreement with the experimental facts. The actual binding energy of the hydrogen molecule is about 4.5 eV, and the average separation of the two protons is about 0.75 Å. These values are quite typical for molecules in general; molecular binding energies are of the order of 1–10 eV, and the internuclear separations are of the order of one Ångström unit, i.e., of the order of 10^{-8} cm.

The same "mechanism" which holds a molecule together also holds solids together, and the typical separation between two neighboring atoms in a solid is also of the order of 1 Å.

31   These estimates now provide us with an understanding of the magnitude of the energy released or absorbed in a *chemical reaction*. The elementary process in a chemical reaction is that two or more different molecules collide and form one or more new molecules. The energy associated with this rearrangement of the atoms into new molecules must be of the order of typical molecular binding energies and hence of the order of 1–10 electron volts per elementary process. The *bulk reaction energies* are therefore of the order of $(1-10) \times N_0$ electron volts/mol, or roughly 20,000–200,000 cal/mol. As an example we may consider the burning of hydrogen gas in an atmosphere of chlorine, according to the reaction

$$H_2 + Cl_2 = 2HCl + (44{,}000 \text{ calories}) \qquad (31a)$$

The order of magnitude is in accordance with our estimate.

32   There is an amusing feature about our macroscopic units which is worth a comment. We have said that the units cm, gm, and sec refer to human characteristics, and we are therefore not surprised that these units are not particularly suited to the discussion of atoms. *One* macroscopic unit seems, however, to be in a special position: namely the *volt* as a unit of potential in that the derived unit electron volt is "just right for atoms." Is this an accident?

The answer is no. Originally the unit volt was chosen such that the EMF of voltaic cells would be of the order of a volt. In fact the EMF of a certain cadmium-mercury standard cell is very close to one volt. We know that the operation of such a cell is based on an electrochemical reaction taking place in the cell, and for every electron leaving the terminal of the battery, one elementary chemical process must have taken place. For every elementary chemical process an energy of, say, X eV is released, and this energy may be converted into mechanical or thermal energy outside the battery. If the EMF of the battery is U we must have $Ue = X$, and, since by the choice of the unit volt, U is of the order

TABLE 30A   *Characteristics of Somewhat Randomly Selected Diatomic Molecules*

Molecule	Distance between Nuclei Å	Dissociation Energy eV
AgH	1.62	2.5
BaO	1.94	4.7
Br$_2$	2.28	1.97
CaO	1.82	5.9
H$_2$	0.75	4.5
HCl	1.27	4.4
HF	0.92	6.4
HgH	1.74	0.38
KCl	2.79	4.42
N$_2$	1.09	9.76
O$_2$	1.20	5.08

of a volt it follows that the typical electrochemical reaction energy is of the order of one electron volt. This explains the mystery of the suitability of the electron volt as a unit of energy in atomic and molecular physics; the volt is in fact an "atomic unit"!

The Most Basic Facts of Nuclear Physics

33 The building blocks of nuclei are protons and neutrons. Protons and neutrons have many important physical properties in common, and they are often regarded as two different *charge states* of a "single" particle called the *nucleon*. The nucleon thus occurs in two versions: the charged version, which is the proton and the neutral version, which is the neutron.†

The number A of nucleons in a nucleus is the *mass number* or *nucleon number*. The number Z of protons is known as the *charge number*, or, when we speak of the corresponding atom, as the *atomic number*.

The masses of the proton and neutron are

$$M_p = (1.00727663 \pm 0.00000008) \text{ amu}$$
$$= (938.256 \pm 0.005) \text{ MeV}/c^2 \qquad (33a)$$
$$M_n = (1.0086654 \pm 0.0000004) \text{ amu}$$
$$= (939.550 \pm 0.005) \text{ MeV}/c^2 \qquad (33b)$$

Consider a nucleus of mass number A and charge number Z. Let its mass be $M(A,Z)$. The quantity

$$\Delta(A,Z) = (ZM_p + (A - Z)M_n) - M(A,Z) \qquad (33c)$$

is called the *mass defect* of the nucleus. This quantity is *positive*, which circumstance has a simple interpretation: the quantity $\Delta(A,Z)c^2$ equals the *binding energy* of the nucleus, or the energy which has to be supplied to break up the nucleus completely into its elementary constituents, i.e., into protons and neutrons.

It is an *empirical fact* that the binding energy per nucleon is *roughly* the same for all stable nuclei, i.e.,

$$\frac{\Delta(A,Z)c^2}{A} \sim 8 \text{ MeV} \qquad (33d)$$

There are some marked exceptions among the very light nuclei, and there is also a slight systematic decrease of the average binding energy as the mass number A increases, as we see in Fig. 33a.

---

† The neutron was discovered by Chadwick in 1932. [J. Chadwick, "The Existence of a Neutron," *Proceedings of the Royal Society* (*London*), ser. A, **136**, 692 (1932).]

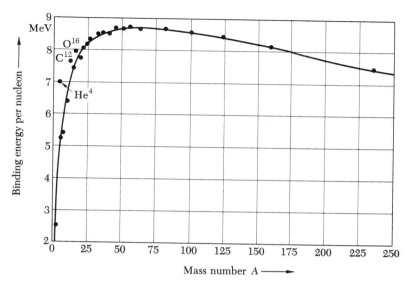

Fig. 33A Graph showing the binding energy per nucleon, $\Delta(A,Z)c^2/A$, versus the mass number A. The dots refer to particular nuclei, some of which are identified. The irregularities occurring for the very lightest nuclei are not well represented by the smooth graph, but for $A > 25$ the curve is an accurate representation of the facts.

The binding energy per nucleon is *roughly* 8 MeV. As the mass number increases, the binding energy per nucleon slowly decreases. This systematic trend is a consequence of the electrostatic energy of repulsion of the protons within the nucleus.

34   The reader should note that the mass values listed in most tables of "nuclear" masses actually refer to the masses of the corresponding neutral *atoms*. If $M(A,Z)$ is the mass of a *nucleus*, and $\overline{M}(A,Z)$ the mass of the corresponding atom, we have

$$\overline{M}(A,Z) = M(A,Z) + Zm - B(Z) \tag{34a}$$

where m is the electronic mass, and where the positive quantity $B(Z)$ expresses the binding energy of all the electrons in the atom.

When we consider the energy balance in a nuclear reaction, it will in most cases make no difference whether we use the true nuclear masses, or the associated atomic masses, because the contributions from the electron masses cancel if we use the latter. The binding energy $B(Z)$ is very small compared with the nuclear binding energy of 8 MeV per nucleon, and it can almost always be neglected.

The reason why the atomic masses are listed instead of the nuclear masses is that the atomic masses can be measured more readily. Through deflection experiments in combined electric and magnetic fields with an instrument developed specifically for this purpose, called the *mass spectrograph*, we can determine the ratios of charge to mass for ions. This work was initiated by J. J. Thomson and F. Aston, and it has led to a precise knowledge of a large number of atomic masses.†

---

† F. W. Aston, "Isotopes and Atomic Weights," *Nature* **105**, 617 (1920). Also F. W. Aston, *Mass Spectra and Isotopes* (Edward Arnold and Company, London, 1942).

Figs. 34A-B A mass spectrometer intended for the analysis of small samples of noble gases from stony meteorites. Here the purpose is not to measure accurate values of atomic masses, but to determine relative abundances of the isotopes of the element (xenon) as it occurs in the meteorite. The data obtained can be used to estimate the age of the meteorite, which is of great interest in our attempts to understand the origin and development of the solar system. For a description of this work, see J. H. Reynolds, "The age of the elements in the solar system," *Scientific American* **203,** 171 (Nov. 1960).

A photograph of the instrument is shown above, and the principle of operation can be understood with reference to the drawing below. The noble gas sample, which is admitted at left into the evacuated glass envelope, is ionized by electron bombardment in the ion source. The ions are accelerated, and deflected by the magnet in the middle. (The pole pieces and coils of the magnet can be seen in the middle of the photograph.) Different isotopes are deflected by different amounts, and by varying the magnetic field strength the current passing through the collector slit at right can be measured for each isotope in turn. The abundance of the isotope is, of course, proportional to the current. The magnetic field is wedge-shaped in order to achieve partial focusing of the ion beams. *(Illustrations by courtesy of Prof. J. D. Reynolds, Berkeley.)*

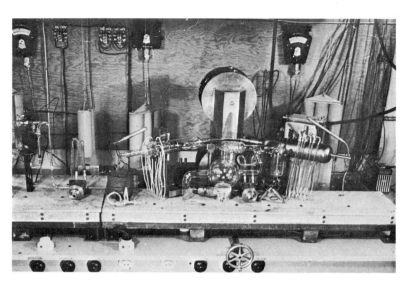

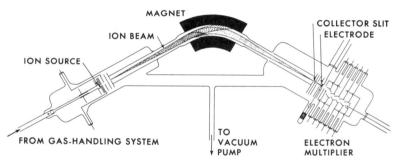

The mass spectrograph can also be used to determine the abundances of the different isotopes in a naturally occurring chemical element, and once these abundances are known we can obtain information about "nuclear" masses from the chemical atomic weights.

Finally we obtain information about nuclear masses from a study of the kinematics of nuclear reactions.

35 The ratio Z/A of charge number to mass number shows a systematic trend as a function of the mass number A. For not too heavy nuclei, say for A less than 50, this ratio is close to $\frac{1}{2}$. As A grows larger the ratio decreases slowly; for the uranium isotope $_{92}U^{238}$ it has the value $Z/A = 0.39$. For very small A we again meet with irregularities; hydrogen, for instance, has three isotopes, $_1H^1$, $_1H^2$ (deuterium), and $_1H^3$ (tritium).

Some nuclei are stable, whereas other nuclei are unstable and decay through the emission of particles or gamma rays. The commonly occurring nuclei are either absolutely stable or else have very long lifetimes; if this were not so they would have decayed at some early stage in the history of the earth and would no longer be present. Nuclei formed in nuclear reactions may have very short lifetimes, of the order of a small fraction of a second. When the lifetime is *very* short we often speak of an *excited state* of a nucleus, especially if the decay takes place through the emission of a gamma ray, in which case A and Z remain unchanged.

At present about 900 nuclei are known, of which about 280 are stable. If we plot these nuclei on a (Z,A)-plane, then the points representing the individual nuclei tend to cluster along a smooth curve, in accordance with what was said before. (See Fig. 35a.) The more removed a nucleus is from the "central curve" the more unstable it tends to be.

36 It has been found experimentally that a nucleus has a quite well-defined size and that it may be regarded as a sphere of nuclear matter of radius

$$r \cong r_0 A^{1/3} \qquad \text{where } r_0 = 1.2 \times 10^{-13} \text{ cm} = 1.2 \text{ fermi} \qquad (36a)$$

(The unit *fermi* $= 10^{-13}$ cm, named in honor of Enrico Fermi, is often employed as a unit of length in elementary particle physics.)

Since the volume of the nucleus is proportional to r^3, and hence, by formula (36a), to the nucleon number A, we conclude that the density of nuclear matter in the different nuclei is approximately constant.

The sizes of the nuclei, as summarized by the formula (36a), have been determined from a variety of experiments.† The most straightforward method consists in the measurement of the effective cross-sectional area which a nucleus presents to a beam of very-high-energy particles in a scattering experiment.

37 Let us now try to say something about the nature of the forces which hold a nucleus together. All our experimental evidence says that

(i) The nuclear force is *not* electromagnetic in nature; compared with the electromagnetic forces the nuclear forces are much stronger.

(ii) The nuclear force is of *short range;* when the separation

---

† R. Hofstadter, "Structure of Nuclei and Nucleons" (Nobel address), *Science* **136**, 1013 (1962).

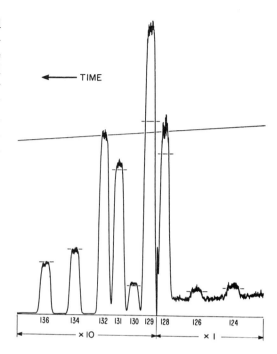

Fig. 34C Mass spectrum recorded with the apparatus shown in Figs. 34A-B, for xenon extracted from a stony meteorite. Graph is taken from J. H. Reynolds, "Determination of the age of the elements," *Physical Review Letters* **4**, 8 (1960). The short horizontal bars show the isotopic abundances for terrestial samples of xenon. As we can see, the meteorite sample is richer in the isotope Xe^{129}. Note that the graph is drawn in two different vertical scales. *(Courtesy of Physical Review Letters.)*

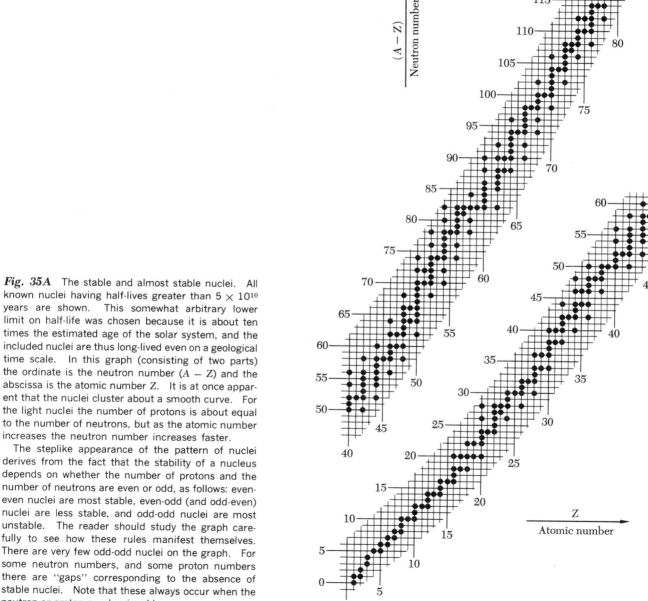

Fig. 35A The stable and almost stable nuclei. All known nuclei having half-lives greater than 5×10^{10} years are shown. This somewhat arbitrary lower limit on half-life was chosen because it is about ten times the estimated age of the solar system, and the included nuclei are thus long-lived even on a geological time scale. In this graph (consisting of two parts) the ordinate is the neutron number $(A - Z)$ and the abscissa is the atomic number Z. It is at once apparent that the nuclei cluster about a smooth curve. For the light nuclei the number of protons is about equal to the number of neutrons, but as the atomic number increases the neutron number increases faster.

The steplike appearance of the pattern of nuclei derives from the fact that the stability of a nucleus depends on whether the number of protons and the number of neutrons are even or odd, as follows: even-even nuclei are most stable, even-odd (and odd-even) nuclei are less stable, and odd-odd nuclei are most unstable. The reader should study the graph carefully to see how these rules manifest themselves. There are very few odd-odd nuclei on the graph. For some neutron numbers, and some proton numbers there are "gaps" corresponding to the absence of stable nuclei. Note that these always occur when the neutron or proton number is odd.

between two nucleons exceeds 10^{-12} cm the specific nuclear force becomes insignificant.

(iii) The specific nuclear force between two protons is the same as that between two neutrons. The nuclear force between two protons is furthermore of the same nature as the nuclear force between a proton and a neutron; we may say that they are actually the same, although this statement does require some qualification.

The evidence for these three statements comes from scattering experiments as well as from the systematic study of stable or radio-active nuclei and their systems of energy levels. In particular the statement about the short range of the nuclear force can be verified as follows. We bombard a nucleus with protons accelerated to high energy in an accelerator, and we study the scattering of the protons from the nucleus. When the proton is far away from the nucleus (i.e., beyond $10^{-11} - 10^{-12}$ cm), the only effective force is the Coulomb repulsion. This repulsion prevents the proton from approaching the nucleus closely enough for the nuclear force to become effective unless the energy of the proton is very high. If the statement about the short range holds we expect that protons (or alpha-particles, as in Rutherford's experiment) of not too high energy will scatter as though the Coulomb repulsion were the only force present. We may thus test the statement (ii) through a detailed analysis of scattering experiments, and the conclusion is as stated.

Since the protons are charged they can also be acted upon by electromagnetic forces, and two protons in a nucleus will certainly exert a Coulomb repulsion on each other. At distances much larger than 10^{-12} cm the electromagnetic forces are for all practical purposes the *only* forces acting, but for smaller distances the nuclear force dominates. Electromagnetic forces do play a role for nuclear structure, but they do not play a dominant role.

In this connection it should be stated very explicitly that *electrons seem to be completely unaffected by the specific nuclear forces;* the only significant forces acting on electrons are the electromagnetic forces.

38 Let us elaborate a bit on the fact that the strong nuclear force is a short-range force. According to our present beliefs the general nature of this force, effective between two nucleons, is reasonably well represented by a potential function $U(r)$ of the form

$$U(r) \simeq C \left(\frac{b}{r}\right) \exp\left(-\frac{r}{b}\right) \tag{38a}$$

Enrico Fermi. Born 1901 in Rome; died 1954. He received his doctorate in 1922 from the Scuola Normale Superiore at Pisa, Italy. In 1926 he became professor of theoretical physics at the University of Rome. Fermi left Italy in 1938, and after holding a professorship at Columbia University he went to the University of Chicago in 1942, where he remained until his death. He received the Nobel prize in 1938.

Fermi's contributions to physics cover an amazingly wide range and no short summary can do justice to his work. From his early work we can mention his invention of the so-called Fermi-Dirac statistics of particles (at the same time, and independently of Dirac who did similar work), and his very successful quantitative theory of beta-decay. His main work was in the fields of nuclear and elementary particle physics. Among the many topics he studied we can mention artificial radioactivity, slow neutrons, nuclear fission and chain reactions, and the pion-nucleon interaction. Fermi was one of the very few physicists who have done outstanding work both in theory and in experiment. *(Photograph by courtesy of Professor E. Segrè, Berkeley.)*

provided that the separation r is larger than 10^{-13} cm.† The constant b is a measure of the range of the force; it has the value $b = 1.4 \times 10^{-13}$ cm. The constant C expresses the strength of the force. The nature of the force at distances smaller than 10^{-13} cm is much more complicated and very poorly understood at present.

We emphasize that the potential function $U(r)$ does not describe the interaction between two nucleons *precisely*, but it does represent the most important feature of this interaction which is that *the potential falls off exponentially with distance.*

Let us see what this really means. At the distance $r = b$ we have $U(b) = C/e$. (This constant is roughly of the order of 10 MeV.) At the distance $r = 10b = 1.4 \times 10^{-12}$ cm the potential equals $U(10b) = (0.1C) \exp(-10) \sim 5 \times 10^{-6}C$. At the distance $r = 100b = 1.4 \times 10^{-11}$ cm the potential equals $U(100b) = (0.01C) \times \exp(-100) \sim 10^{-45}C$. We can conclude from this numerical exercise that when the separation between the two nucleons exceeds 10^{-11} cm, then the nuclear force is *utterly* negligible. For all practical purposes there is no nuclear force beyond the distance quoted.

The reader should think carefully about this. At first sight the expression (38a) might resemble the Coulomb potential. However, the exponential factor makes all the difference in the world. Our numerical exercise was intended to impress the reader with this fact.

The specific nuclear force between nuclei in molecules and solids is thus nonexistent for all practical purposes, and this is the kind of situation in which the electromagnetic forces have a chance to play a dominant role. At *small* distances, $r \sim 10^{-13}$ cm, the specific nuclear force is considerably stronger than the electromagnetic forces, and the latter are relegated to play a secondary role. That this is the case is immediately obvious from the fact that nuclei exist. The electrostatic forces of repulsion try to disperse the charged particles in a nucleus, but the nuclear forces try to keep them together, and the nuclear forces win; they are stronger.

39   Since the typical binding energy of a nucleus is of the order of 8 MeV per nucleon, we can expect that transmutations of nuclei will involve energies roughly of the order of 1 MeV. The energies with which material particles and photons (gamma rays) are emitted from nuclei do in fact range typically from, say, 100 keV to 10 MeV.

---

† In Chap. 9 we shall present a theoretical explanation for this form of the potential function.

The energies involved in nuclear reactions are therefore of an entirely different order of magnitude than the energies involved in chemical reactions, and we can readily understand why nuclei are not affected by chemical processes. From the standpoint of chemistry and atomic physics, nuclei are just small, hard, massive, indivisible, charged spheres.

In our discussion of atoms we concluded that the wavelength of an optical photon is large compared to the size of the atom. It is worth noting that the situation is similar in nuclear physics. Consider a gamma ray of energy 1 MeV, which is a typical nuclear transition energy. The corresponding wavelength is 1.2×10^{-10} cm = 1200 fermi, and it is thus large compared with the typical nuclear size.

Gravitational and Electromagnetic Forces

40 We should now explain why we neglect gravitational forces in our discussion of atoms, molecules, and nuclei. For this purpose we compute the ratio of the gravitational force to the electrostatic force between two protons. This ratio is independent of the distance between the two protons, and we find

$$\frac{M_p{}^2 G/r^2}{e^2/r^2} = \frac{M_p{}^2 G}{e^2} = 8.1 \times 10^{-37} \qquad (40a)$$

where we have inserted the value $G = 6.67 \times 10^{-8}$ dyne cm^2 gm^{-2} for the constant of gravitation.

The ratio of the strengths of the two forces is thus very, very small, and in the presence of electromagnetic interactions we expect the effects of gravitation to be completely negligible. Gravitational forces can play a role only if all other known forces are ineffective, hence only between (large) electrically neutral bodies separated by distances large compared to the typical atomic distances.

Einstein's Theory of General Relativity is a purely *geometric* theory of gravitation. It is a theory of great beauty and internal coherence. In spite of many attempts by Einstein and others it has not been possible, to date, to incorporate the other forces of nature in a natural way into this theory. The phenomenon of gravitation therefore stands quite apart from the interactions which govern the structure of matter on the microscopic scale; gravity appears to be totally irrelevant to microphysics, which is the reason why gravitation is neglected in this book. The reader should note that the ratio given in equation (40a) is nothing but the ratio of the gravitational constant in natural microscopic units to the fine structure constant. We have no place for a number of such small size in our present theories of quantum physics. We can perhaps hope that sometime in the future a connecting link may be found between the apparently unconnected subjects of microphysics and gravitation, but at present we have no hints of how the gap might be bridged.

41 Let us now consider the strength of the electrostatic field at a distance of a Bohr radius a_0 from a proton. Since a_0 is of the order of 10^{-8} cm, and since the electrostatic potential energy of the electron in the hydrogen atom is of the order of 10 eV, we see that this field is of the order of 10^{11} volts/meter, or precisely

$$E_{\text{atom}} = 5.14 \times 10^{11} \text{ volts/meter} \qquad (41a)$$

Compared with the strongest macroscopically realizable electrostatic fields, which are of the order of magnitude of 10^7 volts/meter, this is a very strong field. We conclude, first of all, that the effects of external electric fields which we can produce in the laboratory will be small on atoms and molecules and completely negligible on nuclei. Nevertheless the effects are observable; an electric field will split each of the spectral lines of an atom into several lines of approximately the same frequency. This phenomenon is known as the *Stark effect*.

The circumstance that the electrostatic fields acting on the electrons in an atom are large compared with the electrostatic fields which can be realized macroscopically in the laboratory can be readily understood as follows. It is an important feature of the electrostatic field (as described by Maxwell's equations) that if such a field is maintained in any empty region in space, then the field strength must assume its largest value at some point on the conductors. The conductors are, however, built of atoms, and if the field strength on the conductor should become comparable to the field strength holding the atoms together then the conductor would begin to disintegrate. The estimate (41a) is therefore an absolute upper limit on realizable macroscopic electrostatic fields; in actual practice electrical breakdown occurs long before this upper limit is achieved.

42 Similar considerations apply to macroscopic magnetostatic fields. The fields which we can realize in the laboratory must necessarily be weak in the sense that their effect on the structure of atoms cannot be very marked. A magnetic field also produces a splitting of a spectral line into several components. This phenomenon is known as the *Zeeman effect*.

To set an upper limit on realizable magnetic fields we may determine the magnetic field which produces the same energy density as an electric field of the order of 10^{11} volts/meter; this magnetic field is of the order of 10^7 gauss. Steady magnetic fields of strength up to 50,000 gauss can be produced easily in the laboratory, and fields in the neighborhood of 10^6 gauss can be produced for short time intervals. Consideration of the stresses on the current-carrying conductors which produce the fields, and which must not exceed the limits set by the forces holding the atoms and solids together, tells us that we cannot possibly produce static fields in excess of 10^7 gauss.

43 If we look upon the strength of macroscopic fields from the standpoint of *natural field strengths in quantum electrodynamics*, we can conclude that even the electric fields in the atoms are very weak. We might define the natural unit of field strength (electric or magnetic) as the field which produces an energy density in space of (one electron rest energy)/(Compton wavelength of the electron)^3. This unit of electric field strength is equal to 4.0×10^{17} volts/meter and the corresponding unit for the magnetic field is equal to 1.3×10^{13} gauss. The theory of quantum electrodynamics predicts marked deviations from Maxwell's equations in the vacuum at these field strengths. In particular the principle of superposition no longer holds, and the electromagnetic fields cannot be described by linear equations. Actually quantum electrodynamics also predicts very small deviations from linearity at the very weak fields which can be realized in the laboratory. These deviations are, however, so fantastically small that they are of no practical importance macroscopically, and in fact they have not been detected in macroscopic experiments to date. The smallness of the macroscopic fields when measured in natural units, which ultimately can be traced back to the smallness of the fine structure constant α, provides us with a certain understanding of why the linear Maxwell equations are so accurate in practice.

Concerning Numerical Work

44 Let us say something about the numerical evaluation of a theoretical expression for some physical quantity. The reader may feel that nothing need be said about this subject; numerical work is a necessary evil (especially in homework problems), and we learn no physics from such arithmetical exercises. This is not quite true. There are "bad" numerical evaluations and there are "good" numerical evaluations. To carry out a good numerical evaluation requires *physical insight*. Let us consider an example to show the difference between "good" and "bad" work. In the study of the finer details of the spectrum of the hydrogen atom it is found that the spectral lines, which in a measurement of poor resolution appear to be single, upon higher resolution actually consist of several very closely spaced lines. We say that the spectrum has a *fine structure*. In the theoretical study on fine structure we encounter an energy, E_f, which characterizes the typical separation between two of these closely spaced lines, and this energy is given theoretically by the expression

$$E_f = \frac{e^8 m}{32 \hbar^4 c^2} \tag{44a}$$

Now it would definitely be "bad" to evaluate E_f by inserting the values of the constants occurring in the expression (44a) directly from Table 2A. First of all it would be very tedious since we have to compute e^8 and $\hbar^4$. Secondly, the formula (44a) is very untransparent; before we have carried out the computation we cannot "see" what the magnitude of this energy is, and the expression as it stands shows us nothing about the physical nature of the effect. Suppose, however, that we first group the constants in Eq. (44a) into factors which have a recognizable significance as follows:

$$E_f = \frac{1}{16} \left(\frac{e^2}{\hbar c} \right)^4 \left(\frac{1}{2} mc^2 \right) = \frac{1}{16} \alpha^2 \left(\frac{1}{2} \alpha^2 mc^2 \right) = \frac{\alpha^2}{16} R_\infty \tag{44b}$$

If we look at the expression to the extreme right we see that the *magnitude* of the fine structure separation E_f is now quite transparent; it is a small correction to the gross structure of relative order of about 10^{-5}. If we wish to compute the energy E_f in electron volts the computation is now simple; we have to multiply 13.6 eV by the constant $\alpha^2/16$. It is, therefore, clear that the grouping of the factors, as in Eq. (44b), leads to a simplification of the purely numerical work. However, the expression (44b) also gives us some insight into the nature of the effect. In a purely nonrelativistic theoretical treatment of the hydrogen atom (in the approximation of an infinite proton mass), and with the neglect of effects due to the intrinsic magnetic moment of the electron, there would be no fine structure. To see this we have to remember that in such a theory only the constants e, m, and $\hbar$, but not c, can appear. The ionization energy R_∞ is in fact independent of c. The fine structure constant, which appears in the expression for E_f, is, however, inversely proportional to c, and in the nonrelativistic appproximation, in which $c = \infty$, we obtain $E_f = 0$. We may, therefore, regard E_f as a relativistic correction to the gross structure. We then expect that the magnitude of this correction should be of the order of $(v/c)^2 R_\infty$, where v is the velocity of the electron. We have estimated the velocity v, and we have found that $(v/c) \sim \alpha$ and hence we are led to an expression like the one in (44b). The fine structure in hydrogen is therefore a relativistic effect.

45 The name "fine structure constant" for the constant α arose historically in connection with Sommerfeld's work on the fine structure of hydrogen; the constant α was first recognized as an important constant just in the expression (44b). At the time when Bohr presented his theory of the hydrogen spectrum it did not appear natural to write the hydrogen ionization energy R_∞ in the form

$$R_\infty = \frac{1}{2}\alpha^2 mc^2 \tag{45a}$$

as we have written it. It was rather written as

$$R_\infty = \frac{e^4 m}{2\hbar^2} \tag{45b}$$

and for this reason α was not called the "gross structure constant," which would be more appropriate. The expression (45a) must be regarded as a "better" expression for R_∞ as it gives us a much better insight into the nature of atoms. As we have explained, α is the fundamental coupling constant between the electromagnetic field and the elementary charge. Atoms are "loosely bound structures," with "slowly" moving electrons, because α is small compared to unity. For this reason a nonrelativistic theory leads to a good approximation. The relativistic corrections are of the order of $(v/c)^2$, hence of order α^2.

46 We hope that this example shows something of the spirit in which numerical work should be approached. We always try to recognize physically significant combinations of the constants in our expressions, and we carry out a grouping of the factors, or terms, before we evaluate anything numerically. This grouping clearly requires insight; unless we understand the nature of the phenomenon we cannot do it in a natural and meaningful way.

Our homework problems in this book are *not* intended to be merely exercises in arithmetic. They are intended to familiarize the reader with the orders of magnitude of quantum physics, and they are intended to teach the reader how to apply the ideas discussed in the text to concrete physical situations.

Advanced Topic: The Fundamental Constants of Nature†

47 Let us speculate about the following interesting question: How many independent fundamental constants of nature are there really?

The idea behind this question is as follows. Our present physical theories imply definite relationships between the parameters characterizing physical systems. The ionization energy of hydrogen, for example, can be expressed theoretically in terms of the constants m, e, and h, or, if we like, in terms of the constants m, c, and α. If we already know the constants m, e, and h, we can *predict* the ionization potential, and then check our theory by comparing the prediction with the experimental result. In the same sense a very large number of other physical parameters are "understood theoretically": they can be expressed in terms of a few *fundamental* constants.

The phrase "understood theoretically" should here be interpreted in a very generous sense. We regard a constant as understood theoretically whenever we can formulate a definite equation which *in principle* determines the constant, irrespective of whether our limited mathematical abilities actually suffice for the computation of the numerical value of the constant.

Our classification of physical parameters into fundamental constants and derived constants is in principle quite arbitrary. In practice we pick as our fundamental constants those parameters which occur in a particularly "simple" way in our equations, and which have a reasonably transparent physical interpretation. It is clearly more reasonable to regard the fine structure constant as a fundamental constant, and the ionization energy of hydrogen as a derived constant, than to reverse the status of these two parameters.

A set of independent fundamental constants is thus a set of suitably selected physical parameters which are not related to each other theoretically. We have no understanding of their numerical magnitudes: each one of these constants must be determined empirically. Our question is concerned with the *maximum* number of independent constants, i.e., with the number of constants that must be known before all other physical parameters can be computed (predicted).

It is clear that our question makes sense only with reference to our present physical theories. A constant declared to be purely empirical today might be "explained" tomorrow, within the framework of a new theory.

---

† Can be omitted in a first reading.

48   To explore the situation at present, let us try to list a number of fundamental constants:

(i) The fine structure constant:

$$\alpha = \frac{e^2}{\hbar c} \cong \frac{1}{137}$$

(ii) The ratio of the electron mass to the proton mass:

$$\beta = \frac{m}{M_p} \cong \frac{1}{1836}$$

(iii) The constant of gravitation in natural atomic units:

$$\gamma = \frac{(M_p{}^2 G)/(\hbar/M_p c)}{(M_p c^2)} = 5.902 \times 10^{-39}$$

(iv) A constant which characterizes the strength of the so-called *weak interactions*, which are responsible for the beta-decay of many nuclei.   According to our present beliefs the weak interactions have nothing to do with the strong nuclear force, electromagnetism, or gravity.   All phenomena involving weak interactions seem to be manifestations of a fundamental (universal) interaction in nature, characterized by *one* single coupling constant.   The strength of this interaction is about 10^{-14} times smaller than the strength of the nuclear force.

(v) The ratio $m/m_\mu \sim 1/200$, of the electron mass to the muon mass.   The *muon* (or *mu-meson*), is an elementary particle which does not seem to differ from an electron in any other way than in its larger mass.   The role of the muon in the scheme of things is almost completely unknown.

(vi) Finally, we need some constants which describe the *strong interactions*, which in particular include the strong nuclear forces. The theoretical situation is very unclear, and we do not know how many independent constants there are in this category.   We may consider the following two:

$$s_1 = \frac{\text{(mass of the pi-meson)}}{\text{(mass of the proton)}} \cong 0.15$$

$$s_2 = \frac{B_D}{M_p c^2} \cong 2.35 \times 10^{-3}$$

where $B_D = 2.23$ MeV is the binding energy of the deuteron.

We have arbitrarily picked the constant s_2, because of its immediate physical significance, as *one* possible constant to describe the strength of the nuclear force.   There is nothing very fundamental about this number, but we can imagine that it does give us a measure of the strength of the force. In other words, we believe that all other nuclear binding energies can, in principle, be expressed in terms of s_2 and s_1. In this case we really have to take an extremely generous view of what we mean by "understanding things theoretically." We *do not* know, in this case, what the "correct equation" is, and our pious hope that such an equation, involving only s_1 and s_2, exists, may be totally unfounded.

The truth is that at the time of this writing we cannot really compute the ratios of the masses of such particles as the *K*-mesons, the nucleons, the lambda-particle, etc. We do not have a fundamental theory for doing this, and we should, perhaps, add all these mass ratios to our list. On the other hand, it is possible that a theory may one day emerge through which we can compute the masses of some, or maybe all, of the strongly interacting particles. According to the most extremely optimistic view the "correct" theory will be such that strong interaction physics will contain *no* empirical constant; everything will be computed, including the numbers s_1 and s_2. At this time, however, the question of the number of constants characterizing the strong interactions must be regarded as completely open.

49   We have not included in our list a very remarkable *empirically* determined constant, namely the ratio of the charge of the electron to the charge of the proton. According to an experiment performed by J. G. King in 1960, this ratio equals -1 to an accuracy of one part in 10^{20}. King similarly measured the ratio of the charge of the helium nucleus to the charge of the proton and found that this ratio equals 2, to a similar accuracy.† These results strongly support the idea that the charge of any particle must be an integral multiple of the electronic charge. There is much evidence in favor of this idea, although this evidence is, in most cases, not as stringent as in King's measurements. Physicists have, in fact, believed in the "quantization of charge" for a long time. There is, however, no theoretical understanding of *why* all charges should be integral multiples of the electronic charge.

Why, then, did we not include the constant (-1 ± 10^{-20}) in our list? Because our theories are such that we would find it very upsetting if this constant is not actually equal to -1. We can calmly contemplate the possibility that the constants actually included in the list would be slightly different; they are empirical

---

† We are here stating a reasonable inference. What King actually did was to establish that the hydrogen molecule and the helium atom are *neutral* to the stated accuracy. [J. G. King, "Search for a small charge carried by molecules," *Physical Review Letters* **5**, 562 (1960).]

constants in just this sense. It would not upset quantum electrodynamics if the fine structure constant were one per cent larger; our laws of nature, as we know them, would not have to be changed in any essential point. The situation is different with regard to the quantization of charge; the *structure* of our theory depends on this principle.

50 Quantum electrodynamics, as a theory of atoms, molecules, and matter in bulk, in essence contains only *two* fundamental empirical constants, namely α and $\beta = m/M_p$. What we mean by this statement is that we believe that we know, in principle, the dependence of all physical quantities in this realm of physics on these two constants. Properties of the different atomic nuclei enter only through the *integers* Z and A, and the other physical characteristics of nuclei have only a "small" effect on atoms, molecules, and matter in bulk.

Our statement is, therefore, a simplification of the true situation, but it is interesting to pursue this idea. At first sight it may appear wrong, since the number of "fundamental constants" listed in Table 2A is certainly larger than two. It should be noted, however, that the constants listed are based on completely arbitrary units (human units) and that their numerical values have no absolute significance at all.

In trying to understand the properties of matter in bulk it is therefore essential that we distinguish between the fundamental physical quantities and those that depend on our arbitrary units. Consider for instance the velocity of sound in a crystal. To ask for this velocity in cm/sec is not a "fundamental" question because the answer depends on the arbitrary definition of cm and sec. The most clear-cut theoretical question is to ask for the ratio of the velocity of sound, c_s, to the velocity of light; *this* quantity is clearly independent of all macroscopic standards. We believe firmly that this number is, in principle, computable within quantum electrodynamics.

51 To understand the true meaning of the constants listed in Table 2A let us consider the definitions of our macroscopic system of units.

The *kilogram* has been defined, through international agreement, as the mass of a certain chunk of metal kept in Paris. To indicate that we mean just this chunk we shall denote the unit by $(kg)_P$, the "Paris kilogram." The *gram* is defined as $(gm)_P = (kg)_P/1000$.

This chunk of metal contains a certain number of nucleons, say n_1 nucleons. The precise value of n_1 is unknown, but we could determine it in principle by counting. If we now assume that we can compute, within nuclear physics, and within the theory of strong interactions, the ratio of every nuclear mass to the proton mass, then we can write the mass of the chunk in Paris in the form

$$(\text{kg})_P = n_1 c_1 M_p = n_1 c_1 \beta^{-1} m \tag{51a}$$

where c_1 is a constant, close to unity, which we determine through our computation. Strictly speaking it does depend on α and β, but only weakly. The number n_1, although not known precisely, is a numerical constant selected through international agreement; it is the number of nucleons in the chunk.

52 For the *meter* there exist, or rather have existed, two standards. In the old standard the meter is the distance between two notches on a certain metal rod kept in Paris; we shall denote this meter by $(\text{m})_P$, the "Paris meter." The new standard is "atomic" in nature, and the corresponding meter, which we denote $(\text{m})_a$, the "atomic meter," is defined as a certain multiple of the wavelength of a definite orange line in the spectrum of krypton, the multiple being $n_2 = 1{,}650{,}763.73$, by international agreement.

The wavelength of the orange krypton line is something which we can compute in principle (but not in practice), and we can write this wavelength in the form

$$\lambda = c_2 \alpha^{-2} \left(\frac{\hbar}{mc} \right) \tag{52a}$$

where c_2 is a constant which depends only very weakly on α and β. In a first approximation it is a purely numerical constant, and if we should master the mathematics of atomic physics we could find this number.

The atomic meter can thus be written

$$(\text{m})_a = n_2 c_2 \alpha^{-2} \left(\frac{\hbar}{mc} \right) \tag{52b}$$

53 The adoption of an "atomic standard" of time seems imminent, although the *second* is presently defined astronomically. Let us, however, get ahead of history and assume that the atomic standard *has* been adopted and that the second has been defined in terms of a certain transition frequency in the cesium atom which lies in the radio-frequency region. This frequency, which can be interpreted as the precession frequency of the spin of the cesium nucleus in the field of the orbital electrons, has been measured very accurately and found to have the value (in terms of the astronomical second)

$$\frac{1}{T_0} = \nu_0 = 9{,}192{,}631{,}770 \pm 10 \text{ cycles/sec} \qquad (53a)$$

The precision of this number is characteristic of the precision which can be achieved in radio-frequency measurements. The theoretical expression for this frequency is of the form (according to quantum electrodynamics)

$$\nu_0 = c_3 \alpha^4 \beta \left(\frac{mc^2}{\hbar} \right) \qquad (53b)$$

where c_3 is a numerical constant almost independent of α and β; it can be evaluated in principle but not in practice, provided we are given certain data concerning the cesium nucleus. Suppose, therefore, that we *define* the second, denoted $(\text{sec})_a$, or the "atomic second," as follows:

$$(\text{sec})_a = (9{,}192{,}631{,}770) T_0 = n_3 c_3^{-1} \alpha^{-4} \beta^{-1} \left(\frac{\hbar}{mc^2} \right) \qquad (53c)$$

where T_0 is the period of this atomic oscillation, and where the number $n_3 = 9{,}192{,}631{,}770$ is thus determined through international agreement.

54   Consider finally the old standard of length, the Paris meter $(\text{m})_P$. It is defined as the distance between two notches on a metal rod, and it is therefore equal to the length of a certain string of atoms. The number, n_4, of atoms in this string is, in a sense, determined by international agreement (although the number is not accurately known), since n_4 is the number of atoms between the two notches. The separation between two neighboring atoms in the metal rod is something which we can compute in principle; this separation is of the form $c_4 a_0$, where a_0 is the Bohr radius, and where c_4 is a constant which depends only very weakly on α and β. We can therefore write the Paris meter in the form

$$(\text{m})_P = n_4 c_4 \alpha^{-1} \left(\frac{\hbar}{mc} \right) \qquad (54a)$$

This standard of length has been abandoned for obvious technical reasons; the separation between the two notches is a poorly defined quantity. Two optical wavelengths can be compared with greater accuracy, and there is then no reason why we should try to express these wavelengths in terms of the length of the metal rod.

55   Our discussion shows the true nature of the macroscopic standards. They are defined in terms of somewhat arbitrarily se-

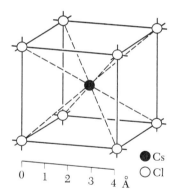

To remind the reader that the spacing between atoms in any solid is of the order of the Bohr radius a_0. The above figure shows the crystal structure of cesium chloride. This kind of lattice is called a body-centered cubic lattice. We can describe it by saying that the chlorine atoms form a cubical lattice, with one cesium atom at the center of each cube. Notice that this structure is different from the one of sodium chloride, shown in Fig. 30A, Chap. 1.

The chemical formula for cesium chloride is CsCl, and the lattice contains an equal number of cesium and chlorine atoms. The reader should convince himself that this is implied by the figure, although it might at first sight seem as if there were more chlorine atoms than cesium atoms in the lattice.

lected "atomic parameters," and in terms of the numbers n_1, n_2, and n_3 which are selected through international agreement. (As we said n_1 is actually not accurately known; it is defined implicitly.) We note the following:

(i) To measure an optical wavelength amounts to comparing this wavelength with the wavelength of the orange krypton line. This can be done very accurately, and for this reason optical wavelengths are accurately known quantities. The Rydberg constant $\tilde{R}_\infty$ is in essence an optical wave number, and that is why this constant is known with such accuracy. Our most accurate length measurements are nothing but measurements of ratios of optical wavelengths. These data have *potential* theoretical significance; if we could master the theory of atomic spectra sufficiently well to enable us to predict these wavelength ratios to a comparable accuracy, we could carry out meaningful comparisons between theory and experiment. Our computational abilities are, however, very limited, and for this reason the actual theoretical significance of the wavelength measurements is limited.

(ii) Two frequencies in the radio-frequency region can be compared very accurately. If we measure an atomic or molecular frequency in this region we, in effect, compare it with the cesium frequency.

(iii) To measure the velocity of light amounts to comparing the *frequency* associated with the orange krypton line with the cesium frequency. It is, therefore, not a measurement of a "fundamental physical constant," but rather an evaluation of our arbitrary standard of length in terms of our arbitrary standard of time.

56 Consider the expressions (51a), (52b), (53c) and (54a). They give us *theoretical* expressions for the macroscopic standards in terms of (i) the numbers n_1, n_2, n_3, and n_4, determined through international agreement; (ii) the basic standards m, $\hbar/mc$, and $\hbar/mc^2$ of quantum electrodynamics; and (iii) the quantities c_1, c_2, c_3, and c_4, which we believe we can compute *in principle*.

Even if we cannot compute the quantities c_1, c_2, c_3, and c_4 accurately in practice, we know that they are, to a first approximation, just purely numerical parameters, independent of α and β. If we could really *compute* these numbers it would mean that we could *compute* the value of the velocity of light, in units of $(\mathrm{m})_a/(\sec)_a$.

Our theoretical expressions for the macroscopic standards enable us to deal with the following question: What would the world be like if our constants of nature were *slightly* different? This means: what would the world look like if the two empirical constants α and β were *slightly* different? This is an interesting question because it tests our understanding of the role of α and β in the world. We leave it to the reader to speculate about this question; he should return to this problem after he has finished reading this book.

57   Suppose we ask the question, why is the size of an atom of the order of 10^{-10} meters: why are atoms as small as they are? This sounds like a metaphysical question, but it really is not. Suppose we reformulate the question as follows: why is the linear size of a human about 10^{10} a_0?   That is what the question amounts to, because the meter was so defined that the linear size of a human is about a meter.   We can answer this question roughly if we can explain the number of atoms in a human, and this problem is not in principle beyond the reach of physics.   It would be absurd to try to compute this number precisely, but we ought to be able to estimate it to within a factor of, say, 10^6.   (If we knew more about biology and related subjects, that is.)   We leave these frivolous speculations to the reader.   We mentioned this issue only to illustrate how *all* the properties of the macroscopic world we live in ultimately depend on the properties of the elementary particles and their interactions.

References for Further Study

Among extensive tables of physical constants we mention the following:
1)   *Handbook of Chemistry and Physics* (Chemical Rubber Publishing Co., Cleveland, Ohio).   New editions have appeared yearly.   The latest editions incorporate the changed scale of atomic masses resulting from the selection of C^{12} as the new standard.
2)   *American Institute of Physics Handbook* (McGraw-Hill Book Company, New York, 1957).
3)   For a very interesting account of the struggles of mankind in the determination of the constants of physics, we recommend Cohen, Crowe and DuMond: *The Fundamental Constants of Physics* (Interscience Publishers, Inc., New York, 1957).
4)   For a critical survey of the fundamental constants, see E. R. Cohen and J. W. M. DuMond: "Our Knowledge of the Fundamental Constants of Physics and Chemistry in 1965," *Reviews of Modern Physics* **37**, 537 (1965).
5)   With reference to our discussion in Sec. 57 the reader may be interested in reading: "Gulliver was a bad biologist," by Florence Moog, *Scientific American*, Nov. 1948, p. 52.

Problems

1   In 1903, P. Curie and Laborde studied the heat emission of radium. They found that 1 gram of pure radium (we now know that the isotope

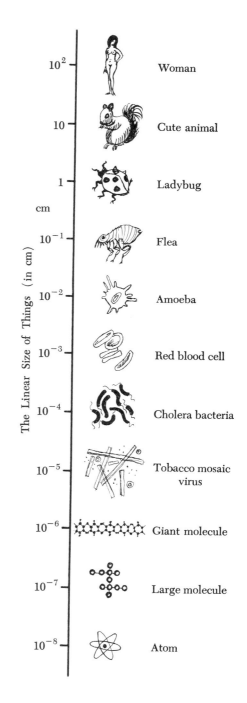

The Linear Size of Things (in cm)

10^2	Woman
10	Cute animal
1	Ladybug
cm	
10^{-1}	Flea
10^{-2}	Amoeba
10^{-3}	Red blood cell
10^{-4}	Cholera bacteria
10^{-5}	Tobacco mosaic virus
10^{-6}	Giant molecule
10^{-7}	Large molecule
10^{-8}	Atom

$_{88}$Ra^{226} is involved) emits about 100 cal/hr. From this and from the known half-life compute the approximate energy in MeV with which the emitted alpha-particles emerge. In the experiment of Curie and Laborde these particles were captured within the source and the calorimeter, and their kinetic energy therefore was converted to heat energy. (The half-life is 1622 years.)

2 (*a*) The radium nucleus has a *positive* mass defect, but nevertheless it is unstable and decays. How is this possible? Is not the necessary and sufficient condition for stability that the mass defect be positive? Explain in detail.

(*b*) The radium isotope referred to above is $_{88}$Ra^{226}; this is the isotope discovered by P. and M. Curie. It decays through the emission of alpha-particles, which are nothing but helium nuclei, $_2$He^4.

We might be led to believe that only stable nuclei or long-lived isotopes occur naturally since any short-lived isotopes would have decayed in geological times. Now, measured against the age of the earth the half-life of 1622 years is not particularly large; it is, rather, quite small. How do you then explain the natural occurrence of radium?

3 In connection with the decay of a radioactive nucleus like Ra^{226} we note a remarkable circumstance; the lifetime is "unreasonably long." Try to form, from the fundamental constants of nuclear physics and electrodynamics, a "natural time" and find this in seconds. No matter how you play with your constants (nicely, that is), you must admit that Ra^{226} lives *much* too long. Clearly here we encounter a problem which we must try to solve later; it is actually possible to account for the observed phenomena, and the reason for the long lifetime (or if you like, the reason for the decay) is an interesting quantum-mechanical effect, known as the *tunneling effect.*

4 The sun radiates energy from its surface at the rate of 3.86×10^{26} watts. Before the development of nuclear physics it was something of a problem to explain where this huge amount of energy comes from. Let us try to make some simple estimates.

The sun is believed to be at least 4 billion years old. The mass of the sun is 1.98×10^{30} kg.

(*a*) What fraction of the sun's mass has to be converted into radiant energy per year in order to account for the power radiated? You will find that this number is quite consistent with the idea that the sun has not changed much during its lifetime, i.e., during the last 4 billion years.

(*b*) Rule out chemical reactions as the source of energy.

(*c*) Do you know of any nuclear process which might take place in the interior of the sun, and which might provide us with the explanation of where the energy comes from? Consult some introductory astronomy book, and convince yourself through some simple estimates that your explanation is plausible, or at least not in violent contradiction with the facts.

5 We have said that the density of nuclear matter, i.e., the density of the "substance" inside a nucleus, is roughly the same for all nuclei. Give this density in macroscopic units, i.e., in gm/cm^3.

6 (*a*) With reference to the discussion in Sec. 17, estimate the mean energy and the mean velocity of a nitrogen molecule in nitrogen gas held at room temperature. The nitrogen molecule consists of two nitrogen atoms. (Give energy in eV.)

(*b*) Under atmospheric pressure, and at room temperature, 1 mole of nitrogen gas (or *any* gas) occupies a volume of 22.4 liters. Estimate the number of collisions that a nitrogen molecule will undergo per second, assuming that a nitrogen molecule is of "typical molecular size." Compare this collision frequency with a typical optical frequency.

7 One of the lines in the hydrogen spectrum has the wavelength 4861.320 Å. It was discovered by H. Urey in 1932 that this line has a faint companion located at 4859.975 Å [See *Phys. Rev.* **39**, 164 (1932); **40**, 1 (1932).] The explanation is that ordinary hydrogen is not isotopically pure, but a mixture of two isotopes, $_1H^1$ and $_1H^2 = D$. The atom of the heavier isotope, deuterium, occurs only to about 0.015 percent, and this isotope is responsible for the faint line mentioned.

In the study of the hydrogen spectrum one may, to a first approximation, neglect the motion of the nucleus. Suppose now that we try to take the nuclear motion into account. It is then no longer the nucleus which stays fixed but rather the center of mass of the nucleus and the electron. A theory which takes the nuclear motion into account therefore predicts slightly shifted locations of the spectral lines relative to the predictions in a theory in which the nucleus is infinitely heavy, and the amount naturally depends on the actual mass of the nucleus (in our case the mass of the proton or deuteron).

Try to formulate a simple theory which explains the ratio of the two wavelengths given. Use the two wavelengths to compute the ratio of the deuteron mass to the proton mass, and compare your result with the ratio which can be obtained from a table of nuclear masses.

8 Singly ionized helium, i.e., a helium atom with *one* electron removed is, like the hydrogen atom, a system consisting of a single electron moving around a nucleus. We may therefore expect that the spectral lines emitted by singly ionized helium are entirely analogous to the spectral lines emitted by the hydrogen atom. The two systems are, however, not identical; the helium nucleus carries two elementary charges whereas the hydrogen nucleus (proton) carries only one. In view of what has been said in this chapter it should be possible to find the consequences for the spectrum of the increased central charge in singly ionized helium as compared to hydrogen, and it should, therefore, also be possible to predict the wavelength of any line emitted by singly ionized helium, given the wavelength of the corresponding line in hydrogen. In other words, it is possible, without a detailed theory of atomic structure, to find the *ratios* of the corresponding wavelengths.

One of the visible hydrogen lines has the wavelength 6562.99 Å. What is the wavelength of the corresponding line emitted by singly ionized helium? Does this line lie in the visible region?

We may here assume that both nuclei are infinitely heavy. This example teaches us that primitive dimensional arguments, like the one given in Sec. 27, can sometimes be employed to make precise, quantitative predictions.

9 Suppose an alpha-particle collides head on with a nucleus of charge number Z and mass number A. Derive an expression for the energy, in MeV, which the alpha-particle must have in order to just reach the surface of the nucleus, as a function of the mass number A. For simplicity we shall assume that the nucleus remains stationary in the collision; that $A = 2Z$; and that the alpha-particle is a point charge without any size. If the alpha-particle does not reach the surface, the short-range nuclear forces cannot come into play, and the collision event proceeds as if only the electrostatic forces were effective. The energy which you have computed is, therefore, very roughly, a characteristic energy above which the outcome of the scattering event will begin to differ significantly from the prediction in which only the electrostatic forces are taken into account.

10 In this problem we shall consider the electrostatic energy of repulsion in a nucleus. Since the density of nuclear matter is roughly a constant we will assume that a nucleus is a uniformly charged sphere. This model is reasonable for a not too light nucleus.

(*a*) Show that for a nucleus of mass number A, and charge number Z, the electrostatic energy U_e is given by

$$U_e \cong A^{5/3} \left(\frac{Z}{A}\right)^2 \times (0.7 \, \text{MeV}) \tag{i}$$

Let us furthermore assume that the number of neutrons equals the number of protons, i.e., that $A \cong 2Z$. From (i) we then obtain an expression for the *electrostatic energy per nucleon*, namely

$$\frac{U_e}{A} \sim A^{2/3} \times (0.17 \, \text{MeV}) \tag{ii}$$

This energy should be compared with the average binding energy of a nucleon which is about 8 MeV. We thus see that for a not too large A the electrostatic energy per nucleon is quite small. It grows, however, with A, and this circumstance provides the explanation for the systematic trend mentioned in Sec. 33. The nature of the specific nuclear force is such that if it were to act alone, then the most stable nuclei would have roughly the same number of neutrons and protons. Since, however, electromagnetic forces are also effective, the net effect of the simultaneous presence of the two kinds of force is that nuclei with an excess of neutrons are favored, and this tendency for neutron excess increases with mass number A.

(b) As a check on this picture of the nucleus we consider the following. The (unstable) fluorine isotope $_9F^{17}$ and the oxygen isotope $_8O^{17}$ differ in mass by $M(17;9) - M(17;8) = 3.0 \times 10^{-3}$ amu. We note that the first nucleus has 9 protons and 8 neutrons whereas the second has 8 protons and 9 neutrons. We may, in other words, obtain one from the other by interchanging the neutrons and the protons. We say that we have a pair of *mirror nuclei*.

We maintained in the text that neutrons and protons are quite similar physically, and if this is true we would expect the mass *defects* for the two nuclei mentioned to be equal. However, the proton and the neutron do differ in their charge, and the same can be said about the two mirror nuclei. Assuming now that the two mirror nuclei are otherwise equal, except for the charge, we may try to account for the difference in mass defect on the basis of the electrostatic energy of repulsion. Carry out a computation to see how well these ideas check.

11 Some of the heaviest known nuclei can decay spontaneously by *fission*. In this process the nucleus splits into two approximately equal parts, with an energy release of about 200 MeV per fission. Fission can also be induced by neutron bombardment. The nucleus absorbs the incident neutron, and is left in an excited state from which it subsequently decays by fission. The uranium isotope U^{235} is an example of a nucleus which readily undergoes fission after neutron absorption. Since heavy elements have an excess of neutrons over protons compared with elements in the middle of the periodic table, the fission process will lead to the emission of a few neutrons. These circumstances make a chain reaction possible: neutrons emitted in fission induce more fissionable nuclei to undergo fission, which leads to the emission of more neutrons, etc. Nuclear reactors and the (fission) atomic bomb work on this principle.

(a) Estimate the energy which would be released (in calories, and kilowatt-hours) if 1 gm of U^{235} underwent complete fission, and compare this energy with the energy which is released in a typical chemical reaction involving 1 gm of matter.

(b) A small chunk of U^{235} metal will not explode spontaneously, whereas a large chunk will. How do you explain this?

(c) To study the origin of the energy released in fission consider, on the basis of the relation (i) in Prob. 10, the electrostatic energy of the nucleus (say U^{235}) before fission and the total electrostatic energy of the fragments. Obviously some of the electrostatic energy will be released. Estimate this energy and compare with the value 200 MeV per fission.

12 The mass of two deuterium nuclei is larger than the mass of the alpha-particle (= the nucleus $_2He^4$). (See Table 4A for *atomic* masses.)

(*a*) Compute the energy released if 1 gm of deuterium undergoes *fusion* to form helium, and compare this with the energy released in fission.

(*b*) Why is it that a container filled with deuterium does not explode spontaneously?

13 Let us assume that the electron is a classical point-particle, and let us assume that an electron in an atom moves in an orbit in a plane perpendicular to the z-axis, and in such a way that its angular momentum is constant and equal to $\hbar$.

(*a*) What is the effective magnetic moment of the electron? We shall call this magnetic moment *one Bohr magneton*.

(*b*) What is the difference in energy (in eV) in the two situations when a magnetic moment of 1 Bohr magneton points in the same, and in the opposite, directions of a magnetic field of 1000 gauss?

(*c*) Suppose that in an iron crystal there is a magnetic moment of 1 Bohr magneton located at the position of every atom in the crystal, and suppose furthermore that all the magnetic moments point in the same direction. Is the resulting magnetization similar in magnitude to the magnetization encountered in a saturated ferromagnet?

We are here interested in estimating the magnitude of the magnetic moments which we can expect to find in atoms. Our naive classical model of atomic magnetism should not be taken too seriously. It turns out, however, that the Bohr magneton is indeed typical for atoms. In the complete quantum-mechanical discussion of atomic magnetism one recognizes two contributions to the magnetic moment. One is due to the "orbital motion" of the electron, and it is analogous to our classical magnetic moment. The other is due to the *spin* of the electron; an electron also possesses an intrinsic angular momentum analogous to the angular momentum of a small billiard ball spinning around an axis passing through its center. The magnitude of this spin angular momentum is $\hbar/2$, and the corresponding magnetic moment is very closely one Bohr magneton.

The purpose of the estimate in part (*c*) of the problem is to find out whether we can *hope* to explain ferromagnetism in terms of magnetic moments of atoms. The result of this estimate is encouraging. It must be mentioned, however, that ferromagnetism is a complicated phenomenon, and our simple estimate does not tell the whole story.

14†   In Secs. 51–56 we discussed the "atomic nature" of some of our macroscopic standards of measurement.

Suppose, then, that at present our standards have been compared and adjusted so that $(m)_P = (m)_a$, and so that the basic atomic constants e, m, M_p, c and $\hbar$ have the values given in Table 2A in terms of these standards. Suppose furthermore that at 1 A.M., on May 30, 1988, the constants α and β change suddenly to

$$\alpha' = \alpha(1 + u), \qquad \beta' = \beta(1 + w)$$

and then remain constant at their new numerical values. We shall assume that the numbers u and w are small, say of the order of 1 percent; otherwise the change in the world order may be too drastic. This natural catastrophe will certainly be noticed, and for some time (after recovering from the initial shock) physicists will be busy with the remeasurement of their sacred constants. Let us denote quantities after the catastrophe by primes.

  (*a*) Find $(m)'_P/(m)'_a$.

  (*b*) What are the new values of the mass of the electron and the mass of the proton? [In $(gm)'_P$.]

  (*c*) What is the new value of the velocity of light, c', in units of $(m)'_a/(sec)'_a$?

  (*d*) What is the new value $\hbar'$ of Planck's constant?

  (*e*) What is the new value of the electronic charge in electrostatic units, and what is the new value in coulombs?

  (*f*) What will be the density of copper [in $(gm)'_P/(cm^3)'_a$] after the catastrophe?

———————

† This problem refers to an advanced topic.

Chapter 3

Energy Levels

Chapter 3   Energy Levels

4742.5—4728.6 A.

Wave-length	Element	Arc	Spk.,[Dis.]	R	Wave-length	Element
4742.589	Mo	–	10	–	4737.642	Sc I
4742.549	Er	3 w	–	–	4737.626	U
4742.5	bh Sc	5	–	Me	4737.561	Pt I
4742.481	Sm	3	–	–	4737.350	Cr
4742.392	Nd	4	–	–	4737.282	Ce
4742.333	U	10	3	–	4737.1	bh C
4742.325	Pr	7	–	–	4737.05	Tl II
4742.266	Th	4 l	2	–	4736.965	Zr
4742.25	Se I	–	[500]	Rd	4736.958	Sm
4742.227	Sm	2	–	–	4736.945	Er
4742.110	Ti I	15	1	–	4736.9	bh Z
4742.04	Ho	10	3	Ex	4736.79	Dy
4741.997	Er	3 w	–	–	4736.782	Ca
4741.937	Ge II	–	50	–	4736.780	Fe
4741.922	Sr I	30	–	ISn	4736.688	Pr
4741.78	Cd II	–	3	Vs	4736.637	Mc
4741.775	Eu	10 W	–	–	4736.608	Eu
4741.726	Sm II	80	–	–	4736.6	Rt
4741.71	O II	–	[20]	Fl	4736.491	Cl
4741.539	Dy	3	2	–	4736.490	S
4741.533	Fe I	12	1	S	4736.30	T
4741.520	W	12	2	–	4736.203	f
4741.503	Pr	30	–	–	4736.151	f
4741.404	Yt I	2	3	–	4736.116	
4741.398	Er	20	–	–	4736.089	
4741.282	U	1	2	–	4736.062	
4741.269	Ru	4	–	–	4735.94	
4741.10	Tm	3	–	Me	4735.93	
4741.018	Sc I	100	60 h	–	4735.848	
4741.005	Pr	6	–	–	4735.847	
4740.97	Se II	–	[600]	Bi	4735.77	
4740.928	Dy	3	2	–	4735.76	
4740.68	Cl I	3	[10]	Ks	4735.66	
4740.614	Cb	3	3	–	4735.49	
4740.524	Eu	500	2	–	4735.4	
4740.517	Th	20	15	–	4735.3	
4740.5	bh Zr	8	–	L	4735.3	
4740.40	Cl II	–	[150]	Ks	4735.2	
4740.359	Mo	5	5	–	4735.	
4740.331	Ru	7	–			

Fig. 1A A very small portion of a table of wavelengths: Massachusetts Institute of Technology *Wavelength Tables*, compiled under the direction of G. R. Harrison (MIT Press, Cambridge, Mass., 1939). This table, which consists of 429 pages, lists more than 100,000 spectral lines between 10,000 Å and 2,000 Å. Each page has three columns, and the lines are listed in order of decreasing wavelength. The chemical element corresponding to each line is identified, and some data on method of excitation and on intensities are given.

It is customary to list wavelengths in the visible region as measured in air, whereas wavelengths in the ultraviolet region refer to vacuum. In the visible region we have approximately: $\lambda_{vac} = 1.0003\ \lambda_{air}$. *(Courtesy of MIT Press.)*

Term Schemes

1 The fact that each chemical element is associated with a unique optical spectrum is one of the striking aspects of nature. This feature of nature is furthermore very general: not only do atoms have characteristic spectra, but so do molecules and nuclei as well. These objects emit and absorb electromagnetic radiation at certain definite frequencies, which range from the radio frequency region (for molecules) up to the region of very short wavelength X-rays, or gamma rays (for nuclei). Historically the optical spectra of the elements were discovered first, by G. R. Kirchhoff and R. Bunsen in the middle of the nineteenth century, whereas the radio-frequency spectra of molecules, and the gamma ray spectra of nuclei, were discovered much later, during this century.

We interpret the spectra in terms of *energy levels* of atoms, molecules and nuclei. Through our study of spectra we become aware of an extremely important property of composite systems, which is this: with each such system there is associated a set of energy levels, or *stationary states*, characteristic of the system. We find these levels in "small" systems, such as atoms, molecules and nuclei, in which cases the levels manifest themselves very directly in the spectra which we observe. We also find these levels in "large systems," such as solids, liquids and gases. At first it might not occur to us that there is a relationship between the emission and absorption of gamma rays by a nucleus, and the vibrations of a quartz crystal in some electronic device, but there is.

2 In this chapter we shall study energy levels in "small" systems. We shall discuss some of the relevant experimental facts, and we shall try to understand some aspects of what is observed on the basis of very simple theoretical ideas. In this chapter we shall not attempt to explain *why* energy levels occur, but rather accept this feature of nature as a basic empirical fact. In Chap. 8 we shall face the challenge of explaining the levels, and we shall see how they can be understood on the basis of quantum mechanics.

Our order of presentation is actually somewhat analogous to the historical development of the subject, in the sense that many of the features of *atomic* spectra discussed in this chapter had been discovered long before a satisfactory theory of atomic structure (i.e., quantum mechanics) had been created. Our account is, however, not *really* historical. We want to discuss the empirical facts concerning energy levels more generally, and therefore we shall also discuss nuclei, although the properties of nuclei became known only much later.

3   Some remarkable regularities in atomic spectra were noticed early.   As an example we mention the *Ritz combination principle,* according to which the wave numbers of many spectral lines of an element equal differences, or *sums,* of wave numbers of other *pairs* of lines.   For instance, in a certain element[†] the following lines have been observed: $\tilde{\nu}_1 = 82258.27$ cm^{-1}; $\tilde{\nu}_2 = 97491.28$ cm^{-1}, and $\tilde{\nu}_5 = 15232.97$ cm^{-1}.   We have $\tilde{\nu}_2 - \tilde{\nu}_1 = 15233.01$ cm^{-1}, which is so close to $\tilde{\nu}_5$ that we would hardly be willing to believe that this agreement is a mere "accident," especially not since the same feature occurs for other lines of the same element as well as for the lines of many other elements.

A more general principle was discovered later.   The wave number $\tilde{\nu}$ of *any* line emitted by an atom can be expressed as a difference $\tilde{\nu} = T' - T''$ of two *spectral terms* T' and T''.   Each atom is characterized by a *set* of such terms (expressed as wave numbers), known as the *term system* of the atom.

This principle contains the Ritz combination principle.   Suppose that three spectral lines are associated with three terms, as follows:

$$\tilde{\nu}_{12} = T_1 - T_2, \quad \tilde{\nu}_{13} = T_1 - T_3, \quad \tilde{\nu}_{23} = T_2 - T_3 \qquad (3a)$$

We then have

$$\tilde{\nu}_{23} = (T_1 - T_3) - (T_1 - T_2) = \tilde{\nu}_{13} - \tilde{\nu}_{12} \qquad (3b)$$

which is an example of the combination principle.

4   Today we interpret a spectral term as corresponding to an energy level of the atom, and the term system is therefore interpreted as a manifestation of a set of energy levels characteristic of the atom in question.   This idea was first formulated by Niels Bohr in his paper on the hydrogen atom.[‡]

Let us consider the matter in the light of what we already know about the quantum nature of electromagnetic radiation.   A light quantum, or photon, of frequency ν, and hence wave number $\tilde{\nu} = \nu/c$, carries the energy $E = h\nu = (hc)\,\tilde{\nu}$.   This energy is the difference between the two energies $E' = (hc)T'$ and $E'' = (hc)T''$ if the wave number is the difference between the two terms T' and

---

† We shall not here reveal the identity of the atom, because that would take all the fun out of Prob. 1 at the end of this chapter.

‡ N. Bohr, *Philosophical Magazine* **26**, 1 (1913).

Fig. 1B   The spectrum of hydrogen.   (Wavelengths in Ångströms.)   The appearance of this spectrum in the visible region is at first sight not particularly dramatic.   The wavelengths of hydrogen are, however, of very great interest.   Since the hydrogen atom is the simplest possible atom it plays the role of a probing stone for all theories of atoms: this spectrum *must* be explained.   That Bohr could account for these lines was a spectacular advance in our understanding of nature.   Modern quantum mechanics can account for everything visible on this plate, and much more, and the history of the theory of the hydrogen atom is indeed a dramatic chapter in the annals of physics.   (*Spectrum photographed by Dr. D. Goorvitch, Berkeley, for this book.*)

T″. The terms can therefore be alternatively expressed as energies, wave numbers, or frequencies, since these quantities are always related through the constants *h* and *c*. In view of this we can say that a table of spectral terms is a table of "energy levels." As we shall see this mode of expression has real physical significance: it is more than a mere change in terminology.

5   In some elementary accounts of atomic spectra and atomic structure the matter is stated somewhat as follows, in the form of two theoretical postulates:

I. "An atom can exist only in certain definite stationary states of internal motion. These states form a discrete set, and every state is characterized by a definite value of the total energy."

II. "When an atom emits or absorbs electromagnetic radiation it jumps from one stationary state to another. If the atom jumps from an upper state of energy E_u to a lower state of energy E_l, (hence we have $E_u > E_l$) a photon will be emitted, and the frequency ω of the emitted photon is given by†

$$h\nu = \hbar\omega = E_u - E_l \qquad\qquad (5a)$$

The inverse of the emission process is the absorption of a photon of frequency ω, in which case the atom jumps from the lower state to the upper state."

Now we should note immediately that if the above postulates are interpreted literally, then the first postulate is manifestly false. The "upper states" cannot be absolutely stationary or stable at all since the atom indeed decays from these states spontaneously. On a *macroscopic* time scale this decay is very rapid: we may quote a time of 10^{-8} seconds as an order of magnitude estimate which describes typical lifetimes of excited states of atoms. We should note, however, that this lifetime is fairly long on an *atomic* time scale. The frequency of an optical photon is of the order of 10^{14} sec^{-1}, and the corresponding period is thus much shorter than the typical lifetime of an excited state.

About the second postulate we may say that it is not very informative: we are left totally in the dark as to what it means that the atom "jumps" from one state to another. Some writers would actually not use the word "jump" at all, but would rather say that "the atom makes a *transition* from one state to another." This mode of expression undoubtedly sounds more learned, but it is

---

† As we explained in Sec. 8, Chap. 2, both ν and the associated quantity $\omega = 2\pi\nu$ are called "frequency." Similarly both h and $\hbar = h/2\pi$ are called "Planck's constant." In the following we will mostly use ω and $\hbar$, because the author likes them better than ν and h.

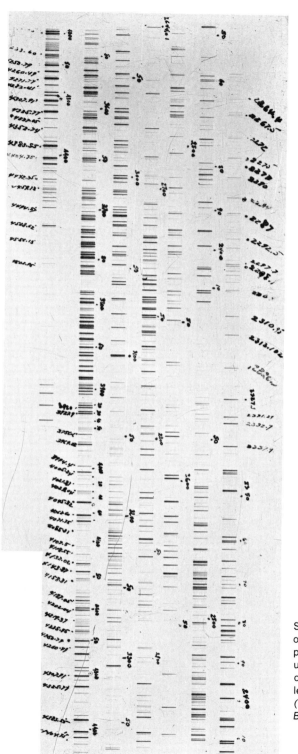

Several portions of the spectrum of iron photographed on the same glass plate. Wavelengths written on the plate are in Ångströms. The purpose of this particular photograph was not to measure the wavelengths of iron, but rather to use these well-known wavelengths to calibrate the quartz prism spectrograph. *(Photograph by courtesy of Professor S. P. Davis, Berkeley.)*

Spectrum of helium (long lines) superimposed on spectrum of iron (short lines). The numbers on the plate give some of the helium wavelengths in Ångströms. The contrast between the complexity of the iron spectrum and the simplicity of the helium spectrum is striking. *(Photograph by courtesy of Professor S. P. Davis, Berkeley.)*

hardly more informative. What precisely takes place when the atom makes the transition?

In spite of what we have just said the reader should not reject our two postulates as meaningless statements: they should be looked upon as a *first approximation* to the description of very complex phenomena, and as such they serve a useful purpose.

6 To account for all the observed spectral lines in an atom (or molecule or nucleus) we try to construct a term system, or energy level system, for the atom, by which we understand a listing of a number of energy levels E_0, E_1, E_2, . . . , etc., such that every observed spectral line corresponds to a transition between two energy levels of the term system.

The level system so constructed is often represented graphically in the form of a *term scheme*, shown schematically in Fig. 6A. The horizontal lines indicate four energy levels of the system. The vertical lines between the levels indicate possible transitions, and the arrow-heads indicate whether the transition is upward (absorption), or downward (emission). The six possible transition frequencies are tabulated below the figure. It is common practice to draw the term scheme against a linear vertical energy scale, and the transition frequencies are thus directly proportional to the lengths of the arrows (or lines) between the levels.

As the figure suggests a comparatively small number of terms describe a much larger number of lines: the number of *pairs* of levels which can be picked from n levels equals $n(n-1)/2$. It should be mentioned, however, that we do not in general observe spectral lines corresponding to transitions between *every* possible pair of levels, and in this respect Fig. 6A is a bit misleading. We shall discuss this important point later.

To fully appreciate the degree of order which this procedure brings to the study of spectra we only need to look at some of the more complicated atomic spectra, or better still at molecular band spectra. (See Figs. 6B, and the other spectra in this chapter.) The latter kind of spectrum characteristically shows a number of *bands*, which when studied with very high resolution are found to consist of an enormous number of closely spaced spectral lines. At first sight a molecular band spectrum appears to be hopelessly complex. Nevertheless it has been found possible to bring order into this complexity: in many cases we can construct term schemes and account for every single observed line.

7 Consider again Fig. 6A and let us suppose that the figure shows the term scheme of an atom, in which case the level separations are typically of the order of an electron volt.

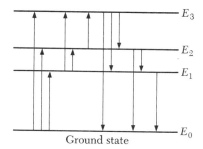

Fig. 6A Term scheme showing four energy levels, and transitions between them. The possible transition frequencies are:

$$\omega_{30} = \frac{(E_3 - E_0)}{\hbar} \qquad\qquad \omega_{31} = \frac{(E_3 - E_1)}{\hbar}$$

$$\omega_{20} = \frac{(E_2 - E_0)}{\hbar} \qquad\qquad \omega_{21} = \frac{(E_2 - E_1)}{\hbar}$$

$$\omega_{10} = \frac{(E_1 - E_0)}{\hbar} \qquad\qquad \omega_{32} = \frac{(E_3 - E_2)}{\hbar}$$

Suppose we study the *absorption spectrum* of the atom. Using light which originates in a source with a continuous spectral distribution, we observe the absorption lines in the light after it has passed through a layer of a (monatomic) gas of the atoms in question. Suppose furthermore that the gas is reasonably cool, say at room temperature. We will then observe absorption lines at the frequencies ω_{30}, ω_{20} and ω_{10}, but not at the remaining three frequencies. The explanation for this is very simple: the overwhelming majority of the atoms in the gas are in the ground state, and we will, therefore, only observe transitions from the ground state to one of the higher states.

As the temperature is raised the probability that an atom is found in one of the excited states increases. In Volume V of this series† we shall learn that in a gas held at temperature T the ratio of the number of atoms in the n:th excited state to the number of atoms in the ground state is given by

$$\frac{N_n}{N_0} = \exp\left(-\frac{E_n - E_0}{kT}\right) \tag{7a}$$

At "room temperature," for which $kT \simeq (1/40)$ eV, this ratio is a *negligibly* small number. It follows that a cool gas will not emit

† Berkeley Physics Course, Vol. V, *Statistical Physics*.

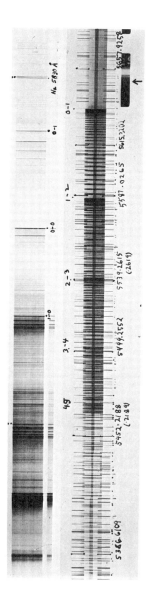

Fig. 6B Parts of the spectrum of the molecule C_2, taken at two different dispersions. The spectrum at left is taken at low dispersion, and shows the "bands" characteristic of molecular spectra. The spectrum at right is taken with much higher dispersion (see the wavelengths in Ångströms on the plate). The lines which make up the bands are here seen very clearly. *(Photograph by courtesy of Professor S. P. Davis, Berkeley.)*

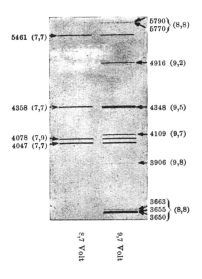

Fig. 8A The spectrum of the mercury atom when excited by electron collisions, for two different electron energies. This photograph is taken from G. Hertz, "Über die Anregung von Spektrallinien durch Elektronenstoss, I," *Zeitschrift für Physik* **22,** 18 (1924).

When the electron energy is increased from 8.7 eV (left spectrum) to 9.7 eV (right spectrum) a whole set of new lines appears, of which there is no trace in the left spectrum. The numbers within parentheses in the figure show the electron energies at which the lines first appear, and the numbers without parentheses are the wavelengths in Ångströms. *(Courtesy of Springer Verlag.)*

Fig. 8B Greatly simplified term scheme of the neutral mercury atom, showing the levels involved in the transitions seen in Fig. 8A. The numbers at left are the energies of the levels expressed as wave numbers. The corresponding energies in eV are shown at right. Note that this term scheme is *not* drawn to scale. Levels for which numerical data are omitted are indicated at right. The numbers on the transition lines are wavelengths in Ångströms. All transitions to the ground state are in the ultraviolet region. Two of these transitions are shown in the diagram (with wavelengths in parentheses). These lines are not visible in the spectra in Fig. 8A. The ionization limit lies at 84184 cm⁻¹ (corresponding to 10.4 eV).

(visible) light, unless the atoms are excited by some other (external) means.

8 If we study the *emission spectrum* of a gas of the atoms, excited for instance by an electrical discharge, we might observe all the spectral lines indicated. If an atom, originally in its ground state, collides with an energetic electron the electron may transfer part of its energy to the atom. This causes the atom to jump to one of the higher states from which it subsequently decays to a lower level with the emission of light. It is self-evident that this process cannot take place unless the electron has sufficient energy to raise the atom to one of the excited states. If the energy of the electron is less than $(E_1 - E_0)$ then the electron can only undergo an *elastic* collision with the atom. If the energy is higher, inelastic collisions leading to light emission become possible.

There is an obvious experimental test of this picture, and indeed of the general ideas underlying our postulates of Sec. 5. We simply vary the energy of the electrons used in exciting the atoms, and as the energy increases new emission lines ought to appear. Fig. 8A shows some results in such an experiment, for a gas of mercury atoms. As we see the appearance of the emission spectrum changes

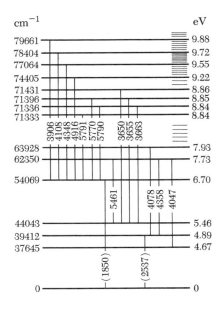

in the expected manner, and the changes can be accounted for on the basis of the term scheme shown in Fig. 8B.

9   The graph in Fig. 9A shows the results of a similar experiment. Mercury atoms, in the form of a gas at low pressure, are excited by electron bombardment.   The excited atoms decay back to the ground state by emission of photons, and the presence of these photons (in particular the ultraviolet photons) is observed through the photo-electric current which they give rise to when they fall on an iron electrode.   As the energy of the bombarding electrons is increased new levels can be excited, and as a consequence of this new transitions become possible.   At each new energy level of the mercury atom the rate of increase of the number of photons emitted with the electron energy increases abruptly, and the slope of the graph will thus show discontinuities at these energies.   The location of these discontinuities in Fig. 9A should be compared with the levels shown in the term scheme in Fig. 8B.

It is difficult to determine the energy of the bombarding electrons precisely, but nevertheless measurements such as these can clearly be very useful in the determination of the level system of an atom. The curve in Fig. 9A establishes the approximate location of many of the levels.   These data are supplemented by accurate wavelength measurements of the emission lines, and since we can observe at what electron energies a line first appears (if we employ that method of excitation by electron bombardment), we get information about the levels involved in the transitions.   Additional information is obtained by study of the absorption spectrum: in this case we know that the ground state must be the lower level.

These and many other methods have been employed in the past, and are still employed, in the collection of an enormous amount of data on atomic spectra and atomic levels.

10   The phenomenon of *fluorescence* is easily understood on the basis of Fig. 6A.   A photon of energy $(E_3 - E_0)$ is absorbed by the atom in its ground state, and the atom accordingly makes a transition to the level of energy E_3.   It may decay from this level via the other levels, and we will then observe photons of all the frequencies listed in Fig. 6A.

On the basis of this picture we can immediately understand Stokes' rule: the frequency of the light emitted in fluorescence cannot exceed the frequency of the exciting light.   This rule holds quite generally, although exceptions can occur if some of the atoms which absorb the exciting light are not originally in the ground state.

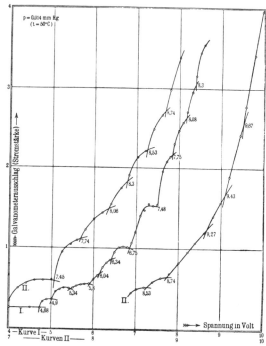

Fig. 9A   The excitation of mercury atoms by electron collisions.   Figure is taken from J. Franck and E. Einsporn, ''Über die Anregungspotentiale des Quecksilberdampfes,'' *Zeitschrift für Physik* **2**, 18 (1920).

The abscissa shows the energy of the electrons (in two different scales) and the ordinate is a measure of the light emitted by the mercury atoms.   (See explanation in text.)   As the electron energy increases new levels will be excited, and for each new level the slope of the curve changes abruptly because new transitions involving additional photons become possible.

The atoms are in the form of mercury gas, at a pressure of 0.014 mm Hg and a temperature of 50°C. (*Courtesy of Springer Verlag.*)

In the paper[†] in which Einstein discussed the photoelectric effect he also discussed Stokes' rule, from the standpoint of the photon picture. At that time the idea of energy levels had not yet been born, but the rule can also be understood if we assume that the energy of the emitted quantum must derive from the energy of an absorbed quantum.

11   At a certain energy above the energy of the ground state the atom will become ionized. This energy is the smallest energy at which an electron and the singly ionized atom can exist completely separated from each other. At this energy, and above, the "atom" no longer exists as an atom, but we may still consider the system consisting of the singly ionized atom and the electron. This system can clearly have *any* energy we like above the ionization energy. The set of possible energies of the *system* therefore consists of a set of discrete energy levels *below* the ionization energy, and of a continuum *above* this energy. This situation is shown schematically in Fig. 11A. The shaded area above the ionization level E_i represents the continuum.

The vertical line to the left represents the transition from the ground state to the energy E' in the continuum through the absorption of a photon of energy $(E' - E_0)$. This process is the photoelectric effect for a single atom. The ejected electron will appear with the *kinetic* energy $(E' - E_i)$.

The inverse of the photo-ionization process (photo-electric effect) is the *radiative recombination* of the electron and the singly ionized atom. This process is shown by the vertical line to the right in Fig. 11A. An electron of kinetic energy $(E'' - E_i)$ collides with the ion (at rest), and the system "jumps" down to the level of energy E_2 with the emission of a photon of energy $(E'' - E_2)$. From this level the atom continues down to the ground state via the first excited state, as indicated by the arrows. In each transition in this *cascade* a photon of the appropriate frequency is emitted.

In atomic physics the ionization level is often assigned the energy value zero, and the bound states thus all have negative energies. Other conventions are possible, and suitable, depending on the circumstances. In nuclear physics we often assign the value zero to the ground state of the nucleus. We should note that the choice of the zero-point is purely a matter of convention.

12   We have so far considered only atoms in the light of our two postulates. The notion of energy levels and transitions between

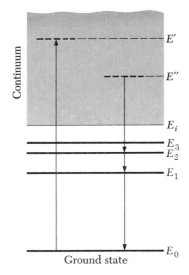

Fig. 11A Term scheme showing discrete energy levels and the continuum (in gray) above the ionization level. Transitions between the discrete levels, and to and from the continuum, are indicated by vertical arrows. The dotted horizontal lines embedded in the continuum do not symbolize energy levels of the *atom*, but rather two particular energies in a continuous range of energies within which the system consisting of an electron and an ion may be found.

† A. Einstein, *Annalen der Physik* **17**, 132 (1905).

these levels is, however, of very general applicability, and we can discuss molecules and nuclei in the same manner. Consider an arbitrary system of particles, of any number and kind. The ionization level, or dissociation level, above which the possible energies of the system form a continuum, is the lowest energy at which the system can exist in two separate parts, at a large distance from each other. Below this energy we may encounter a number of discrete energy levels corresponding to bound states of the system. (This characterization, which is in the spirit of our two postulates, must be qualified if we wish to be very precise.)

As an example of a term scheme in nuclear physics we consider the term scheme of the deuteron, shown in Fig. 12A. The deuteron has no discrete excited states. The binding energy, B, of the deuteron is $B = 2.23$ MeV, which means that the continuum begins at the energy B above the ground state. Above this energy the "deuteron" no longer is a deuteron, but is a system consisting of a neutron and a proton separated from each other.

The vertical arrow in the figure represents the photo-disintegration of the deuteron. A photon of energy $E_{ph} > B$ brings about the dissociation of the deuteron into a proton and a neutron of combined kinetic energy ($E_{ph} - B$). This process, which has been studied experimentally in great detail, is clearly completely analogous to the photo-ionization of an atom, which we discussed in the preceding section. The inverse of this process is the radiative capture of a neutron by a proton.

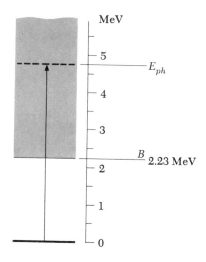

Fig. 12A Term scheme of the proton-neutron system, showing the ground state of the deuteron, and the continuum which begins at the dissociation energy 2.23 MeV above the ground state. The arrow symbolizes the photodisintegration of the deuteron.

13 The author hopes that these introductory remarks on term schemes have convinced the reader that our two postulates are indeed useful in our study of the structure of atoms, molecules and nuclei. We are led to organize our observational material on spectra with the help of term schemes. The important part of the second postulate is clearly the relation (5a). The statement about the "jumping atom" is *not intended* to describe the details of the emission or absorption processes: it is merely a picturesque way of saying that something goes on.

Through custom the word "jump" has become well-established as a colloquial term in quantum physics. In the opinion of the author it has not been a happy choice of term, and one may surmise that it has caused much needless suffering in the study of physics. The word is dangerous because of its connotations: when we say that a "system jumps from one state to another" it seems to imply that the process has a certain abrupt and discontinuous character, and the mental picture which we accordingly form may mislead us seriously.

The Finite Widths of Energy Levels

14 In our discussion so far we have not run into any difficulties with the "jump-picture," and the reason for this is simply that we have really not made any use of it: we have merely used equation (5a). Let us now consider a situation where we may get into trouble if we take the "jump-picture" too literally.

A photon of frequency ω_0 is incident on an atom originally in its ground state. The frequency ω_0 just happens to correspond to the transition energy of the atom from the ground state to one of the excited states, and the atom accordingly absorbs the photon and makes the jump. Eventually it jumps down to the ground state again, with the re-emission of a photon of frequency ω_0. This photon may be emitted in any direction, and the atom will thus scatter incident light of the right frequency ω_0. Suppose, however, that the incident light is *not* of the right frequency ω_0, but is, instead, of a frequency ω very slightly different from ω_0. Will the atom then scatter light? The answer is yes. One finds, experimentally, that as we let the incident frequency ω vary from a value below ω_0 the effectiveness of the atom as a scatterer changes: it first increases to a sharp maximum at $\omega = \omega_0$ and then decreases again. Somehow photons of the wrong frequency can also induce the "jumps": experiments tell us that this is so. We may furthermore ask what the frequency of the scattered radiation will be if the incident frequency is $\omega \neq \omega_0$. The "jump-picture" might suggest to us that this frequency should be the "correct" one, namely ω_0, which is not what is found experimentally: the re-emitted frequency is actually ω, as we expect on the basis of energy conservation (and the photon picture).

In the discussion of this phenomenon, known as *resonance fluorescence*, the term "jump" can hardly be regarded as appropriate: it could even lead us seriously astray.

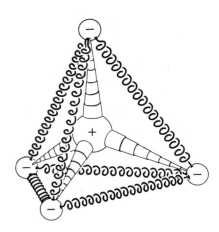

Fig. 15A A mechanical model of an atom, which is helpful in understanding resonance fluorescence. If this contraption is excited by a kick (say in a collision with an electron), it will oscillate, and since the electrons are charged, electromagnetic radiation will be emitted at the resonant frequencies of the system. The motion is necessarily damped, because the system loses energy through radiation.

Under the influence of an incident electromagnetic wave, the atom will undergo forced oscillations at the frequency of the incident wave, and hence will emit radiation at the same frequency. This is the phenomenon of resonance fluorescence.

15 The observed facts can be understood very easily in terms of another model. Let us regard the atom as a mechanical system in which the electrons are bound to the nucleus by springs. Such a system will have a number of resonant frequencies, one of which is the frequency ω_0. In the ground state of the atom this system is at rest, but an incident electromagnetic wave will excite oscillations in the system. As a result the oscillating electrons will radiate an electromagnetic wave of the *same* frequency as the incident wave. The amplitude of the oscillations will be larger the closer we are to the resonant frequency ω_0 and the effectiveness of the atom as a scatterer will clearly be largest when the incident frequency is equal to the frequency ω_0. Furthermore, and this is most important, the radiated wave will stand in a definite phase relationship to the incident wave, and it will accordingly interfere with the incident wave in a very definite way which the "jump-picture" could not very well explain. The most serious defect of the "jump-picture" in this case is that it dissects the scattering process in a manner which in no way corresponds to reality: the scattering process should be regarded as a single coherent process and not as being made up of two jumps in which the photon emitted in the second jump stands in no definite phase relationship to the photon absorbed in the first jump.

Whether the re-radiated wave is coherent with the incident wave or not is something which can be checked experimentally, and the evidence is conclusively in favor of the oscillator model which predicts coherence.

16 Our discussion of resonance fluorescence suggests to us a new interpretation of the energy levels of atoms, nuclei and molecules: the energy level differences correspond to frequencies at which the system can resonate. *The energy level differences are resonances.*

Of course we should not take any mechanical model with springs and levers seriously: that would be obvious nonsense. The reason why such admittedly wrong models can nevertheless work so well in describing such phenomena as resonance fluorescence is simply that many aspects of a resonance phenomenon do not depend on the details of the model: all that counts is the system of resonant frequencies (with the associated damping constants) and the nature of the couplings of the various resonant modes to the external source of excitation.

17 Suppose, now, that we try to determine the energy of a level above the ground state of an atom by determining the frequency of photons which can cause transitions from the ground state to the excited state. We try, in other words, to determine the frequency at which the atom resonates. There is, however, no *unique* such frequency: the atom responds throughout some small frequency *interval*. We might, of course, say that the "correct" frequency, which defines the energy of the level, is the frequency ω_0 at which the response is a maximum. The fact remains, however, that the atom also responds in the immediate neighborhood of ω_0, and the line in the absorption spectrum of the atom cannot, therefore, be absolutely sharp: it has a *finite width*. That this is so is an experimental fact: the lines in an absorption spectrum have finite widths.

We can then ask the question: how about the spectral lines

emitted by the atom? Do they also have finite widths? The answer is yes. The width of an emission line is the same as the width of the corresponding absorption line. (We must mention here that the lines of optical spectra, such as we observe them in practice, are broadened because of several different effects. We are here concerned with the width of a spectral line emitted, or absorbed, by an isolated atom, originally at rest with respect to the observer. This width is an intrinsic property of the atom. Let us forget temporarily about all other causes of broadening: we will discuss these causes later in this chapter.)

What does it mean that an emission line has a finite width? It literally means what it says: if we photograph the line with a spectrograph of *extremely* high resolution, we find that the line has a finite width. The frequency of the emitted light is not precisely ω_0, but we also find frequencies in the immediate neighborhood of ω_0.

18 Since the location of an energy level is determined through observations on emission and absorption lines, and since these lines always have a finite width, we must conclude that the energy of an excited state cannot be a precisely defined quantity. If we believe in the photon picture, and in the principle of conservation of energy, we are forced to this conclusion. Our first postulate in Sec. 5 is therefore not literally true. *The energy levels above the ground state have a finite width.*

Suppose that we determine the energy of a particular excited state in an atom (or molecule, or nucleus) by observing the absorption line connecting the excited state with the ground state. If the response of the atom is a maximum at the frequency ω_0, we can assign to the excited state the *mean energy* $E = E_0 + \hbar\omega_0$, where E_0 is the energy of the ground state. If the width of the spectral line is $\Delta\omega$, (defined in some suitable manner), we say that the width of the excited level is $\Delta E = \hbar\Delta\omega$. Once we recognize that energy levels are of finite width there is not much point in using the clumsy term "mean energy": we simply talk about the "energy" of the level with the understanding that this energy refers to a suitably defined mean energy.

19 The nature of the simplifying assumption underlying our first postulate can be very well illustrated by an example from classical mechanics. Consider a pendulum set in motion, and then left to swing by itself. We assume that the frictional forces (the most important of which is air resistance) are small, but not zero, so that the pendulum may execute several hundred oscillations before its energy of oscillation has diminished to $1/e$ times its original value. (The time required for this is the "mean-life of the oscillatory state.") Let the time interval between two successive outswings to the right be one second.

Suppose now that someone asks us for the frequency of the pendulum. Without much reflection we would answer that the frequency is one per second. This is certainly a reasonable answer, but strictly speaking it is wrong: by "frequency" we understand the repetition rate of a *periodic* phenomenon. The motion of the pendulum is, however, only approximately periodic since the amplitude of oscillation does diminish as time goes on. The frequency of a *damped* harmonic motion is not precisely defined, although it may, for all practical purposes, be very well defined indeed.

An atom emitting radiation is in some respects analogous to a damped pendulum. The emission process does not go on forever, and this must mean that the "oscillation inside the atom" is a damped oscillation. There is not, therefore, a *precisely* defined frequency, because the oscillatory phenomenon is not strictly periodic. The electromagnetic radiation emitted by that "something which is oscillating inside the atom" is thus not monochromatic. The emitted line has a finite width.

20   If we think about Fig. 19A it might occur to us that the smaller the damping, the better is the frequency defined, and we might conjecture that perhaps the uncertainty $\Delta\omega$ in the frequency is inversely proportional to the mean-life τ.

To investigate this question we shall consider the emission and scattering of light by an atom in the spirit of the "oscillator model" of Sec. 15. We assume that only two states are involved: the ground state and an excited state of energy $\hbar\omega_0$ above the ground state.

We first consider the atom by itself, just after it has been excited. We denote by $A(t)$ the amplitude of whatever it is that oscillates inside the atom, and we shall assume the time dependence

$$A(t) = A \exp\left(-i\omega_0 t - \frac{t}{2\tau}\right) \qquad (20a)$$

where A is a constant. This is the time dependence of the amplitude of a damped harmonic oscillator of mean frequency ω_0, in the complex representation.

Since the oscillation phenomenon involves charged particles we expect that electromagnetic radiation (of mean frequency ω_0) will be emitted, and the time dependence of the amplitude of the

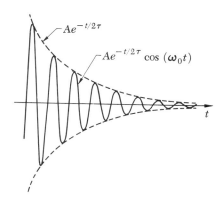

Fig. 19A   An exponentially damped oscillatory process, showing the amplitude as a function of time. Since the process is not strictly periodic in time it is wrong to say that the frequency of the oscillation is ω_0, because the concept of frequency refers to a *periodic* phenomenon. If the damping is not too large it is fair to say that the frequency is *approximately* ω_0. It is intuitively clear that the smaller the damping, i.e., the smaller the decrease in amplitude for two successive maxima, the better is the frequency defined.

emitted wave must be of the same form (20a). The *intensity I(t)* of the emitted radiation is proportional to the absolute square of the amplitude:

$$I(t) = C|A(t)|^2 = C|A|^2 \exp\left(-\frac{t}{\tau}\right) \tag{20b}$$

where C is some constant. We can therefore write

$$I(t) = I(0) \exp\left(-\frac{t}{\tau}\right) \tag{20c}$$

We have chosen to write the exponential decay factor in (20a) in the form $\exp(-t/2\tau)$ because we wanted to have the factor $\exp(-t/\tau)$ in the expression for the intensity. It is clearly a matter of convention how we write this factor, i.e., how we define τ. With our definition τ is the time it takes for the *intensity* of the radiation to decrease by the factor $1/e$. Since τ is a measure of the duration of the process, we can interpret τ as the *mean-life of the excited state*. "Most of the decay takes place within a time of the order τ."

21   The oscillator-amplitude $A(t)$ given by (20a) satisfies the first-order differential equation:

$$\frac{dA(t)}{dt} + \left(i\omega_0 + \frac{1}{2\tau}\right)A(t) = 0 \tag{21a}$$

This *homogeneous* differential equation describes the oscillator in the absence of any external influences. Suppose now that monochromatic light, of frequency ω, is incident on the oscillator. The equation (21a) must then be modified by the addition of a term describing the harmonically varying applied driving force. The resulting *inhomogeneous* differential equation for the oscillator is then of the form

$$\frac{dA(t)}{dt} + \left(i\omega_0 + \frac{1}{2\tau}\right)A(t) = F \exp(-i\omega t) \tag{21b}$$

where F is a constant which describes the magnitude of the driving force.

The differential equation (21b) has the steady-state solution (ignoring transients)

$$A(t) = \frac{iF \exp(-i\omega t)}{(\omega - \omega_0) + i/2\tau} \tag{21c}$$

which corresponds to an oscillation with constant amplitude at the applied frequency ω.

The intensity of the radiation emitted by the oscillator is proportional to the absolute square of $A(t)$. The emission from the *driven* oscillator is observed as scattered radiation, and the amount of scattering is proportional to the intensity. Let us denote the total amount of radiation scattered per unit time, for unit amplitude of the incident radiation, by $S(\omega)$, where ω is the incident frequency. In view of (21c) we may then write

$$S(\omega) \text{ proportional to } \left| \frac{1}{(\omega - \omega_0) + i/2\tau} \right|^2$$

or

$$S(\omega) = S(\omega_0) \frac{(1/2\tau)^2}{(\omega - \omega_0)^2 + (1/2\tau)^2} \qquad (21d)$$

where $S(\omega_0)$ is the amount of scattering "at resonance," i.e., when $\omega = \omega_0$.

A schematic plot of $S(\omega)$ versus ω is shown in Fig. 21A.

22   The function $S(\omega)$ expresses "the intensity of response" of the system under an external perturbation at the frequency ω. *This kind of resonant response is a very general phenomenon in quantum physics, and it is by no means restricted to the interaction of light with atoms.* We find the same resonant response when we study the scattering of material particles, such as protons, of a well-defined energy, from a nucleus, or the scattering of pions from a proton. One might well say that a quasi-stable energy level of a quantum-mechanical system "exists" in *precisely* the sense that the system exhibits a resonant response, as given by equation (21d), at the appropriate frequency.

In nuclear physics the resonance formula (21d) is known as the *Breit-Wigner one-level resonance formula*, after G. Breit and E. P. Wigner.

23   Let us now note an important feature of the resonance formula (21d). We consider the frequency ω at which the response is one-half of the response at maximum and we find

$$\omega = \omega_0 \pm \frac{1}{2\tau} \qquad (23a)$$

The width of the resonance curve (see Fig. 21A) at half-maximum is accordingly given by

$$\Delta\omega = \frac{1}{\tau} \qquad (23b)$$

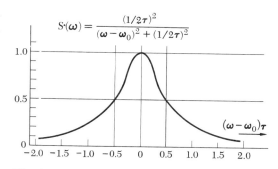

Fig. 21A The Universal Resonance Curve. It describes the response of *any* linear (or approximately linear) system to a sinusoidally varying external force in the neighborhood of a resonant frequency provided no other resonant frequency is close by.

(Two bell-shaped curves play a particularly important role in physics: the resonance curve and the gaussian curve. As usually drawn they may look very similar. It must be remembered, however, that the gaussian falls off *very* rapidly outside the central region, whereas the resonance curve has a long "tail.")

This agrees with our conjecture, in Sec. 20, about the relationship between the uncertainty in the frequency and the mean-life of the excited state.

Since we can define the width of the (excited) energy level by $\Delta E = \hbar \, \Delta\omega$, we immediately derive from (23b) the very important relation

$$\Delta E = \frac{\hbar}{\tau} \qquad (23c)$$

which gives the uncertainty ΔE of the energy of a level in terms of the mean-life τ of the state. The longer the state lives, the better is the energy defined.

24 The reader may have strong doubts that a simple differential equation such as (21b) can really describe such a complicated phenomenon as the interaction between light and an atom. Actually it does not, but the point is that we do not try to describe *every* aspect of the interaction, but merely the response of the atom to (almost) monochromatic light of a frequency in the *immediate* neighborhood of the resonant frequency ω_0, corresponding to a transition from the ground state to an excited state. The formula (21d) describes only a single resonance, and if there are several, as is always the case with atoms, molecules and nuclei, then the theory must be modified. The formula (21d) can be expected to hold with good accuracy in a neighborhood of the resonance line, far away from all other resonances.

To present the complete story of radiative transitions would take us too far, and we must be satisfied by our somewhat vague theory. The essence of the matter is clearly that *something* oscillates, and that this something is charged, and that the response (in amplitude) to an external perturbation is linear.

25 Let us next consider the width of a line emitted in a transition between two *excited* states. This situation is shown schematically in Fig. 25A. The widths of the energy levels are represented (in a greatly exaggerated fashion) by the thickness of the horizontal lines. We consider a cascade of two transitions: from the second excited state to the first excited state followed by a transition from the first excited state to the ground state. The width of the line (of frequency ω_{10}) emitted in the second transition is $\Delta\omega_{10} = \Delta E_1/\hbar$.

We may also ask for the uncertainty in the *sum* of the two frequencies emitted in the cascade from a *single* atom. If we denote the sum of the two frequencies by $\omega_{20} = \omega_{21} + \omega_{10}$ we will have

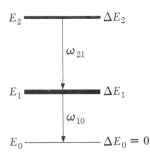

Fig. 25A Schematic term scheme to illustrate discussion in Sec. 25. The width of the line (of mean frequency ω_{21}) emitted in the transition from the upper excited state to the lower excited state depends on the widths of *both* levels, and we have: $\Delta\omega_{21} = (\Delta E_2 + \Delta E_1)/\hbar$.

$\Delta\omega_{20} = \Delta E_2/\hbar$. This result follows from the principle of energy conservation: the uncertainty in the total available energy is clearly the same as the uncertainty in the energy of the second excited level.

From this we can surmise that the width of the line (of frequency ω_{21}) emitted in the first transition is given by $\Delta\omega_{21} = (\Delta E_2 + \Delta E_1)/\hbar$, and if the first excited state is broad the emitted line will be broad even if the width of the second excited state is very narrow (corresponding to a long lifetime). The width of the first excited level introduces an uncertainty in the way in which the total available energy is partitioned between the two emitted photons.

The results which we have just presented, and which are based on energy conservation and on the idea of a finite width of the energy levels, are certainly extremely plausible. Our discussion is, however, not rigorous, but it suffices for us to understand the qualitative features of the problem, and the important point is that the width of an emitted line must depend on the widths of *both* levels involved.

26　Consider again the relation $\Delta\omega = 1/\tau$. Since the frequency is inversely proportional to the wavelength λ, the fractional uncertainty in the wavelength equals the fractional uncertainty in the frequency, and we have

$$\frac{\Delta\lambda}{\lambda} = \frac{\Delta\omega}{\omega} = \frac{1}{\omega\tau} \tag{26a}$$

For optical transitions in atoms the quantity $\omega\tau$ is always very large. The frequency $\nu = \omega/2\pi$ is of the order of 5×10^{14}/sec, whereas the mean-life is of the order of $\tau \sim 10^{-7} - 10^{-8}$ sec, and the fractional uncertainty in the wavelength (or frequency) is thus of the order of $\Delta\lambda/\lambda \sim 10^{-7}$, which is a very small quantity. The resulting width of the spectral line is known as the *natural line-width:* it is an intrinsic property of the atom (i.e., of the levels involved in the transition.)

Further Discussion of Levels and Term Schemes

27　Let us now look at a number of typical term schemes. They have been constructed on the basis of actual measurements, interpreted within the framework of quantum mechanics. We should look at them with proper respect: each diagram, or rather the associated table of wavelengths, is the fruit of a considerable amount of human labor.

44　　　　　　　　　　　　　　　　　　　　JACK SUGAR

TABLE V. — Observed spectral lines of Ce

λ_{air} Å	Intensity	σ (cm⁻¹)	Classification		o − c	λ_{air} Å
4623.197	20	21624.00	101354₂	−122978°₃	− 0.03	4356.835
4616.233	60	21656.62	103612₄	−125269°₃	+ 0.01	4346.353
4613.803	60	21668.02	21849°₃	− 43517₄	+ 0.03	4344.025
4612.528	2	21674.01	101354₂	−123028°₂	− 0.05	4339.205
4612.384	4	21674.69	101354₂	−123029°₁	+ 0.03	4336.143
4610.723	30	21682.50	103612₁	−125295°₀	− 0.03	4335.515
4599.803	1	21733.97				4331.168
4582.264	200	21817.16	103351₄	−125168°₃	0.00	4327.503
4576.904	300	21842.71	103351₄	−125193°₄	+ 0.01	4321.384
4575.494	3	21849.44	0₄	− 21849°₃	− 0.03	4314.767
4570.430	2	21873.65				4309.634
4568.802	20	21881.44	103351₄	−125232°₄	− 0.02	4304.710
4551.460	60	21964.81	103231₂	−125196°₃	+ 0.01	4300.970
4544.250	100	21999.66	103231₂	−125230°₂	− 0.01	4296.170
4536.526	1	22037.12	103231₂	−125268°₁	− 0.05	4289.790
4536.330	10	22038.07	103231₂	−125269°₁	+ 0.01	4287.787
4535.726	1000	22041.01	21476°₄	− 43517₄	+ 0.01	4285.507
4527.861	6	22079.29	103079₄	−125158°₅	− 0.01	4284.777
4526.655	4	22085.17	103079₄	−125164°₄	− 0.02	4282.307
4525.931	2	22088.71	103079₄	−125168°₅	+ 0.01	4280.497
4525.330	100	22091.64	100814₄	−122905°₄	+ 0.03	4271.27
4524.689	10	22094.77	100814₄	−122908°₃	− 0.04	4264.6
4521.924	1000	22108.28	100814₄	−122922°₅	− 0.01	4247.5
4520.709	3	22114.22	103079₄	−125193°₄	− 0.02	4239
4519.918	10	22118.09	100814₄	−122932°₄	− 0.04	422°
4503.372	10	22199.36	100734₂	−122933°₃	+ 0.02	
4502.825	100	22202.05	70433₂	− 92635°₄	0.0	
4494.689	2	22242.24	100734₂	−122976°₁		
4491.454	100	22258.26	102897₂	−¹		
4490.855	4	22261.23	100°			

Fig. 27A　Portion of a table in a paper by J. Sugar, "Description and Analysis of the Third Spectrum of Cerium (Ce III)," *Journal of The Optical Society of America* **55**, 33 (1965); exhibit taken from page 44. First column shows wavelength in air of observed lines of the doubly ionized cerium atom. Second shows relative intensity of line. Third shows energy of photon, expressed as a wave number. The fourth column shows spectral terms involved, with energies expressed as wave numbers.

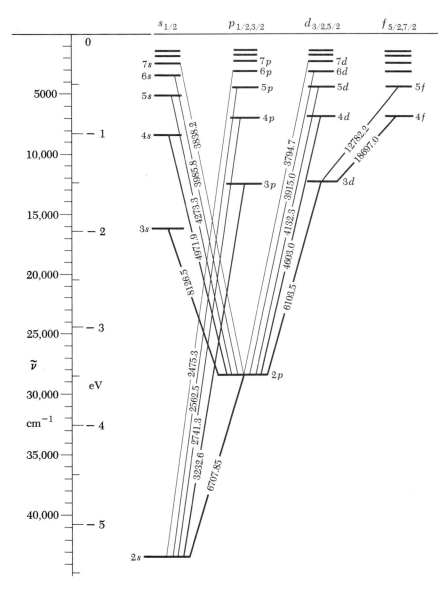

Fig. 28A Term scheme for the neutral lithium atom. The slanted lines represent observed electric dipole transitions. The numbers on these lines are the wavelengths in Ångströms. For other details, see the explanation in the text. Based on a figure in W. Grotrian, *Graphische Darstellung der Spektren von Atomen. . .* , vol. II, p. 15 (Verlag von Julius Springer, Berlin, 1928).

We have drawn our term schemes in the form in which the reader will find them in the literature. The drawing of such schemes, and the labeling of the different energy levels, is governed by a number of conventions of long standing. For greater realism we wanted to adhere to these conventions, even if we cannot here explain every detail of the drawings. The reader may want to object that we should not show *anything* in the diagrams that we are not now prepared to explain theoretically. Such an attitude, carried to its logical conclusion, would forbid us to consider term schemes at all, before we have shown theoretically that energy levels do occur. Our purpose in this chapter is, however, to discuss some aspects of physical systems, given the empirical fact that energy levels do occur. It is also an historical fact that term schemes for atoms, of which the one shown in Fig. 28A is a typical example, were constructed on the basis of spectroscopic measurements *before* the full significance of the details of the schemes was understood: i.e., *before* the discovery of quantum mechanics.

28 The energy levels of a quantum-mechanical system are labeled by a number of *quantum numbers*. These numbers are the numerical values of important physical parameters which occur in the quantum-mechanical description of the system. We shall discuss the physical interpretation of some of these quantum numbers in connection with the term schemes. The reader is, however, under no obligation to understand and remember in detail all the labels on the levels shown.

Figure 28A shows the term scheme of the neutral lithium atom. The energy scale at the left expresses the energy both in electron volts and in equivalent wave numbers. The horizontal bars represent the energy levels. The lines connecting the levels represent observed electromagnetic transitions between the levels, and the numbers attached to these lines give the wavelengths, in Ångström units, of the spectral lines. Particularly prominent lines in the spectrum are indicated by thick lines between the levels.

The energy levels in the term scheme in Fig. 28A have been arranged into columns. Four of these are shown, labeled by the letters *s*, *p*, *d* and *f*. Actually the lithium atom has more levels, which we would arrange into columns to the right of the ones shown, but these levels lie close to the ionization level, and they do not contribute to the visible spectrum of lithium.

We notice that the spectral lines indicated in Fig. 28A obey an interesting rule: the transitions are between two levels in *neighboring* columns. Actually the transitions shown in Fig. 28A do not include all possible transitions. Quantum mechanics predicts that there will also be transitions from the *s*-column, or the *d*-column, to the 3*p*-level; from the *p*-column to the 3*s*-level; from the *p*-column, or the *f*-column, to the 3*d*-level, and so on. Many of these transitions have actually been observed but we have not shown them in the diagram in order not to overload the figure. These additional transitions, which lie in the infrared region, also obey the rule mentioned, namely that the transitions are between levels

62

Fe III—Continued

Authors	Config.	Desig.	J	Level	Interval
$z\ ^3P_2$	$3d^3(a\ ^4P)4p$	$z\ ^3P^\circ$	2	*119697.64*	
$^3P_1$			1	*119982.26*	-284.62
$^3P_0$			0	*120179.95*	-197.69
$y\ ^5F_1$	$3d^3(a\ ^4D)4p$	$y\ ^5F^\circ$	1	*120697.10*	
$^5F_2$			2	*120826.17*	129.07
$^5F_3$			3	*121008.78*	182.61
$^5F_4$			4	*121241.67*	232.89
$^5F_5$			5	*121468.82*	227.15
$z\ ^3G_3$	$3d^3(a\ ^4G)4p$	$z\ ^3G^\circ$	3	*121919.74*	
$^3G_4$			4	*121941.29*	21.55
$^3G_5$			5	*121949.62*	8.33
$z\ ^3D_3$	$3d^3(a\ ^4P)4p$	$z\ ^3D^\circ$	3	*122346.61*	
$^3D_2$			2	*122628.34*	-281.73
$^3D_1$			1	*122843.03*	-214.69
$y\ ^5D_4$	$3d^3(a\ ^4D)4p$	$y\ ^5D^\circ$	4	*122944.15*	
$^5D_3$			3	*122829.55*	114.60
$^5D_2$			2	*122898.84*	-69.29
$^5D_1$			1	*122921.37*	-22.53
$^5D_0$			0	*123455.92*	-534.55
$y\ ^5P_1$	$3d^3(a\ ^4D)4p$	$z\ ^5P^\circ$	1	*123552.95*	
$^5P_2$			2	*123697.18*	144.23
$^5P_3$			3	*123750.39*	53.21
$y\ ^3D_3$	$3d^3(a\ ^4D)4p$	$y\ ^3D^\circ$	3	*124854.04*	
$^3D_2$			2	*124903.92*	-49.88
$^3D_1$			1	*124954.88*	-50.96
$y\ ^3F_4$	$3d^3(a\ ^4D)4p$	$y\ ^3F^\circ$	4	*125443.58*	
$^3F_3$			3	*125637.98*	-194.40
$^3F_2$			2	*125672.83*	-34.85
$z\ ^3S_1$	$3d^3(a\ ^4P)4p$	$z\ ^3S^\circ$	1	*126390.57*	

Fig. 28B Term schemes (diagrams) are useful for an overall view, but extensive sets of accurate data are best presented in tables. Above is shown a portion of a table of energy levels in doubly ionized iron. The energies, measured from the ground state, are expressed in wave numbers, cm⁻¹ (fifth column). The column labeled *J* gives the angular momentum of the state. The first three columns show various designations of the levels, which need not be explained here.

Exhibit taken from C. E. Moore, *Atomic Energy Levels*, vol. II, p. 62. (Circular of the National Bureau of Standards 467, U.S. Government Printing Office, Washington, 1952).

in neighboring columns. This rule is an interesting example of a *selection rule*, which tells us that only certain pairs of levels can be involved in a transition. The empirical support for this rule is strikingly apparent when we look at the observed lines shown in Fig. 28A. We note in particular the absence of transitions between the 3s- and 2s-levels; between the 3p- and 2p-levels, and so on. Because this selection rule governs the spectrum of the lithium atom, it appears natural to arrange the levels into columns as we have done.

29 The selection rule mentioned is a striking feature of the spectrum of the lithium atom. Can we explain it theoretically? The answer is yes: we can understand this phenomenon completely. The explanation is based on two things: the isotropy of physical space, and the smallness of the fine structure constant $\alpha = e^2/\hbar c \sim 1/137$. We shall not attempt to present the complete explanation in this book, because we cannot assume knowledge of the appropriate mathematical tools, but we shall try to give the reader at least a rough idea of what is involved.

Because of the smallness of the fine-structure constant a certain type of electromagnetic transition plays a dominant role in atomic physics, namely transitions in which the emitted electromagnetic wave has the same symmetry properties as a wave emitted by a small electric dipole oscillator. This we shall actually show later. We call such a wave (or photon) an *electric dipole wave* (or *electric dipole photon*). It can be shown, within the framework of quantum mechanics, that it carries off an amount of angular momentum $\hbar$.

The isotropy of physical space means that there is no distinguished direction in the world: the behavior of an isolated system does not depend on how it is *oriented* in space. Under very general conditions this implies (within quantum mechanics, as well as classical mechanics) that the angular momentum vector of an isolated system is conserved: it does not change with time. This means that if an atom emits an electric dipole photon the angular momentum of the atom before the emission must equal the angular momentum of the atom after the emission *plus* the angular momentum carried away by the dipole photon. This conservation principle implies selection rules because each stationary state of an atom is characterized by a definite value of angular momentum.

30 According to quantum mechanics the square of the angular momentum of an atom (neglecting any angular momentum which might possibly be carried by the nucleus) is of the form

$$J^2 = j(j + 1)\hbar^2 \tag{30a}$$

where j is the *angular momentum quantum number*. The possible values of j are restricted by the rule that $2j$ can be any non-negative integer, $2j = 0, 1, 2, \ldots$, such that $2j$ is even if the atom has an even number of electrons, and odd if the number of electrons is odd. It is common practice to say that a state characterized by the angular momentum quantum number j "has angular momentum j."

Within quantum mechanics it can then be proved that in an electric dipole transition from an initial state of angular momentum j_i to a final state of angular momentum j_f the allowed changes in the angular momentum are governed by the rule:

$$\Delta j = j_f - j_i = -1, \quad 0, \quad \text{or} +1 \tag{30b}$$

This is a rigorous rule which holds for all isolated quantum-mechanical systems, be they atoms, molecules or nuclei, and it follows from the conservation principle discussed in the preceding section. In this book we shall not discuss the theory of angular momentum, and we shall therefore leave the reader's curiosity about the derivation of the relations (30a) and (30b) unsatisfied.

31   The theorem which we have stated is, however, not the whole story of the selection rules which operate for the lithium atom. In atomic physics there is an additional, *approximate*, selection rule governing electric dipole transitions, and it can be stated as follows: in an electric dipole transition the *orbital angular momentum* of the electrons must change by precisely one unit, or

$$\Delta l = l_f - l_i = -1 \quad \text{or} \quad +1 \tag{31a}$$

where the letter l, suitably indexed, denotes the *orbital angular momentum quantum number* of the electrons in the atom. What is the significance of l? This quantum number also has a "classical" interpretation: if we think about the atom in classical terms, then l expresses the magnitude of the angular momentum associated with the *orbital* motion of the electrons. It is a fact that every electron also has an *intrinsic* angular momentum, or *spin*. For an electron the spin angular momentum quantum number has the value $j_{\text{spin}} = \frac{1}{2}$, and we say that "the electron has spin $\frac{1}{2}$." The *total* angular momentum of an electron in an atom is made up of two parts: it is the vector sum of the orbital angular momentum and the spin.

The theoretically possible values of l are all the non-negative integers: $l = 0, 1, 2, 3, 4, \ldots$. The letters s, p, d, f, by which the

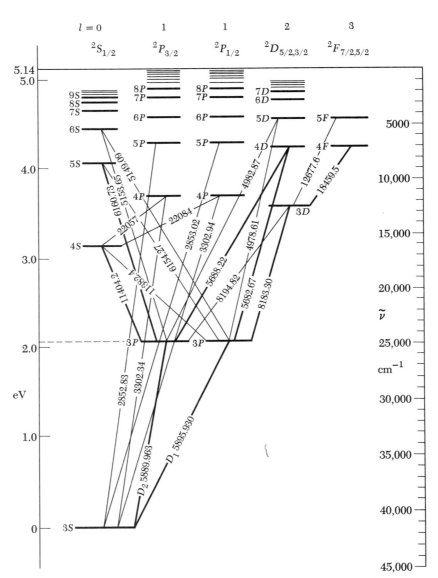

Fig. 32A Term scheme for the neutral sodium atom. The numbers on the slanted lines give wavelengths in Ångströms of observed transitions. *(After Grotrian.)*

columns in Fig. 28A are labeled, are in fact code letters for the orbital angular momentum, as follows: "s" means $l = 0$; "p" means $l = 1$; "d" means $l = 2$ and "f" means $l = 3$. The selection rule which we mentioned in Sec. 28 is equivalent to the selection rule (31a).

It is not always possible to make a definite assignment of orbital angular momentum quantum number to an energy level in an atom, although it so happens that it can be done without ambiguity for an alkali atom, such as lithium. The reason for this is that whereas the total angular momentum is a constant of motion, neither the orbital angular momentum, nor the spin angular momentum is. The levels, in other words, do not, in general, have a definite value of l. This is the sense in which the rule (31a) is only approximately valid. As we have said, it is a good rule for alkali atoms (and for hydrogen).

32   Consider again Fig. 28A. What about j, and the selection rule (30b)? This rule does not quite show up in Fig. 28A because we have exhibited a simplified form of the term scheme. We should actually have drawn the p-, d-, and f-columns double. The subscripts $\frac{1}{2}$, $\frac{3}{2}$, $\frac{5}{2}$ and $\frac{7}{2}$ on the column-labels s, p, d and f indicate the total angular momentum j. For alkali atoms (and hydrogen) the following rules hold: if $l = 0$, then $j = \frac{1}{2}$ (the entire angular momentum is due to the electron spin). For all other values of l, j can have the values $j = l + \frac{1}{2}$ and $j = l - \frac{1}{2}$. (For other atoms the rules are different). The level $2p$ is thus actually double, but the separation in energy between the two levels of the *doublet* is quite small, and within the accuracy of the figure, the levels coincide.

Fig. 32A shows the term scheme of the sodium atom. Sodium is also an alkali atom, and it is quite obvious that its term scheme is, in many respects, similar to the term scheme of lithium. In this case we have drawn the p-column double, but left the d- and f-columns single to save space (and labor). All the transitions shown in Fig. 32A are electric dipole transitions. The transitions responsible for the characteristic yellow light of a sodium lamp are the transitions from the $3p_{1/2}$- and $3p_{3/2}$-levels to the ground state $3s_{1/2}$. "The yellow sodium line" is in fact a doublet.

The reader should ponder the term scheme in Fig. 32A and convince himself that the transitions shown obey the selection rules (30b) and (31a) for j and l, respectively.

33   The energy levels of the helium atom, shown in Fig. 33A form, as it were, two almost completely independent systems: the singlet

90

Na I—Continued

Config.	Desig.	J	Level	Interval
$6f$	$6f$ $^2F^\circ$	$\left\{\begin{array}{l}2\frac{1}{2}\\3\frac{1}{2}\end{array}\right.$	58400.1	
$6h$	$6h$ $^2H^\circ$	$\left\{\begin{array}{l}4\frac{1}{2}\\5\frac{1}{2}\end{array}\right.$	58403.4	
$7p$	$7p$ $^2P^\circ$	$\begin{array}{l}\frac{1}{2}\\1\frac{1}{2}\end{array}$	$\begin{array}{l}58540.40\\58541.14\end{array}$	0.74
$8s$	$8s$ ^2S	$\frac{1}{2}$	38968.35	
$7d$	$7d$ ^2D	$\begin{array}{l}2\frac{1}{2}\\1\frac{1}{2}\end{array}$	$\begin{array}{l}39200.962\\39200.963\end{array}$	-0.001
$7f$	$7f$ $^2F^\circ$	$\left\{\begin{array}{l}2\frac{1}{2}\\3\frac{1}{2}\end{array}\right.$	39209.2	
$8p$	$8p$ $^2P^\circ$	$\begin{array}{l}\frac{1}{2}\\1\frac{1}{2}\end{array}$	$\begin{array}{l}39298.54\\39299.01\end{array}$	0.47
$9s$	$9s$ ^2S	$\frac{1}{2}$	39574.51	
$8d$	$8d$ ^2D	$\left\{\begin{array}{l}2\frac{1}{2}\\1\frac{1}{2}\end{array}\right.$	39729.00	
$8f$	$8f$ $^2F^\circ$	$\left\{\begin{array}{l}2\frac{1}{2}\\3\frac{1}{2}\end{array}\right.$	$[39734.0]$	
$9p$	$9p$ $^2P^\circ$	$\begin{array}{l}\frac{1}{2}\\1\frac{1}{2}\end{array}$	$\begin{array}{l}39794.53\\39795.00\end{array}$	0.47
$10s$	$10s$ ^2S	$\frac{1}{2}$	39983.0	
$9d$	$9d$ ^2D	$\begin{array}{l}2\frac{1}{2}\\1\frac{1}{2}\end{array}$	40090.57	
$9f$	$9f$ $^2F^\circ$	$\left\{\begin{array}{l}2\frac{1}{2}\\3\frac{1}{2}\end{array}\right.$	40093.2	
$10p$	$10p$ $^2P^\circ$	$\left\{\begin{array}{l}\frac{1}{2}\\1\frac{1}{2}\end{array}\right.$	40137.23	

Fig. 32B   A portion of a table of energy levels in the neutral sodium atom. The energies (fourth column) are expressed in wave numbers, cm^{-1}, measured from the ground state. The column labeled J gives the angular momentum of the state.

Exhibit taken from C. E. Moore, *Atomic Energy Levels*, vol. I, p. 90 (Circular of the National Bureau of Standards 467, U.S. Government Printing Office, Washington, 1949).

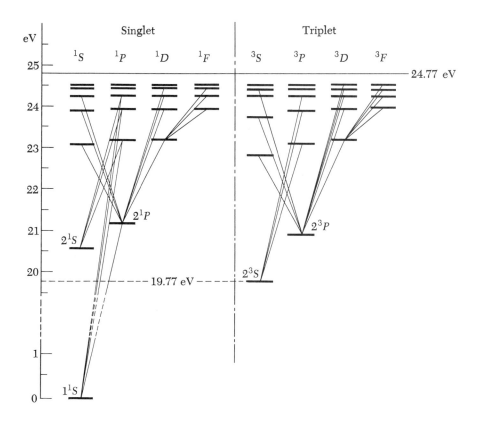

Fig. 33A Term scheme for the neutral helium atom. Note the remarkable separation between the singlet and the triplet systems of levels. In the triplet states the electron spins are parallel, and in the singlet states anti-parallel. There is an obvious correspondence between the triplet and singlet levels, except that the singlet ground state has no analog among the triplet states. This circumstance is a consequence of the Pauli exclusion principle: two electrons whose spins point in the same direction cannot both occupy the lowest level. There is no such restriction if the spins are oppositely directed.

system and the triplet system. The observed spectral lines arise from transitions *within* these systems: from singlet levels to singlet levels, and from triplet levels to triplet levels.

The helium atom is a two-electron atom. In the singlet levels the two electron spins are oppositely directed, whereas in the triplet levels the two electron spins are parallel.

The letters S, P, D, F, . . . , designate the total orbital angular momentum of the electrons. The left superscript 1 or 3 designate the *multiplicity* (singlet or triplet). For the singlet levels the total angular momentum equals the orbital angular momentum. For the triplet levels the total angular momentum j can assume the values $j = l - 1, l, l + 1$, with the provision that we always have $j \geqq 0$. In the triplet system the S-levels are single, and the remaining levels triple. The singlet levels are, of course, single.

34 We note an interesting detail in the term scheme for the

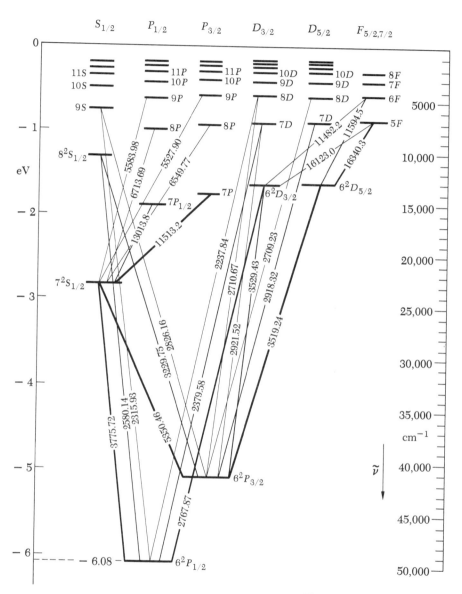

Fig. 34A Term scheme for the neutral thallium atom. The numbers on the slanted lines give the wavelengths in Ångströms of observed transitions. (After Grotrian.)

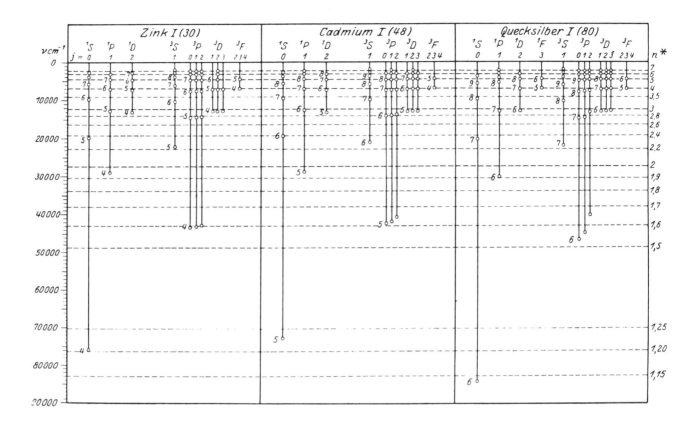

Fig. 35A Term schemes of zinc, cadmium and mercury, shown together to illustrate the fact that chemically similar elements have similar term schemes. This figure is taken from W. Grotrian, *Graphische Darstellung der Spektren von Atomen und Ionen . . . ,* vol. II, *Struktur der Materie,* Band VIII, p. 131 (Verlag von Julius Springer, Berlin, 1928). (*Courtesy of Springer Verlag.*)

thallium atom, Fig. 34A. An atom in the state $7^2S_{1/2}$ can decay *either* to the state $6^2P_{3/2}$, *or* to the ground state $6^2P_{1/2}$. The atom has a choice which way to make the "jump." There are other examples of this feature in the thallium term scheme, as well as in some of the other term schemes shown in this chapter. (The reader should hunt for these examples.) If an excited state can decay in several different ways, each mode of decay occurs with a definite probability. This probability is known as the *branching ratio* for the decay mode in question. That the branching ratios are intrinsic properties of the excited state, i.e., insensitive to *how* the excited state was reached, is an experimental fact.

35 The quite similar term schemes of sodium and lithium, which are both alkali metals, are strikingly different from the term schemes of helium and thallium. An examination of a large number of term schemes would reveal the remarkable fact that chemically similar

1																	2
H 1.0080																	**He** 4.003
3 **Li** 6.940	4 **Be** 9.013											5 **B** 10.82	6 **C** 12.011	7 **N** 14.008	8 **O** 16.000	9 **F** 19.00	10 **Ne** 20.183
11 **Na** 22.991	12 **Mg** 24.32											13 **Al** 26.98	14 **Si** 28.09	15 **P** 30.975	16 **S** 32.066	17 **Cl** 35.457	18 **Ar** 39.944
19 **K** 39.100	20 **Ca** 40.08	21 **Sc** 44.96	22 **Ti** 47.90	23 **V** 50.95	24 **Cr** 52.01	25 **Mn** 54.94	26 **Fe** 55.85	27 **Co** 58.94	28 **Ni** 58.71	29 **Cu** 63.54	30 **Zn** 65.38	31 **Ga** 69.72	32 **Ge** 72.60	33 **As** 74.91	34 **Se** 78.96	35 **Br** 79.916	36 **Kr** 83.80
37 **Rb** 85.48	38 **Sr** 87.63	39 **Y** 88.92	40 **Zr** 91.22	41 **Nb** 92.91	42 **Mo** 95.95	43 **Tc**	44 **Ru** 101.1	45 **Rh** 102.91	46 **Pd** 106.4	47 **Ag** 107.880	48 **Cd** 112.41	49 **In** 114.82	50 **Sn** 118.70	51 **Sb** 121.76	52 **Te** 127.61	53 **I** 126.91	54 **Xe** 131.30
55 **Cs** 132.91	56 **Ba** 137.36	57–71 **La Series**	72 **Hf** 178.50	73 **Ta** 180.95	74 **W** 183.86	75 **Re** 186.22	76 **Os** 190.2	77 **Ir** 192.2	78 **Pt** 195.09	79 **Au** 197.0	80 **Hg** 200.61	81 **Tl** 204.39	82 **Pb** 207.21	83 **Bi** 208.99	84 **Po**	85 **At**	86 **Rn**
87 **Fr**	88 **Ra** 226.03	89–103 **Ac Series**	(104)	(105)	(106)	(107)	(108)										

Lanthanide series	57 **La** 138.92	58 **Ce** 140.13	59 **Pr** 140.92	60 **Nd** 144.27	61 **Pm**	62 **Sm** 150.35	63 **Eu** 152.0	64 **Gd** 157.26	65 **Tb** 158.93	66 **Dy** 162.51	67 **Ho** 164.94	68 **Er** 167.27	69 **Tm** 168.94	70 **Yb** 173.04	71 **Lu** 174.99
Actinide series	89 **Ac** 227.04	90 **Th** 232.05	91 **Pa** 231.05	92 **U** 238.04	93 **Np**	94 **Pu**	95 **Am**	96 **Cm**	97 **Bk**	98 **Cf**	99 **Es**	100 **Fm**	101 **Md**	102 **No**	103 **Lw**

elements have similar term schemes. Figure 35A shows an example of this. The reason for this is that the optical spectrum and the chemical properties of an element are both determined by the electronic configuration in the atom, and in particular by the configuration of the outermost electrons.

The remarkable periodic table of the chemical elements, shown in Fig. 35B, can be understood in terms of the shell structure of atoms. In this rectangular table the elements are arranged in a certain way, in order of increasing atomic number Z, and with elements of similar chemical properties standing in the same column. The number of electrons in an atom equals Z, and as we progress in the table, in the direction of increasing Z, the "shells" are filled with electrons in a regular manner. The chemical properties depend on how the shells are filled. For instance, noble gases will occur in the table when certain shells are completely filled.

Fig. 35B The periodic chart of the elements. The atomic number Z is given above the chemical symbol, and the atomic weight (for reasonably stable elements) below.

Note the lanthanide series (rare earth series) consisting of 15 chemically very similar elements. All these atoms have the same configuration of electrons in the outermost shell. The series arises because inner shells which were "bypassed" are being filled as we progress in the series. On the basis of this picture Bohr predicted that the element of atomic number 72, hafnium, at that time undiscovered, would be chemically similar to zirconium rather than to the rare earths. Hafnium was indeed later found in a zirconium mineral, which was a striking triumph for the theory.

The so-called actinide elements form an analogous series.

The number of electrons which can be accommodated in a shell is determined by the Pauli Exclusion Principle, and this principle is therefore of *decisive* importance for chemistry. These circumstances were, of course, quite unsuspected before Pauli's great discovery.

To explain the details of the periodic table along the lines outlined is a fascinating task which we shall not undertake in this book. This study is best done in connection with a systematic study of atomic spectra and energy levels, and this would be a bit too much in an introductory course. To whet the reader's appetite we show, in Fig. 35C, a portion of a table of electronic configurations of atoms.

36   When the periodic table was first proposed by D. I. Mendelejeff in 1869 neither electrons nor nuclei were known. Mendelejeff therefore did not arrange the elements according to the charge Z, but rather in order of increasing atomic weight. Fortunately this gives the right order, with very few exceptions. The sequence argon–potassium is such an exception: argon has a larger atomic weight than potassium although the chemical properties of these elements (argon is a noble gas and potassium an alkali metal) establish without a trace of doubt that argon *must* come first. From the standpoint of chemistry the order of the elements in the table is quite clear, and on this basis it is thus possible to assign an atomic number Z to each element.

We should mention here that Mendelejeff had the foresight to leave some empty spaces in his table, to accommodate elements not yet discovered.†

37   The realization that the atomic number actually measures the nuclear charge, and hence equals the number of electrons, was an important step forward in atomic theory. The work of H. G. J. Moseley around 1913 was particularly important in settling this question. He systematically measured the wavelengths of X-rays from a large number of elements and was able to show that the wavelengths of analogous lines (in different elements) depend on the atomic number in a very simple way.‡ Let us discuss this question briefly.

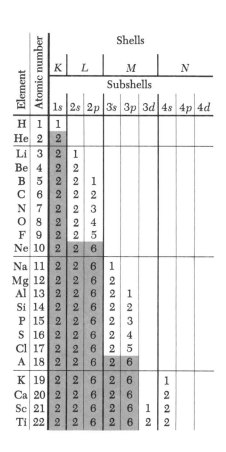

Element	Atomic number	Shells								
		K	L		M			N		
		Subshells								
		1s	2s	2p	3s	3p	3d	4s	4p	4d
H	1	1								
He	2	2								
Li	3	2	1							
Be	4	2	2							
B	5	2	2	1						
C	6	2	2	2						
N	7	2	2	3						
O	8	2	2	4						
F	9	2	2	5						
Ne	10	2	2	6						
Na	11	2	2	6	1					
Mg	12	2	2	6	2					
Al	13	2	2	6	2	1				
Si	14	2	2	6	2	2				
P	15	2	2	6	2	3				
S	16	2	2	6	2	4				
Cl	17	2	2	6	2	5				
A	18	2	2	6	2	6				
K	19	2	2	6	2	6		1		
Ca	20	2	2	6	2	6		2		
Sc	21	2	2	6	2	6	1	2		
Ti	22	2	2	6	2	6	2	2		

Fig. 35C The shell structure of light atoms. The main shells, designated by the letters *K, L, M, N*, . . . , are divided into subshells as shown. The different periods are indicated by the thin horizontal lines. Completed noble gas configurations are shown in gray. For the first three periods the shells are successively filled in a pleasingly regular manner, but beginning with potassium an outer shell is being filled before an inner shell has been completed. This phenomenon also occurs later in the periodic table. It is well understood theoretically.

An *s*-subshell can accommodate 2 electrons, a *p*-shell 6, and a *d*-shell 10 electrons.

† For an account of Mendelejeff's work, and the history of the periodic table, see *The World of the Atom*, Vol. I, edited by H. A. Boorse and L. Motz (Basic Books, Inc., New York, 1966).

‡ H. G. J. Moseley, "The High-Frequency Spectra of the Elements," *Philosophical Magazine* **26**, 1024 (1913), and **27**, 703 (1914).

When an atomic species is bombarded with energetic electrons (whose energy may range up to 100 keV) it is found that short wavelength electromagnetic radiation is emitted in the form of X-rays. It is furthermore found that the spectrum of this radiation consists of a number of sharp lines, which are characteristic of the element in question, superimposed on a continuous background. (See Fig. 23A, Chapter 4, for an experimental curve.) In the spirit of our discussion in Sec. 27, Chapter 2, we assume that the innermost electrons must be involved in the emission of the characteristic lines. The incident electron can knock out an electron from the innermost shell (known as the K-shell), and one of the electrons in an outer shell will subsequently "fall" into the empty "hole." The difference in binding energy will appear in the form of an X-ray photon.

In Sec. 27, Chapter 2, we argued that the binding energy of the innermost electron should be approximately of the form

$$B_K = Z^2 R_\infty \qquad (37a)$$

where $R_\infty = \frac{1}{2}\alpha^2 mc^2$ is the Rydberg constant. We have not presented a theory for what the binding energy in the next shell should be, but let us assume that it is *proportional* to B_K, but smaller. Therefore, if an electron falls into the innermost shell from the next outer shell we expect that the wavelength λ of the emitted photon should be of the form

$$\lambda = \frac{C}{Z^2 \tilde{R}_\infty} \qquad (37b)$$

where C is a constant which depends but weakly on Z. A plot of $\ln(\lambda)$ versus $\ln(Z)$ should therefore be a straight line if these ideas are correct. Such a plot is shown in Fig. 37A, and as we see the experimentally determined wavelengths do fall on a straight line to good accuracy. The constant C is approximately equal to $\frac{4}{3}$, which is what Bohr's theory predicts.

Since the electron which fills the hole can come from a number of different shells, and since the hole might have been created in one of several shells, we expect several characteristic lines. This is in fact what is found. In Fig. 37A we have plotted only one of these lines, involving the same shells in all the atoms.

As we see, the nuclear charge can be determined through such X-ray measurements, and Moseley's work thus led to a new understanding of the meaning of the periodic table.

38 Let us next discuss some aspects of nuclei. The term scheme

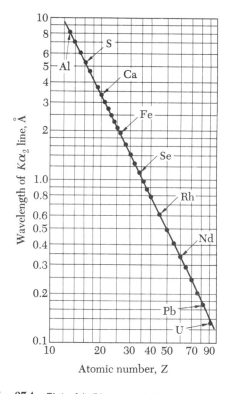

Fig. 37A Plot of $\ln(\lambda)$ versus $\ln(Z)$. Here λ is the wavelength of the so-called $K\alpha_2$-line in the X-ray spectrum of an element of atomic number Z. To the accuracy of the drawing the experimental points all fall on a straight line. Data are available for almost all the elements, although only selected atoms appear on this graph. For a simple theory of the graph, see text.

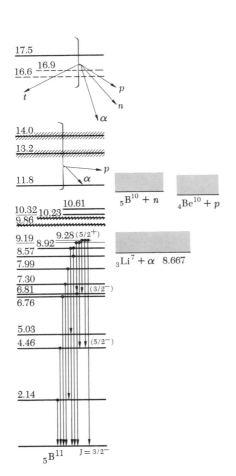

in Fig. 38A shows the nuclear energy levels of the boron isotope $_5B^{11}$, as they have been determined experimentally.

In this figure we have assigned the energy zero to the ground state. The total angular momentum of the ground state is $j = \frac{3}{2}$.

Levels which are particularly broad have been drawn hatched, and the hatching is an approximate measure of the widths.

The dissociation limit of this nucleus is at 8.667 MeV: above this energy the nucleus can dissociate into an alpha-particle and the lithium isotope $_3Li^7$. This mode of dissociation is indicated at the right of the main term scheme. Above an energy of about 11 MeV the boron nucleus can dissociate in two different ways: either into a neutron and the boron isotope $_5B^{10}$, or else into a proton and the beryllium isotope $_4Be^{10}$. These modes of dissociation are likewise indicated at the right of the level scheme for the isotope $_5B^{11}$.

Note, however, that the isotope $_5B^{11}$ has a system of energy levels *above* the dissociation energy 8.667 MeV. Below this energy the nucleus can only emit gamma rays, but above it the nucleus can also emit material particles. (The observed gamma ray transitions in $_5B^{11}$ are indicated by vertical lines.)

As this example shows, we have to be a bit careful about interpreting the "continuum." Levels can very well exist above the dissociation limit. The dissociation energy is merely an energy at which the system can dissociate into two *material* particles. Below this limit the system can still "dissociate," but only into a photon and *one* material particle. If we wish to treat photons on the same footing as material particles we can conclude that the levels above the dissociation limit (which are often called "virtual levels") are not different in principle from the levels below the dissociation limit: all levels above the ground state are unstable. Actually even the ground state may be unstable: consider the ground state of a radioactive nucleus. In our example in Fig. 38A the ground state is stable: the isotope $_5B^{11}$ occurs in the boron found in nature.

39 Two nuclei are said to form a pair of *mirror nuclei* if one can be obtained from the other by changing all protons into neutrons, and vice versa.

As we said in Sec. 37, Chapter 2, the *strong interactions*, which are the dominant interactions in nuclear physics, are believed to be invariant under this change. The proton-proton force is the same as the neutron-neutron force. If this belief is correct, and if there were no other interactions but the strong interactions, then the level systems in two mirror nuclei must be identical.

In Fig. 39A and Fig. 39B we show the experimentally found energy levels of two pairs of mirror nuclei. As we see, it is possible to establish a correspondence between the levels of the pairs.

The energies of the corresponding levels are not, however, identical, as the figures show. The reason for this is that electromagnetic forces are also present, and the electromagnetic forces are *not* invariant under neutron-proton interchange.

40 The term scheme in Fig. 40A explains why the alpha particles emitted by a radioactive nucleus do not always emerge with a single well-defined energy. The figure shows the alpha decay of the bismuth isotope $_{83}Bi^{212}$ to the isotope $_{81}Tl^{208}$ of thallium. The decay takes place from the *ground state* of the *parent nucleus* to one of several excited states, or to the ground state, of the *daughter nucleus*. The term scheme is drawn so that the ground state of the parent nucleus lies 6.2 MeV above the ground state of the daughter: this energy is the maximum kinetic energy with which the alpha particle can be emitted. It is clear that if the decay takes place to an excited state of the daughter, then the alpha particle will emerge with a smaller energy. For the level system shown in the figure, the alpha particle may be emitted with one of five well-defined different energies. The tilted lines show these decays. The numbers within parentheses are the branching ratios for the different decay modes.

If the daughter nucleus is left in one of the excited states it will emit gamma rays, indicated by the vertical lines, and eventually reach the ground state.

For many other alpha-active nuclei the decay always takes place to the ground state of the daughter, because no suitable excited states are available. The alpha particles will then emerge with a single well-defined energy, and there will be no gamma-rays associated with the alpha-decay.

41 By *beta-disintegration* we understand a process in which a nucleus emits an electron or a positron. The simplest process of this kind is the beta-decay of a neutron, which is a phenomenon well established experimentally. The mean-life of the free neutron is 16 minutes. Since the neutron-proton mass difference is $(m_n - m_p) = 1.3$ MeV, we might draw a term scheme like the one shown in Fig. 41A. The oblique line indicates the transition. If only an electron were emitted, it would always be emitted with the *same* energy (about 1.3 MeV), just as is the case in alpha-decay. Experimentally it is found that the electron, in fact, can be emitted

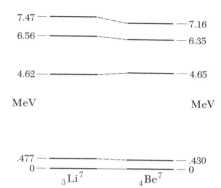

Fig. 39A The lithium and beryllium isotopes of mass number 7 form a pair of mirror nuclei: if the neutrons in the lithium nucleus are changed into protons, and vice versa, we obtain the beryllium nucleus. Mirror nuclei have similar, but not identical, level systems. The difference is an effect of electromagnetism.

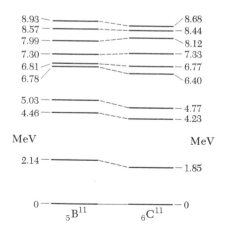

Fig. 39B The boron and carbon isotopes of mass number 11 form another pair of mirror nuclei.

with *any* energy between the rest energy 0.5 MeV and the available energy 1.3 MeV.

The explanation for this is that another particle, in this case the massless anti-neutrino, is also emitted, and the available energy is shared between the electron and the anti-neutrino. The reaction formulas for beta-decay thus read

Electron emission: $_zX^A \rightarrow {}_{z+1}X^A + e^- + \bar{\nu}$

Positron emission: $_zX^A \rightarrow {}_{z-1}X^A + e^+ + \nu$

where X stands for the chemical symbol of the radio-isotope; $e^\pm$ for the positron or electron; ν for the neutrino and $\bar{\nu}$ for the anti-neutrino.

42 The term scheme in Fig. 42A shows the origin of a beta-gamma-cascade emitted by the cobalt isotope $_{27}Co^{60}$. This isotope first beta-decays to an excited state of the nickel isotope $_{28}Ni^{60}$, which lies 2.4 MeV above the ground state. The *maximum* kinetic energy of the emitted electron is 0.3 MeV. The electron may emerge with any energy between zero and this maximum energy. The reaction formula for this part of the process may be written

$$_{27}Co^{60} \rightarrow {}_{28}Ni^{60*} + e^- + \bar{\nu}$$

where the asterisk indicates that the nickel isotope is left in an excited state. It subsequently decays (for all practical purposes immediately) from this state, via another excited state 1.3 MeV above the ground state, to the ground state by emission of gamma rays. The beta-decay is therefore always accompanied by two gamma rays, of energies 1.1 MeV and 1.3 MeV.

The half-life of the cobalt nucleus is 5.3 years, and this cascade process thus provides us with a fairly long-lived source of gamma rays.

Beta-active nuclei frequently have very long half-lives, just like the alpha-emitters. In the case of beta-emitters the reason for this is the intrinsic weakness of the interaction responsible for the beta-decay. This interaction, known as the *weak interaction*, is, roughly estimated, about 10^{14} times weaker than the strong interactions, and hence also considerably weaker than the electromagnetic interaction. The weak interaction is responsible for the (comparatively slow) decay of many fundamental particles, which might be stable but for the weak interaction. Examples of this are the charged pions, the neutron, the muon, the K-mesons and the lambda-hyperon.

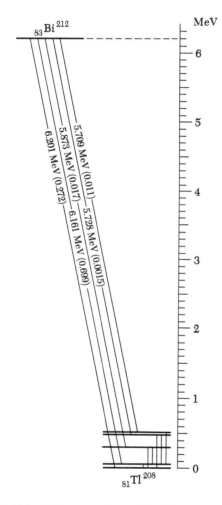

Fig. 40A In the alpha decay of the bismuth isotope $_{83}Bi^{212}$ the daughter nucleus may be left in the ground state, or in any one of four excited states. The alpha particles can accordingly emerge with five different energies. The daughter nucleus decays from the excited states through gamma emission.

Doppler Broadening and Collision Broadening of Spectral Lines

43   Earlier in this chapter we discussed the relationship between the natural linewidth $\Delta\omega$ of a spectral line emitted by an atom, and the mean-lives of the states involved in the transition. In the particular case that the lower state is the ground state we found that

$$\Delta\omega = \frac{1}{\tau} \tag{43a}$$

where τ is the mean-life of the upper state.

In Sec. 26 we quoted typical values of τ for atoms, and we estimated the fractional linewidth to be $\Delta\omega/\omega \sim 10^{-7}$. This is, of course, only a very rough order of magnitude estimate.

Spectral lines of atoms, as observed in nature, are in general much broader than the above estimate. Our theory in Secs. 14–26 applies to an *isolated* atom, originally at rest, but in practice the atoms under study are neither isolated, nor at rest. To study the causes of the additional broadening, let us suppose that we study the emission of light from a gas of atoms, at a temperature T and pressure P. Let the atomic weight be A. The atoms in the gas will move around in a random fashion, and they will incessantly collide with each other.

44   Because of the random thermal motion some of the atoms will move toward the observer, and some of the atoms will move away from the observer. As a result, the spectral line, which is a superposition of the lines emitted by many atoms, will be broadened because of the Doppler effect. For an atom moving toward the observer with velocity v the Doppler shift is given by $\delta\omega/\omega = v/c$. To estimate the amount of the Doppler broadening, $(\Delta\omega/\omega)_D$, we insert the mean velocity v_0 of the atoms in the gas into the formula for the Doppler shift. Actually v_0 is the mean velocity in the direction of observation, which we may take to the 3-axis. In Sec. 17, Chapter 2, we stated that the mean kinetic energy of the atoms and the temperature T are related by

$$E_{\text{kin}} = \tfrac{1}{2}M(v_{01}{}^2 + v_{02}{}^2 + v_{03}{}^2) = \tfrac{3}{2}kT \tag{44a}$$

where $M \cong AM_p$ is the mass of the atom. (M_p is the mass of a proton.) The mean velocities in the three different coordinate directions are clearly equal, and we obtain

$$v_0 = v_{03} = \sqrt{\frac{kT}{AM_p}} \tag{44b}$$

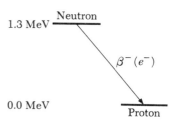

Fig. 41A Term scheme showing the beta-decay of the neutron. The mass of the neutron is 939.55 MeV, and the mass of the proton is 938.25 MeV. Part of the difference of 1.30 MeV, namely 0.50 MeV, appears as the rest mass of the electron, and the remainder appears as kinetic energy of the electron, antineutrino and proton, which result from the decay. The kinetic energy carried by the proton is very small, and most of the available energy is therefore shared between the electron and the antineutrino.

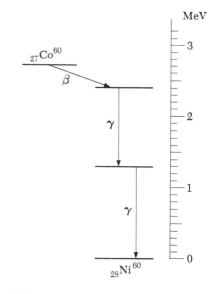

Fig. 42A Term scheme showing beta-gamma cascade emitted by the cobalt isotope $_{27}Co^{60}$. This isotope first beta-decays to an excited state of the nickel isotope $_{28}Ni^{60}$, which lies 2.4 MeV above the ground state. The maximum *kinetic* energy of the electron is 0.3 MeV. The excited state of the nickel isotope subsequently decays through the emission of two gamma rays in rapid succession.

The Doppler broadening is thus given by

$$\left(\frac{\Delta\omega}{\omega}\right)_D \sim \frac{1}{c}\sqrt{\frac{kT}{AM_p}} = (0.52 \times 10^{-5})\sqrt{\frac{1}{A}\left(\frac{T}{293°K}\right)} \qquad (44c)$$

45   The collisions between the atoms also lead to a broadening of the spectral lines. To estimate this effect we shall assume that, for any single atom, there is a time interval τ_c between two successive collisions. The inverse of this time, $1/\tau_c$, is the *collision rate* of the gas. We shall furthermore assume that each collision completely interrupts the emission process. The time τ_c is then the *effective* mean-life of the atoms, and in analogy with the relation (43a) we can assume that the corresponding broadening of the spectral line is given by

$$(\Delta\omega)_c \sim \frac{1}{\tau_c} \qquad (45a)$$

We must now estimate the collision frequency $1/\tau_c$. We regard the atoms as spheres of radius r. Let us focus our attention on one particular atom, immediately after it has undergone a collision. Let its velocity be v. We wish to find the average time τ_c it takes before this atom undergoes its next collision. For an order of magnitude estimate of this time it is permissible to assume that all the other atoms are at rest: for a *precise* value of τ_c we would, of course, have to consider the motion of the other atoms as well. In a small time interval dt our atom travels a distance $v\,dt$. Consider a cylinder, of radius $2r$, with the trajectory of the particle as axis, and centered about this trajectory. The height of this cylinder is then $v\,dt$. If no other atom is found inside the cylinder there will be no collision in the time interval dt. The probability that a collision takes place during the interval equals the probability of finding another atom inside the cylinder. The volume of the cylinder is $4\pi r^2\,v\,dt$, and if n is the average number of atoms per unit volume in the gas, the average number of atoms inside the cylinder will be $4\pi r^2\,nv\,dt$. If this number is small compared with unity it will also give the probability of finding an atom inside the cylinder, and thereby the probability that a collision takes place in the time dt. To estimate τ_c we impose the condition

$$4\pi r^2 n v \tau_c \sim 1 \quad \text{or} \quad \frac{1}{\tau_c} \sim 4\pi r^2 n v \qquad (45b)$$

which says that the mean number of atoms found inside a cylinder of radius $2r$ swept out by an atom in the time τ_c shall be of order unity.

One mole of any gas contains $N_0 \cong 6 \times 10^{23}$ molecules (in our case the molecules are atoms). At a temperature of 273°K, and at a pressure of 1 atm, 1 mole occupies a volume of 22.4 liters. In other words, at this temperature and pressure the number of atoms per unit volume is given by

$$n_0 = \frac{N_0}{(22.4 \text{ liters})} \cong 2.7 \times 10^{19} \text{ atoms/cm}^3 \qquad (45\text{c})$$

The number of atoms per unit volume at any other pressure P, and temperature T, is then given by

$$n = n_0 \left(\frac{P}{1 \text{ atm}}\right)\left(\frac{T}{273°\text{K}}\right)^{-1} \qquad (45\text{d})$$

(This result follows from the equation of state for a gas.)

As a reasonable estimate of the radius r we may take the Bohr radius, $r \cong 0.5 \times 10^{-8}$ cm. We obtain the characteristic velocity v from

$$\frac{Mv^2}{2} = \frac{3}{2}kT \qquad (45\text{e})$$

where $M \cong AM_p$ is the mass of the atom. Combining all the above equations (45) we finally obtain

$$(\Delta\omega)_c \sim \frac{1}{\tau_c} \sim$$

$$(2 \times 10^9 \text{ sec}^{-1}) \times \left(\frac{P}{1 \text{ atm}}\right) \times \sqrt{\frac{1}{A}\left(\frac{273°\text{K}}{T}\right)} \qquad (45\text{f})$$

46   If we now compare the collision broadening, as given by (45f), and the Doppler broadening, as given by (44c), with the broadening due to the finite lifetime of the excited state of an *isolated* atom, we notice that the broadening due to the last mentioned cause will in general be very small compared to the broadening due to the first two causes. The magnitude of the collision broadening diminishes as the pressure is reduced, and at low pressure the Doppler broadening dominates, and is the principal cause of the finite width of the spectral lines. The natural linewidth can only be seen under very special conditions.

We shall not discuss collision broadening and Doppler broadening further. These phenomena, although extremely important in practice, are extraneous to the basic problem of emission and absorption of light by an atom. The author felt it was necessary to discuss them anyway in this context, as the reader may otherwise be left with the impression that the width of the *observed* spectral line is always the natural linewidth.

Advanced Topic: On the Theory of Electromagnetic Transitions†

47   We consider two important questions. Why is it that the mean life of an excited state (in an atom, or nucleus) which is stable against *particle* emission, but unstable against *photon* emission, is long compared with the inverse of the frequency of the photon emitted? Why is it that electric dipole radiation is the most prominent mode of radiation in atomic physics?

Let us try to discuss these questions on the basis of a "semi-classical" electromagnetic theory. This means that our arguments are partly classical and partly quantum mechanical in spirit. The justification of such a simple-minded approach as the one in this chapter lies in its success: we *can* answer the above two questions in a reasonable way.

48   The answer to the first question is: "Because the fine structure constant α is so small." Let us try to see what this means.

First of all, we recall the conclusion which we reached in Secs. 29 and 39 in Chapter 2 that the wavelength of the emitted electromagnetic radiation is in general large compared with the size of the atom or nucleus which emits the radiation. This circumstance has important physical consequences, and it also simplifies the mathematical discussion of radiation phenomena. Let us first suppose that an atom, or nucleus, in an excited state acts like an oscillating electric dipole. Let ω be the frequency of oscillation: this is also the frequency of the emitted light. Let a denote the size of the object. Since the thing that oscillates is one or more elementary charges, we can assume that the electric dipole moment is of the order of magnitude ea. That the object is small compared with the wavelength is expressed by the condition

$$\frac{a\omega}{c} \ll 1 \tag{48a}$$

In Volume III of this series‡ we learned that such an electric dipole emits radiative energy at the rate

$$W = \frac{1}{3c^3}\omega^4\,(ea)^2 \tag{48b}$$

This formula gives the *power* emitted. Since we know that our

---

† Can be omitted in a first reading.
‡ Berkeley Physics Course, Vol. III, *Waves*, Chap. 7.

atom (or nucleus) will only emit a single photon, we are interested in the time τ it takes for the object to emit an amount of energy $\hbar\omega$. This time is given by

$$\frac{1}{\tau} = \frac{W}{\hbar\omega} = \frac{\omega}{3}\left(\frac{e^2}{\hbar c}\right)\left(\frac{a\omega}{c}\right)^2 \tag{48c}$$

or, as an order of magnitude estimate,

$$\frac{1}{\tau} \sim \omega\alpha\left(\frac{a\omega}{c}\right)^2 \tag{48d}$$

We interpret τ as the mean-life of the excited state: this is the time it takes for the excited state to decay through the emission of a photon. Let us consider the dimensionless quantity

$$\omega\tau \sim \frac{1}{\alpha}\left(\frac{a\omega}{c}\right)^{-2} \tag{48e}$$

This quantity is proportional to the number of oscillations which the system has time to perform during the time τ, before it decays. Clearly the excited state is the more stable the larger is the quantity $\omega\tau$. As we see $\omega\tau$ is large for two reasons: it is proportional to the "large" quantity $1/\alpha \cong 137$, and it is proportional to the inverse of the square of the quantity $(a\omega/c)$, and as we have said $(a\omega/c)$ is in general small.

49   For the case of an atom we can take a to be the Bohr radius, $a_0 = (1/\alpha)(\hbar/mc)$. For an optical transition the frequency is of the order of magnitude $\omega \sim \alpha^2 mc^2/\hbar$, and we thus obtain

$$\omega\tau \sim \alpha^{-3}, \qquad \tau \sim \left(\frac{\hbar}{mc^2}\right)\alpha^{-5} \tag{49a}$$

which shows the dependence of τ and $\omega\tau$ on the fine-structure constant. In the optical region the formula predicts mean lives ranging from 10^{-7} to 10^{-9} sec, which is in accordance with observed values.

To obtain a *crude* estimate of the lifetime of an excited state of a nucleus, which can decay through an electric dipole transition, we may take $a = 10^{-13}$ cm. A gamma ray of energy 200 keV has a wavelength of about 6×10^{-10} cm, and we obtain $\tau \sim 10^{-12}$ sec. This estimate, we emphasize, is very crude, but as a rough order of magnitude estimate it agrees with what is observed experimentally. Note that according to (48e) the lifetime is inversely proportional to the cube of the emitted frequency.

We have answered the first of the two questions raised in Sec. 47, and we now understand why excited states which can only decay electromagnetically live long compared with the inverse of the frequency of the light emitted.

50   Let us now turn to the second question, concerning the dominant role of the electric dipole transitions in atoms. To study this issue we must consider the emission rate from a configuration of moving charges which is such that the electric dipole moment vanishes at all times.

Fig. 50A shows a source which emits electric *quadrupole radiation*. The two arrows represent two electric dipoles oscillating at the frequency ω. These dipoles are of the same magnitude, but they are oppositely directed. The distance between the dipoles is a, and they are placed symmetrically with respect to the origin O, which is the center of the "atom." We observe the radiation at the point P, at a large distance r from the atom.

The electric dipole moment of this source is clearly zero. The same is true for the magnetic dipole moment, because we have no circulating currents in the source.

Let us now consider the electric field in a fixed direction, at a very large distance r from the source. This field lies in the plane of the figure, and is perpendicular to the radius vector OP. Let E_1 be the electric field which we would find at P if only dipole 1 were present, located at the origin O. This field is of the form

$$E_1 = \frac{C(\theta)}{r} \exp\left[i\left(\frac{r}{c} - t\right)\omega\right] \tag{50a}$$

where $C(\theta)$ is a function of θ which is proportional to the electric dipole moment. Its precise form need not concern us here.

If both dipoles are now present, as in the figure, then the electric fields due to the two dipoles almost cancel, but not quite, because the distance from P to dipole 1 is $\cong (r + \frac{a}{2}\cos\theta)$, whereas the distance from P to dipole 2 is $\cong (r - \frac{a}{2}\cos\theta)$: consequently the field due to dipole 1 differs in phase from the field due to dipole 2. The electric field, E_2, is therefore given by

$$E_2 = \left\{\frac{C(\theta)}{r} \exp\left[i\left(\frac{r}{c} - t\right)\omega\right]\right\} \times$$
$$\times \left[\exp\left(\frac{ia\omega\cos\theta}{2c}\right) - \exp\left(\frac{-ia\omega\cos\theta}{2c}\right)\right] \tag{50b}$$

51   We shall now make use of our assumption (48a) that $(a\omega/c)$ is very small compared to unity: this assumption is clearly valid for

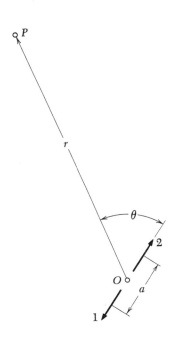

Fig. 50A Schematic picture of electric quadrupole source. The arrows represent two electric dipoles, oscillating with the same frequency ω. They are of equal magnitude, but oppositely directed. The electric, as well as the magnetic, dipole moment of this configuration vanishes, but the electric quadrupole moment does not. If a is small compared with the wavelength λ the rate at which energy is radiated from the system is smaller by a factor $(a/\lambda)^2$ than the rate from a single dipole.

optical transitions in atoms since a could not very well be larger than the typical atomic size. We can, therefore, expand the two exponential functions within the bracket in the right side of (50b), and neglecting all terms of higher order than the first in a we obtain

$$E_2 \cong i\left(\frac{a\omega}{c}\right)(\cos\theta)\,E_1 \tag{51a}$$

where E_1 is given by (50a). The electric field E_2 produced by the electric quadrupole shown in Fig. 50A is therefore everywhere smaller by at least a factor $(a\omega/c)$ than the electric field E_1 produced by a single one of the dipoles "making up the quadrupole." Since the radiation rate is proportional to the square of the electric field we can conclude that the typical rate of electric quadrupole radiation is smaller than the typical rate of electric dipole radiation by a factor $(a\omega/c)^2$. The corresponding lifetimes are then related by

$$\tau_{E2} \sim \left(\frac{a\omega}{c}\right)^{-2}\tau_{E1} \tag{51b}$$

where τ_{E1} stands for the mean-life in electric dipole transitions, and τ_{E2} stands for the mean-life in electric quadrupole transitions.

We have estimated that $(a\omega/c)$ is of the order of α in an atom, and the ratio of the lifetimes thus ranges from $10^{-4} - 10^{-6}$.

Similar considerations apply to nuclei, in which case a is a length characteristic of nuclei, and ω is the emitted frequency. In this case too $(a\omega/c)$ is small, say of the order of 10^{-3} or less.

52 Fig. 52A shows an example of a source with a vanishing *electric* dipole moment, but with a non-vanishing *magnetic* dipole moment. The small arrows again represent (oscillating) electric dipoles, and we can imagine that such a dipole consists of a charge oscillating back and forth along the direction of the arrow. This corresponds to an alternating current along the edges of a square, and the magnetic dipole moment of the system is proportional to the product of the strength of the current and the area of the square.

It is obvious that arguments very similar to those presented in Secs. 50 and 51 apply here too, and we can conclude that

$$\tau_{M1} \sim \left(\frac{a\omega}{c}\right)^{-2}\tau_{E1} \tag{52a}$$

where τ_{M1} stands for the mean-life in magnetic dipole transitions.

53 The classification of the emitted radiation into categories such as electric dipole, magnetic dipole, electric quadrupole, magnetic

Fig. 52A A configuration of oscillating electric dipoles with a vanishing electric dipole moment, and a vanishing electric quadrupole moment, but a non-vanishing magnetic dipole moment. The four arrows represent four electric dipoles of equal magnitude oscillating at the same frequency.

quadrupole, electric octupole, etc., is a classification according to the symmetry properties of the emitted radiation. Each type of radiation is characterized by a distinctive kind of intensity distribution as a function of direction, and by a distinctive polarization pattern. The symmetry pattern of the emitted radiation is, of course, uniquely defined by the symmetry properties of the source, and we may just as well classify the types of radiation in accordance with the properties of the source. An electric dipole emits electric dipole radiation, (abbreviated $E1$), a magnetic dipole emits magnetic dipole radiation, (abbreviated $M1$), an electric quadrupole emits electric quadrupole radiation, (abbreviated $E2$), etc. In term schemes showing electromagnetic transitions in nuclei we often find symbols such as $E1$, $M3$, $E4$, etc., which indicate the nature of the radiation emitted.

Our discussion of electric quadrupole and magnetic dipole radiation can be readily generalized to the study of higher multipoles. To produce an electric octupole we place two electric quadrupoles close to each other, but oppositely oriented, so that the resulting quadrupole moment vanishes. It is easy to understand that the rate of radiation from such a system is smaller than the rate of radiation from a single quadrupole by the factor $(a\omega/c)^2$. Every time we go one step higher in the hierarchy of electric multipoles the characteristic rate drops by a factor of order $(a\omega/c)^2$, where a is a typical linear dimension of the system. Similarly for the magnetic multipoles.

We can thus understand the prominence of electric dipole transitions in atoms. If an excited state can decay in several different ways, one of which is by $E1$-radiation, then it will decay through $E1$-radiation with a very high probability. The other types of radiation may also be present, but the intensities of the spectral lines which do not correspond to $E1$-radiation are *much* smaller than those of the $E1$-lines.

54 When we discussed the selection rules for electric dipole transitions in Secs. 29–31 we said that these rules derive from the principle of conservation of angular momentum. We also mentioned that the origin of the latter principle is the isotropy of physical space. We can therefore describe the selection rules in a seemingly different manner: *the selection rules derive from the isotropy of physical space.* Let us explore this idea a bit.

We have said that the angular momentum quantum number j measures the angular momentum of the state of a system, say an atom. Within the framework of quantum mechanics j has an alternate interpretation: j describes the *rotational symmetry type* of the state. We can say that j describes the appearance of the atom when looked at from all possible directions. For instance, if the atom is in a state such that $j = 0$, then the atom has the same appearance from all directions: $j = 0$ means that the state is spherically symmetric. If $j = 1$, then the state has the same symmetry properties as a vector. The radiation field emitted in an electric dipole transition is an example of such a state *of the photon:* the entire field pattern in space must have the same symmetry properties under rotations as the source, and the source is an electric dipole vector. We have said that an electric dipole photon carries one unit of angular momentum, and this is an example of the general connection between symmetry types and angular momentum. The radiation pattern from an electric quadrupole is characterized by the rotational symmetry quantum number $j = 2$, and correspondingly an electric quadrupole photon carries *two* units of angular momentum. The selection rules for electric quadrupole transitions are therefore different from the selection rules for electric dipole transitions: in a quadrupole transition the angular momentum of the atom can change by as much as two units.

55 In view of the above, all the selection rules governing electromagnetic transitions can be derived from the principle that the rotational symmetry properties of a system are conserved. To illustrate this profound idea we shall prove one particular selection rule, namely that the transition $(j_i = 0)$ to $(j_f = 0)$ is forbidden for all (one-photon) electromagnetic transitions. Differently stated this says that an atom in an excited state which is spherically symmetric, (i.e., $j_i = 0$), cannot decay through the emission of a photon to another state which is also spherically symmetric, (i.e., $j_f = 0$).

We argue as follows: *before* the emission the atom is in a spherically symmetric state. It has the same appearance from all directions. *After* the emission the system, which now consists of the atom in the final state *plus* the emitted electromagnetic wave, must also be in a spherically symmetric state. Originally there was no preferred direction in space, and if physical space is isotropic there cannot be any preferred direction after the emission either. This is what we mean by the preservation of the rotational symmetry properties. Consider now the situation after the emission. If the final state of the *atom* is spherically symmetric, corresponding to $j_f = 0$, we conclude that the emitted electromagnetic wave must also be spherically symmetric: it can have no angular dependence. *Such an electromagnetic wave does not exist, and it follows that the transition cannot take place at all.* It is clear that there cannot be any electric (or magnetic) dipole wave which is spherically

Eugene Paul Wigner. Born 1902 in Budapest, Hungary. Studied in Berlin, and received a doctorate in chemical engineering from the Technische Hochschule in 1925. After spending some time in Berlin and at Göttingen, Wigner came to the United States in 1930. He is now professor of Physics at Princeton University. Wigner was awarded the Nobel prize in 1963.

Wigner's work in theoretical physics covers a remarkably wide range. He has made many important contributions in such diverse fields as atomic physics, theoretical chemistry, solid state physics, nuclear physics, the theory of nuclear reactors, relativity and elementary particle theory. In the author's opinion his most outstanding contribution is his amazingly deep and penetrating analysis of the role of symmetry principles in quantum mechanics. His ideas on this topic are presented in a series of papers (and one book) spanning the time from 1931 to the present. *(Photograph by courtesy of Reviews of Modern Physics.)*

symmetric, because an electric dipole (or magnetic dipole) defines a direction. There cannot be any other spherically symmetric multipole wave either, because at a given instant of time, and at a given position in space, the electric field defines a direction perpendicular to the radius vector. The electric vector at that point and at that instant of time cannot, therefore, remain unchanged under a rotation of the field configuration about the radius vector, and hence the field pattern cannot be spherically symmetric.

56 A transition forbidden by the dipole selection rule can be allowed for quadrupole, or higher multipole transitions. If we examine the term schemes for atoms, shown in this chapter, we see that almost all the excited states can decay to some lower state by electric dipole transitions. The level structure of nuclei is often quite different, and we may find a state just above the ground state which differs in j-value by several units from the ground state. Such an excited state cannot decay by dipole emission, and consequently lives longer. If the difference in j-values is very large, and the energy difference small, the lifetime may be of the order of minutes since the photon emitted is of high multipole order. Such states are called *isomeric states*.

References for Further Study

1) Energy levels of atoms, molecules and nuclei are, of course, discussed in very many texts on these subjects. Among these we mention the following fairly elementary ones:

a) G. Herzberg: *Atomic Spectra and Atomic Structure* (Dover Publications, New York, 1944).

b) H. White: *Introduction to Atomic Spectra* (McGraw-Hill Book Co., New York, 1934).

c) G. Herzberg: *Molecular Spectra and Molecular Structure: I, Spectra of Diatomic Molecules* (D. van Nostrand Co., New York, 1953).

d) D. Halliday: *Introductory Nuclear Physics* (John Wiley and Sons, Inc., New York, 1950).

e) E. Segrè: *Nuclei and Particles* (W. A. Benjamin, New York, 1964).

2) a) Term schemes of many atoms can be found in the book: W. Grotrian: *Graphische Darstellung der Spektren von Atomen und Ionen mit Ein, Zwei und Drei Valenzelektronen*, vol. II (Verlag von Julius Springer, Berlin, 1928).

b) For energy level diagrams of selected nuclei see: F. Ajzenberg and T. Lauritsen: "Energy levels of light nuclei," *Rev. Mod. Phys.* **27**, 77 (1955).

3) For shorter tables relating to spectra and energy levels we refer to:

a) *Handbook of Chemistry and Physics* (Chemical Rubber Publishing Co.).

b) *American Institute of Physics Handbook* (McGraw-Hill Book Co., New York, 1957).

4) There are several articles in the *Scientific American* which the reader can read with profit at this point:

   a) A. L. Bloom: "Optical Pumping," October 1960, p. 72.

   b) H. Lyons: "Atomic Clocks," February 1957, p. 71.

   c) G. E. Pake: "Magnetic Resonance," August 1958, p. 58.

   d) J. P. Gordon: "The Maser," December 1958, p. 42.

   e) A. L. Schawlow: "Advances in Optical Masers," July 1963, p. 34.

   f) S. de Benedetti: "The Mössbauer Effect," April 1960, p. 72.

Problems

1 The following spectral lines were observed (early in this century) for a certain atom:

$$\bar{\nu}_1 = 82258.27 \qquad \bar{\nu}_5 = 15232.97 \qquad \bar{\nu}_8 = 5331.52$$
$$\bar{\nu}_2 = 97491.28 \qquad \bar{\nu}_6 = 20564.57 \qquad \bar{\nu}_9 = 7799.30$$
$$\bar{\nu}_3 = 102822.84 \qquad \bar{\nu}_7 = 23032.31 \qquad \bar{\nu}_{10} = 2469.$$
$$\bar{\nu}_4 = 105290.58$$

where the numbers listed are *wave numbers*, in units of cm^{-1}.

(*a*) Find as many instances as you can which illustrate the Ritz combination principle, i.e., cases in which a wave number can be expressed as a difference of two other wave numbers.

(*b*) Show that all the lines can be regarded as combinations of five terms. Find these (up to a common arbitrary additive constant) and draw a term scheme showing the terms and the transitions which correspond to the above lines.

(*c*) Can you find a *simple* formula for the terms? Does this term scheme occur anywhere in this book?

(*After* you are through with your analysis you may want to peek in a table of wavelengths to identify the atom.)

2 In a study of resonance fluorescence the contents of a quartz vessel C is illuminated by ultraviolet light of wavelength 2537 Å emitted by a mercury lamp (in the lamp an electrical discharge is run through mercury vapor contained in a quartz vessel).

The following facts can be observed:

(*a*) If the vessel C contains mercury vapor, and nothing else, then the gas in C will scatter the incident light very strongly: the atoms in the gas will resonate. The scattered radiation is also of wavelength 2537 Å

(*b*) If the vessel C contains thallium vapor, and nothing else, then C will be transparent to the incident radiation and there will be very little scattering of the incident light.

(*c*) If the vessel C contains *both* thallium and mercury vapor, then C will emit the mercury line 2537 Å, and it will also emit a number of lines characteristic of thallium at the wavelengths 2768 Å, 3230 Å, 3529Å, 3776 Å and 5350 Å. If a glass plate is placed between C and the lamp, none of the above lines will be emitted.

(*d*) Under the conditions described in (*c*) it is found that the thallium line 3776 Å is much broader than the thallium line 2768 Å, and the first mentioned line is, in fact, much broader than could be explained on the basis of the Doppler broadening corresponding to the temperature in the vessel C, and it is also much broader than the same line would be if it were emitted from a discharge tube filled with thallium vapor.

Try to explain all these phenomena. As a hint we refer to the term scheme of thallium, Fig. 34A of this chapter. It is interesting to note that only a few of the lines of thallium are observed in the experiment. The lines 2826 Å and 5584 Å, for example, are notably absent.

3 The lifetime of the $3p_{1/2}$-state in sodium (see Fig. 32A of this chapter) is about 10^{-8} sec. Consider a vessel filled with argon gas at a pressure of 10 mm Hg, and at a temperature of about 200°C. Inside the vessel we have a small speck of sodium, which is heated so that the vessel will contain a small amount of sodium vapor. We observe the absorption line 5896 Å in light from a tungsten filament passing through the vessel. (The heated tungsten filament emits radiation with a continuous spectral distribution.) Estimate:

(*a*) the natural width of the line;

(*b*) the magnitude of the Doppler broadening of the line;

(*c*) the magnitude of the collision broadening of the line.

Express your results in *wave numbers* (also express the frequency of the line in question in wave numbers, cm^{-1}). Compare these widths with the fine-structure separation of the (yellow) sodium lines D_1 and D_2.

(*d*) In the term scheme in Fig. 32A we note a line of wavelength 5688.22 Å. Will we see this line in the absorption experiment described above?

The argon gas in the vessel has no other effect on the process than to establish a pressure and a mean temperature in the vessel. Its presence must be taken into account when we wish to consider the effect of collisions on the absorption line: since the number of sodium atoms in the vessel is extremely small compared to the number of argon atoms the sodium atoms primarily collide with the argon atoms.

4 Let us consider the shape of the spectral lines *emitted* by an atom. We assume that the atoms are present in the light source in the form of a gas. We measure intensity versus frequency with a spectrograph of very high resolution. For some light sources the line may have the appearance shown in the upper figure in the margin, whereas for other constructions of the light source the same spectral line may have the appearance shown in the lower figure in the margin. Furthermore we may note that as a rule only the lines arising in a transition to the ground state show the appearance in the lower

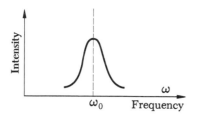

These two figures refer to Prob. 4. The upper figure shows a common appearance of a spectral line (under extremely high resolution) from a gas discharge tube.

Under certain conditions the *same* spectral line, from a similar gas discharge tube, may instead have the appearance shown below.

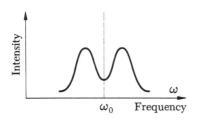

figure. Can you explain these phenomena, and can you explain the physical characteristic of the light sources in which the line can be expected to be of the kind shown in the upper figure?

5   Under the experimental conditions described in Prob. 3, estimate, on the basis of formula (7a) in this chapter, the fraction of sodium atoms which at any given time are in the first excited state. (Assume $T = 200°C$.)

6   (*a*) From the experimental data presented in Fig. 37A, compute the constant C in Eq. (37b).

(*b*) In the study of X-ray emission it is found that in order to make one of the characteristic lines (of frequency ω) appear, the energy E of the bombarding electrons must be quite a bit higher than $\hbar\omega$. For the $K\alpha$-lines, to which Fig. 37A refers, the condition for the appearance of the lines is roughly $E > \frac{4}{3}\hbar\omega$. Why does not the line appear as soon as $E > \hbar\omega$?

7   Although the author cannot assume *any* responsibility for the harmful mental images which might be formed if the reader studies Bohr's planetary model of the atom, he does not want to go as far as to outright *forbid* the reader to consider this model. Bohr assumed that the electron moves in a circular orbit in the hydrogen atom, and in such a way that the angular momentum of the electron is a positive integral multiple of $\hbar$. It is a remarkable accident that this model gives the correct location of all the energy levels to a very high accuracy. Since this model is of considerable historical interest the reader may wish to follow in Bohr's footsteps, and construct the term scheme, and identify the lines shown in Fig. 1B. (The wavelengths noted on the photographic plate are: 4861.3 Å, 4340.5 Å, 4101.7 Å, 3970.1 Å, 3889.1 Å, and 3835.4 Å.)

8   The radioactive nucleus $_{84}Po^{212}$ (formerly called ThC′) emits alpha-particles of several different energies. In this case the explanation is not as given in Fig. 40A of this chapter. Find out what the explanation might be. Draw a term scheme to illustrate your ideas and your acquired knowledge. Label correctly the states corresponding to the different nuclei involved.

9   Fig. 38A is a simplified version of a diagram appearing in an article by F. Ajzenberg and T. Lauritsen, *Reviews of Modern Physics* **27** (1955), p. 107, Fig. 15. Study the original figure. Note that above the line labeled $Li^7 + \alpha$, there is shown a curve with a number of maxima. These coincide with some of the levels of the nucleus B^{11}. This curve shows the results of some actual measurements. Explain in detail the significance of this curve, and discuss the measurements on which it is based.

To the right in the original figure, we furthermore note a horizontal bar labeled $B^{11} + p - p'$, and above this a short horizontal bar labeled 15.6.

This latter bar is joined to some of the levels of B^{11} by lines with arrowheads. This feature of the diagram also refers to some measurements. Discuss these measurements, and explain what the arrows represent.

10 Consider an experiment in which a beam of atoms moves parallel to a screen with a narrow slit in it. The slit is perpendicular to the direction of the beam. For simplicity we assume that all the atoms in the beam have the same velocity v. Some of the atoms are brought to an excited state at some point before they fly past the slit. Let x be the distance between the slit and the point at which the atoms are excited. The atoms can decay from the excited state to the ground state through the emission of a photon, of frequency ω. Let τ be the mean life of the excited state. We study the light emerging through the slit.

(*a*) How does the intensity of the light emerging through the slit depend on the distance x? Motivate your answer.

(*b*) Suppose that we let the light emerging through the slit fall upon a photocell, and suppose that we determine the retarding potential at which the photocell ceases to register. State, and motivate, your expectations as to how this retarding potential will depend on the distance x. It does not matter whether you come up with the right answer or not: the important thing is that you *think* about the problem and formulate a definite prediction based on your present knowledge.

11† It is interesting to study the angular distribution (*in intensity*) of the electric quadrupole radiation emitted by the source shown in Fig. 50A and to compare this angular distribution with the one we would observe for a *single* electric dipole. The intensity is proportional to the square of the electric field. Show that the emitted intensity as a function of the direction of observation is given by

$$I_{E1}(\theta) = A \sin^2(\theta)$$

for the case of an electric dipole, and by

$$I_{E2}(\theta) = B \sin^2(2\theta)$$

for the case of the electric quadrupole shown in Fig. 50A. Here A and B are constants. The intensity is independent of the azimuthal angle. This example indicates how the different kinds of multipole radiation can be distinguished from each other by their characteristic intensity patterns.

---

† This problem refers to an advanced topic: Sec. 50.

Chapter 4

Photons

Chapter 4   *Photons*

The Photon as a Particle

1 In this chapter and the next we shall explore both the particle and wave aspects of such fundamental entities as the photon, the electron, the proton, the neutron, and the other elementary particles found in nature. We shall look at some basic experimental facts and try to obtain a preliminary consistent picture of what is observed. In many instances, the outcome of a particular experiment may suggest a new experiment: when this is the case, we shall try to make a prediction and then study what has actually been observed. Our approach is one of experimentation with ideas, and we should be careful not to commit ourselves too firmly to any specific model yet: let us see how things work out.

2 We may properly begin with the study of photons. The photons are the "quanta" of the electromagnetic field: we know that almost monochromatic radiation of frequency ω comes in packets carrying an energy $E = \hbar\omega$. The most immediate evidence for this comes from the study of the photoelectric effect, but there are, as we shall see, other observations which lead to the same conclusion. Considered all together, these observations imply that the relation $E = \hbar\omega$ must hold over a very wide range of frequencies. We shall now make the (daring) extrapolation that this relation between the energy of the packet and the frequency is completely general, and holds for *all* photons.

3 We ask the following question: suppose that we have a packet of electromagnetic radiation of frequency ω traveling in some direction, with the velocity of light c. Will this packet also carry momentum, and, if so, what is the magnitude of the momentum? If the packet, which we call the photon, has some particle properties, we would expect that it does carry momentum, and we may think about experiments in which we can directly measure the momentum.

In Vol. III† of this series we learned that for a monochromatic electromagnetic wave traveling in a well-defined direction the energy E and the momentum p are related by $p = E/c$, where the momentum is in the direction of propagation. This is what clas-

---

† Berkeley Physics Course, Vol. III, *Waves*, Chap. 7.

sical electromagnetic theory predicts, and it is reasonable to expect that the same relation holds for the electromagnetic quanta.

4 It is instructive to derive the relation between energy and momentum from a different point of view. Let us therefore pretend that we do not yet know that $p = E/c$, but that we do believe that the relation $E = \hbar\omega$ is universally valid. This means, in particular, that this relation holds in *every* inertial frame. The principle of special relativity implies that if we can find general relations between energy, momentum, frequency and direction of propagation, which hold for *all* photons in *one* inertial frame, then these same relations must hold in *every* inertial frame. The requirement of relativistic invariance therefore introduces a constraint of the possible relations between the physical quantities mentioned, and the idea of our argument is to make use of this constraint to find an expression for the momentum **p** of a photon.

Let the photon travel, in one inertial frame, in the direction of the positive x-axis. We shall regard the photon as a particle of energy $E = \hbar\omega$, and of an unknown momentum **p**. For reasons of symmetry **p** must be in the direction of the x-axis. Consider now the same situation as observed in another inertial frame, the "primed frame," moving with a uniform velocity v along the positive x-axis with respect to the "unprimed frame." An observer in the primed frame sees a photon of frequency ω', carrying an energy $E' = \hbar\omega'$, and a momentum **p'**. Since $c > v$ the photon in the primed frame will travel in the direction of the positive x'-axis. Furthermore we conclude (on the basis of symmetry) that in *both* frames the momenta must be directed along the direction of motion of the photon. We can therefore omit the vector notation for the momenta, and simply write p and p' for the x- and x'-components, the other components being zero.

5 Let us recall two results on Lorentz transformations from Vol. I of this series.† The first of these is the formula for the longitudinal Doppler shift, which relates the frequencies ω and ω' according to

$$\omega' = \omega \sqrt{\frac{c - v}{c + v}} \tag{5a}$$

The second is the relativistic transformation law for the energy

---

† Berkeley Physics Course, Vol. I, *Mechanics*. The formula for the longitudinal Doppler shift was derived in Chap. 11, and the transformation law for energy and momentum was derived in Chap. 12.

and the momentum of a particle. Acording to this law the energy E' is given by

$$E' = \frac{E - vp}{\sqrt{1 - (v/c)^2}} \tag{5b}$$

If we now make use of our assumption

$$E = \hbar\omega, \qquad E' = \hbar\omega' \tag{5c}$$

to eliminate E and E' from (5b), and then eliminate ω' from the resulting equation by using (5a), we obtain

$$\hbar\omega \sqrt{\frac{c - v}{c + v}} = \frac{\hbar\omega - vp}{\sqrt{1 - (v/c)^2}}$$

This equation can immediately be solved for p, and we obtain

$$p = \frac{\hbar\omega}{c} \tag{5d}$$

or

$$p = \frac{E}{c} \tag{5e}$$

These relations of course hold in *all* inertial frames, because there was nothing special about our "unprimed frame." In particular they hold in our "primed frame." As we said, the relation (5e) can be derived within classical electromagnetic theory. The relation (5d) is definitely quantum mechanical: it says that a light quantum of frequency ω always carries the momentum $\hbar\omega/c$. This relation, of course, follows immediately from (5e) and (5c), and conversely (5c) follows from (5d) and (5e).

6   The rest mass, m_{ph}, of the photon is zero. In Vol. I we derived a general relation between rest mass, energy and momentum, which when applied to this particular case reads

$$(m_{\text{ph}}c^2)^2 = E^2 - (cp)^2 \tag{6a}$$

In view of (5e) the right side of this equation vanishes, and we accordingly have $m_{\text{ph}} = 0$.

At first sight this result might appear slightly peculiar: since the photon has some particle properties it ought to have a mass when observed in its rest frame. However, *there is no inertial frame in which the photon is at rest:* electromagnetic radiation propagates with the velocity c in *every* inertial frame. A photon at rest is therefore a meaningless concept.

One might well argue that an object which can never be at rest should not be called a "particle." It has, however, become established custom to talk about "mass-less particles," of which the photon and the neutrino are examples, and we should conform to this practice. Ultimately it is purely a matter of taste how we define the word "particle." It is clearly convenient to treat photons and neutrinos on the same footing as the massive particles. On the other hand it should be very strongly emphasized that a photon is no billiard ball: it merely has *some* properties in common with billiard balls.

7 Let us next consider some thought experiments in which we try to see whether the particle picture of the photon is consistent with some results obtainable from classical electromagnetic theory. In this manner we can further familiarize ourselves with the idea that packets of electromagnetic radiation have particle properties.

A word of explanation is in order. When we talk about "particle properties" here, we mean the properties which particles are supposed to possess within classical physics. Actually the word "particle" is of course nowadays used as a common name for such objects as photons, electrons, protons, neutrons, etc. Strictly speaking, "particle properties" are therefore all those characteristics which are shared by these objects. In particular, it is a property of a real physical particle that it can behave like a wave. At this point of our discussion, we are, however, trying to find out what the properties of the *real* particles are, and one aspect of this study is to try to see to what extent the real particles behave like the *imaginary* "classical particles."

8 Consider a stationary source of light, emitting photons of frequency ω. We let this light be incident perpendicularly on a perfect mirror, at rest in the rest frame of the light source.

Classical electromagnetic theory predicts that the reflected light will also be of frequency ω, and that the flux of energy against the mirror is the same as the flux of energy away from the mirror.

Furthermore, classical electromagnetic theory predicts that the incident radiation will exert a pressure on the mirror, namely the radiation pressure. If we assume that the intensity of the radiation is uniform over the mirror, this pressure P is given by

$$P = W \qquad (8a)$$

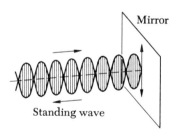

Fig. 8A Reflection of light from a mirror (with a perfectly conducting surface) according to the wave picture. A standing wave builds up in front of the mirror, and currents are induced in the surface. The wave exerts a force on the mirror through the interaction of the magnetic field in the wave with the induced currents. For normal incidence the radiation pressure P is given by $P = W$, where W is the energy density in front of the mirror.

where W is the energy density of the radiation field in the immediate neighborhood of the reflecting surface.

Suppose now that Φ is the flux of incident radiation: i.e., Φ is the amount of energy flowing against the mirror per unit time through a unit area perpendicular to the direction of incidence. If we similarly let Φ' denote the flux of the reflected radiation we must have $\Phi = \Phi'$. In unit time the radiation travels a distance c, and the density of energy W must then be given by

$$W = \frac{\Phi}{c} + \frac{\Phi'}{c} = \frac{2\Phi}{c} \tag{8b}$$

where the first term gives the energy density due to the incident radiation, and the second term gives the energy density due to the outflowing radiation. The flux and the radiation pressure are thus related by

$$P = \frac{2\Phi}{c} \tag{8c}$$

which we obtain by combining (8a) and (8b).

9 Let us now look upon this situation from the standpoint of the photon picture. In this picture there is a flux of, say, N photons per unit time through a unit area against the mirror. Each photon carries an energy $E = \hbar\omega$ and a momentum $p = \hbar\omega/c$. After colliding with the mirror each photon has its momentum reversed (regard the mirror as being infinitely heavy since it remains at rest), and each photon therefore transfers an amount of momentum $2p$ to the mirror: in this picture the radiation pressure arises from the bombardment of the mirror by the photons.

The radiation pressure P is equal to the amount of momentum transferred per unit time to a unit area of the mirror, and we thus have

$$P = 2Np = \frac{2N\hbar\omega}{c} \tag{9a}$$

On the other hand the energy flux Φ is simply given by

$$\Phi = N\hbar\omega \tag{9b}$$

and the energy density (since each photon travels with the velocity of light) is given by

$$W = \frac{2N\hbar\omega}{c} \tag{9c}$$

If we combine the formulas (9a)–(9c) we recover the relations

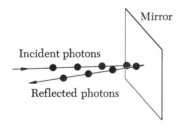

Fig. 9A Reflection of light from a mirror according to the particle picture. The radiation pressure arises when the photons collide with the mirror, and have their momenta reversed (for the case of normal incidence). The relation between radiation pressure and energy density is the same as in the wave theory. (See Fig. 8A.)

(8a)–(8c), which means that for the case considered the photon picture is consistent with the wave picture.

10   Let us next consider the following situation: a source of light is stationary in the laboratory. It emits photons of frequency ω, and these photons are incident perpendicularly on a perfect mirror moving away from the source with the *small* velocity v. We shall assume that the mass M of the mirror is very large. (We let v be small and M large so as to be able to discuss the problem non-relativistically.)

From the standpoint of the photon picture, let us consider what happens when a single photon collides with the mirror. Before the collision the photon has an energy E and a momentum $p = E/c$, and after the collision the photon has the energy E' and the momentum $p' = E'/c$. The conditions for conservation of energy and momentum read:

$$p + Mv = -p' + Mv' \qquad \text{(momentum)} \qquad \text{(10a)}$$

$$E + \tfrac{1}{2}Mv^2 = E' + \tfrac{1}{2}Mv'^2 \qquad \text{(energy)} \qquad \text{(10b)}$$

Here we have taken into account the fact that the mirror may have a (slightly) different velocity v' after the collision: the direction of velocity will, however, remain unchanged. The reflected photon will travel in the opposite direction, and hence the term $-p'$ in (10a).

Let the frequency of the reflected photon be $\omega' = E'/\hbar$. We may rewrite equations (10a) and (10b) in the form

$$\frac{\hbar\omega}{c} + Mv = -\frac{\hbar\omega'}{c} + Mv' \qquad \text{(momentum)} \qquad \text{(10c)}$$

$$\hbar\omega + \tfrac{1}{2}Mv^2 = \hbar\omega' + \tfrac{1}{2}Mv'^2 \qquad \text{(energy)} \qquad \text{(10d)}$$

Eliminating v' from these two equations we obtain

$$\hbar(\omega - \omega') = \left(\frac{v}{c}\right)\hbar(\omega + \omega') + \frac{1}{2M}\left(\frac{\hbar}{c}\right)^2(\omega + \omega')^2 \qquad \text{(10e)}$$

We consider the limiting case of an infinitely heavy mirror, in which case the second term in the right-hand side of (10e) drops out, and we obtain

$$\omega' = \omega\left(1 - \frac{v}{c}\right)\Big/\left(1 + \frac{v}{c}\right) \qquad \text{(10f)}$$

Since we assumed that v/c is small we may expand (10f) in powers of v/c, and, retaining only the linear terms, we obtain for the reflected frequency the approximate expression

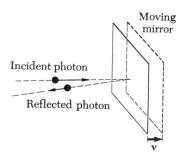

Fig. 10A   The laws governing elastic collisions predict that the energy E' of the reflected photon will be smaller than the energy E of the incident photon, if the mirror moves away from the source. From the relations $E = \hbar\omega$ and $E' = \hbar\omega'$ we can find the shift in frequency. Assuming that the mirror is infinitely heavy we obtain the same result as from the wave picture. (See Fig. 12A.)

$$\omega' \cong \omega \left(1 - \frac{2v}{c}\right) \qquad (10\text{g})$$

11 Let us also consider the intensity of the reflected radiation. For this purpose we imagine that the observer is located at a plane fixed in the laboratory, and parallel to the mirror. Let there be a flux of N photons per unit time through a unit area of this plane towards the mirror, and let the returning flux be N' photons per unit time through a unit area. We shall assume that the light source is of large lateral extent, and that all the photons travel strictly perpendicularly to this plane. We claim that

$$N' = N \left(1 - \frac{2v}{c}\right) \qquad (11\text{a})$$

To see this we argue as follows: let the incident photons passing through a unit area of the plane of observation be equally spaced in time. The interval between the passage of two successive photons is thus $1/N$. Let a given photon return at a time t: the *next* photon must, however, travel a larger distance since the mirror has meanwhile moved by the amount v/N, and it will return at the time $t + 1/N + 2(v/c)/N$. The spacing in time between the returning photons is thus $1/N' = (1/N)(1 + 2v/c)$ which, for a small v/c leads to the approximate expression (11a).

Now the *intensity* of the beams of photons, i.e., the flux of energy per unit area per unit time, is given by $\Phi = \hbar\omega N$ for the incident beam, and $\Phi' = \hbar\omega' N'$ for the reflected beam, and we thus conclude that the intensities are related by the (approximate) formula

$$\Phi' = \Phi \left(1 - \frac{4v}{c}\right) \qquad (11\text{b})$$

We have been led to two interesting results: the frequency of the reflected photon is changed in accordance with (10g) and the intensity Φ' of the reflected beam is related to the intensity Φ of the incident beam by (11b). Can we obtain the same results through classical electromagnetic theory?†

12 On the basis of the wave theory we argue as follows: to an observer stationary in the laboratory the reflected light appears to come from a "source behind the mirror," i.e., from the mirror image of the light source. This mirror image moves with a velocity v

---

† Of course we can. This game is not really necessary, but it is instructive. An alternative way of discussing problems of this kind is to make a transformation to the rest frame of the mirror, and back again.

with respect to the mirror, and the mirror itself moves with a velocity v with respect to the stationary observer. Since v is small we can employ the non-relativistic law for addition of velocities, and we conclude that the image of the light source appears to move away from the observer with the velocity $2v$. The frequency must therefore be Doppler shifted, and the reflected frequency ω' will be given (in the non-relativistic approximation) by $\omega' = \omega(1 - 2v/c)$, which is in accordance with (10g).

13   Let us next consider the intensity. In Vol. II of this series† we have discussed the transformation laws for the electromagnetic fields under Lorentz transformations. Let E and B be the amplitudes of the electric and magnetic fields of the wave in the rest frame of the source. The fields E and B are perpendicular to the direction of propagation. We denote the corresponding amplitudes in the frame in which the source moves away from the observer with the velocity v' by E' and B'. For a plane linearly polarized wave we in fact have $E = B$ and $E' = B'$. The transformation laws then say that the primed and unprimed amplitudes are related by

$$E' = E\sqrt{\frac{c - v'}{c + v'}} \qquad (13a)$$

The intensity (i.e., the flux of energy) is in this case proportional to the square of the amplitude, and we accordingly have

$$\Phi' = \Phi\left(\frac{c - v'}{c + v'}\right) \qquad (13b)$$

where Φ is the intensity in the rest frame of the source, and where Φ' is the intensity in the frame in which the source moves away from the observer with velocity v'. If we now write $v' = 2v$, and expand the right side of (13b) in powers of v/c, assuming that this quantity is small, we recover the expression (11b) in the linear approximation.

We see that the particle picture leads to conclusions identical with those which can be drawn from the wave picture, i.e., from classical electromagnetic theory.

14   Finally we remark that we can account for the net flow of energy through the "plane of observation" towards the mirror: since the reflected radiation is of lower intensity than the incident radiation there will be a non-zero net flow. Where does the energy

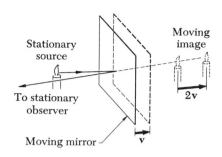

Fig. 12A Light from a stationary source reflected from a moving mirror appears to come from a moving source: the image moves with twice the velocity of the mirror. The wave theory accordingly predicts that the frequency of the reflected light will be Doppler shifted. (Imagine, for simplicity, that the figure shows a monochromatic candle.)

---

† Berkeley Physics Course, Vol. II, *Electricity and Magnetism*, Chap. 6, Sec. 7.

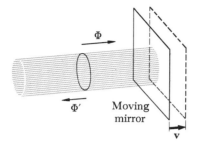

Fig. 14A The intensity, i.e., the flux of energy per unit area per unit time, of light reflected from a mirror moving away from the source and observer is smaller than the incident intensity. The radiation pressure does work on the mirror, and the volume filled with radiative energy increases.

Both the particle picture and the wave picture correctly account for the energy balance.

go? Since the mirror is moving, the radiation pressure will perform work on the mirror: this work accounts for *half* the net flux. The other half goes toward building up the electromagnetic field in the space between the mirror and the plane of observation: since the volume of this space increases steadily but the energy density remains constant, energy has to be supplied at a steady rate. In the photon picture we would rather say that the number of photons in transit between the mirror and the plane of observation increases uniformly because the distance increases. The reader should carry out the indicated very simple computations in detail to convince himself that the energy flow is balanced.

15 Let us next consider an example which teaches us the need for caution. An extremely monochromatic beam of light of frequency ω_0 (as may be obtained using a *laser* as a light source) is incident perpendicularly on a mirror which vibrates with the frequency ω_m in the direction of the beam. We wish to find the frequency of the reflected light.

On the basis of a naive particle picture, one might argue as follows: if the photon happens to hit the mirror at an instant when the velocity of the mirror is v, away from the source, then the reflected photon will be of frequency $\omega = \omega_0(1 - 2v/c)$, in accordance with our earlier discussion. The photons arrive randomly at the mirror, and therefore we will encounter a *continuum* of frequencies ranging from $\omega_0(1 - 2v_0/c)$ to $\omega_0(1 + 2v_0/c)$ in the reflected light: the spectral distribution of the initially almost monochromatic light will be broadened. In the above formula v_0 is the maximum velocity of the mirror.

16 On the basis of the classical wave picture, we arrive at a different conclusion. The reflected light is the product of two periodic processes, and we therefore expect that the frequencies observed in the reflected beam will be *combination frequencies* formed from the two frequencies ω_0 and ω_m. A careful study of this problem, on the basis of classical electromagnetic theory, shows that the frequencies expected in the reflected light will form a discrete set of the form $\omega = \omega_0 + n\omega_m$, where n is any integer (positive, negative, or zero). The *intensities* associated with these different frequencies will, for a physically realistic case in which the velocity of the mirror is small compared with c, be largest for small values of n.

The author hopes that the above result appears plausible to the reader. We shall not study the general case here, but we can increase the credibility of our statement by considering a special case. Suppose that ω_0 is, in fact, an integral multiple of ω_m. In this case, the entire process giving rise to the reflected beam is strictly periodic, with period $2\pi/\omega_m$. After the time $2\pi/\omega_m$ everything repeats itself. This clearly implies that the electric field observed in the reflected beam must also be a periodic function of the time, with period $2\pi/\omega_m$. The frequencies observed in the reflected beam must therefore be integral multiples of the frequency ω_m, which is consistent with the statement that the frequencies are of the form $\omega = \omega_0 + n\omega_m$. It is certainly plausible that the intensities associated with the various frequencies are largest for frequencies in the neighborhood of the frequency ω_0. (To see this, consider what must happen in the limit when the amplitude tends to zero.) In any case, it is clear that we cannot expect to observe a *continuum* of frequencies, as predicted by the naive particle picture.

The frequencies predicted by the classical wave theory correspond to what is actually observed. Experiments of this kind have been performed with the light source itself vibrating. In one such experiment, performed by Ruby and Bolef, the "light source" consisted of gamma-emitting Fe^{57} nuclei located on the surface of an oscillating quartz crystal. Several of the predicted frequencies were observed in this experiment, as shown in Fig. 16A.

17 The apparent sharp contradiction between the predictions of the wave theory and the particle theory can be blunted if we note that our particle theory was in this case outrageously naive. We assumed that the reflection takes place *suddenly*, as if the photon were a *point particle*, without any spatial extent. This assumption is not justified: the wave train has a finite length which is inversely proportional to the precision with which the frequency is defined. We can easily estimate the length of the wave train, on the basis of our discussion in Sec. 23, Chap. 3, of the relation between the uncertainty $\Delta\omega_0$ of the frequency and the duration τ of the emission process. We concluded that

$$\tau \cong \frac{1}{\Delta\omega_0} \qquad (17a)$$

The length L of the wave train (in space) is then

$$L = c\tau \cong \frac{c}{\Delta\omega_0} \qquad (17b)$$

and we see that if the frequency is very well defined then it is certainly not correct to regard the photon as a point particle.

We can also state the matter as follows: suppose that $\omega_m \gg \Delta\omega_0$. The time which the photon "spends" at the vibrating mirror is

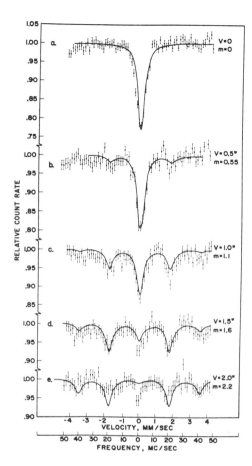

Fig. 16A Graphs showing frequency spectrum of gamma-rays emitted from a vibrating source of excited Fe^{57} nuclei. The different curves correspond to different amplitudes of oscillation, at the same vibrational frequency of 20 Mc/sec. The dips in the curves show the emitted spectral lines. As we see, we have lines at the central frequency, and at ± 20 Mc/sec and ± 40 Mc/sec away from the central frequency.

The curves actually show the transmission rate of gamma rays through a *uniformly* moving absorber containing Fe^{57} nuclei in the ground state, versus the velocity of the absorber. With the source at rest we have strong absorption at zero velocity. With the source oscillating we have strong absorption at those velocities at which the Doppler shifted emitted lines coincide with the resonance line in Fe^{57}.

Figure taken from S. L. Ruby and D. I. Bolef, "Acoustically modulated γ rays from Fe^{57}," *Physical Review Letters* **5**, 5 (1960) *(Courtesy Physical Review Letters.)*

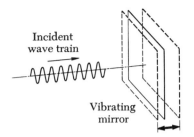

Fig. 17A It is wrong to describe the interaction of a photon with a vibrating mirror as if the photon would collide with the mirror at one definite instant of time: the photon is not a point particle. The wave picture is more appropriate in this case. The length of the wave train, and hence the duration of the collision process, is inversely proportional to the precision with which the frequency of the photon is defined. A strictly monochromatic photon is infinitely long. If the frequency of vibration of the mirror is ω_m, and the frequency of the incident light is ω_0, then the frequencies found in the reflected light are of the form $(\omega_0 + n\omega_m)$, where n is any integer.

then larger than the period of oscillation of the mirror, and it is clear that we cannot imagine that the photon is reflected from the mirror at an instant when the mirror has a definite velocity v. The reflection takes place over a time interval during which the mirror has time to perform several complete oscillations.

The Compton Effect, Bremsstrahlung;
Pair Creation and Annihilation

18 Let us now turn to an experiment in which the energy and momentum of a photon can be studied, namely A. H. Compton's experiment in which the collision of a photon with an electron is observed. Figure 18A shows this collision schematically.

A photon of frequency ω collides with an electron, of mass m, originally at rest. After the collision there emerges a photon of frequency ω' at an angle θ relative to the incident direction. The electron recoils in the collision, and it emerges with an energy E_e at an angle φ with respect to the incident direction.

Energy and momentum can be conserved only if the whole process takes place in a plane, and in this plane (say the plane of the figure) the conservation laws read:

$$\hbar\omega + mc^2 - \hbar\omega' = E_e \qquad \text{(energy)} \qquad \text{(18a)}$$

$$\mathbf{p} - \mathbf{p}' = \mathbf{p}_e \qquad \text{(momentum)} \qquad \text{(18b)}$$

If we now subtract the square of the second equation from the square of the first divided by c we obtain

$$\frac{1}{c^2}(\hbar\omega + mc^2 - \hbar\omega')^2 - (\mathbf{p} - \mathbf{p}')^2 = \frac{E_e^2}{c^2} - p_e^2 = m^2c^2 \qquad \text{(18c)}$$

Since

$$p = \frac{\hbar\omega}{c}; \quad p' = \frac{\hbar\omega'}{c} \quad \text{and} \quad \mathbf{p} \cdot \mathbf{p}' = pp' \cos\theta \qquad \text{(18d)}$$

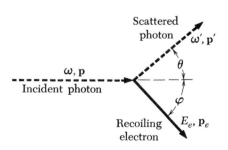

Fig. 18A To illustrate the kinematics of Compton scattering, in which a photon collides with an electron originally at rest. The conservation laws of energy and momentum imply a unique frequency ω', and momentum $\mathbf{p}'$, of the scattered photon as a function of the scattering angle θ.

we can solve (18c) for ω', and we obtain

$$\omega' = \frac{\omega}{1 + (\hbar\omega/mc^2)(1 - \cos\theta)} \qquad \text{(18e)}$$

19 If we introduce the wavelengths $\lambda = 2\pi c/\omega$, $\lambda' = 2\pi c/\omega'$, we may write (18e) in the alternative form

$$\lambda' = \lambda + 2\pi(\hbar/mc)(1 - \cos\theta) \qquad \text{(19a)}$$

The quantity $2\pi(\hbar/mc) = h/mc$ is known as the *Compton wave-*

length of the particle, in this case the Compton wavelength of the electron: $h/mc = 2.43 \times 10^{-10}$ cm $= 2.43 \times 10^{-2}$ Å.

The wavelength of the scattered radiation is longer than the wavelength of the incident radiation, or, what amounts to the same, the frequency of the scattered radiation is smaller than the incident frequency: this has to be so since some energy is transferred to the electron. By inspection of (18e) we see that the fractional change in frequency is very small if the quantity $(\hbar\omega/mc^2) \cong (\hbar\omega)/(0.5$ MeV$)$ is small: we must therefore go to hard X-ray energies before we can see an appreciable effect. We might have concluded the same thing by inspection of (19a): the fractional change in wavelength is small as long as the Compton wavelength is small compared to the incident wavelength.

20   The scattering phenomenon which we have just discussed was observed experimentally by A. H. Compton in 1922.[†] It seems that he was led to his experiment by a previous observation by Barkla that when hard X-rays are scattered at large angles by a solid material, the scattered rays seemed to consist of two components: one component having properties identical with the incident radiation but the other component being different, which difference manifested itself through a difference in the rate at which this radiation was absorbed by intervening media. On the basis of the wave picture we can readily understand the occurrence of the first component. The incident electromagnetic waves, i.e., the incident X-rays, set the electrons bound in the atoms in oscillation at the same frequency ω as the frequency of the wave, and these oscillating electrons will then emit electromagnetic radiation in all directions at the frequency ω. In this process, the state of the atom is only temporarily disturbed, and the electrons are not ejected. We may expect that it is mostly the *tightly* bound electrons which will give rise to this kind of scattering.

Some electrons in an atom are, however, very loosely bound, with binding energies of the order of 10–100 eV, and such electrons could conceivably be ejected in the scattering process. In Compton's experiment X-rays from an X-ray tube with a molybdenum target, operated at a voltage of about 50,000 volts, were scattered at various angles from graphite. The wavelength of the incident radiation was the so-called Mo K-radiation of wavelength 0.7 Å, corresponding to an energy of about 20,000 eV. This energy is

_____

† A. H. Compton, "The Spectrum of Scattered X-rays," *Physical Review* **22**, 409 (1923). For Compton's theoretical analysis, see "A Quantum Theory of the Scattering of X-rays by Light Elements," *The Physical Review* **21**, 483 (1923).

very large compared to the binding energy of the outermost electrons in the carbon atom: actually it is large compared to the binding energies of all the electrons. Under these circumstances we can expect that the scattering process will proceed in a manner very similar to the situation when the electrons are not bound at all, and the analysis of Sec. 18 then applies. Compton, in fact, found that the wavelength of the scattered radiation contained a second component with a wavelength λ' which depend on the angle of scattering in accordance with the formula (19a). (See Fig. 20A).

In later experiments by Compton and others, the recoiling electron was also detected, and it was possible to show that the recoiling electron and the scattered photon are associated with each other, and that momentum and energy are conserved in the process.†

21 Let us now assess the significance of the observations with regard to the Compton effect. First of all, we may note that a classical packet of electromagnetic radiation would also be expected to scatter from an electron, and the phenomenon of scattering as such therefore does not require quantum mechanics for its explanation. However, the specific relation (18e) between the frequency of the scattered radiation and the angle of scattering *does* depend on Planck's constant, and in a way that lends strong support to the photon picture. We should note that we derived the formula (18e) under the assumption that a *whole* photon is scattered, and not a third, or a fifth, of a photon: if only a fifth of a photon were scattered the conservation laws would give a quite different result. The importance of the Compton effect is therefore that the observational results give further support for the universality of the relation $E = \hbar\omega$. The photons cannot be "split" in Compton's experiment: a photon of frequency ω *always* carries the energy $\hbar\omega$ and the momentum $\hbar\omega/c$.

In optical experiments with the photo-cell (in the visible, or ultraviolet region) we can check the relation $E = \hbar\omega$ only over a very limited frequency range. The study of the Compton effect extends this range to the hard X-ray region. Of course, if we believe firmly in special relativity, which we do, we conclude that the relation is completely general as we argued in the beginning of this chapter. Nevertheless any experiment which directly tests this relation in a new frequency region is worthwhile: we test the

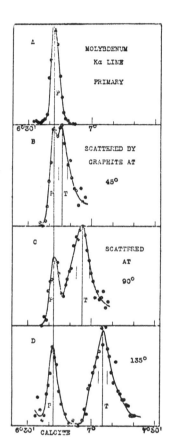

Fig. 20A Graph taken from Compton's paper [*Phys. Rev.* **22**, 409 (1923)], showing spectrum of scattered radiation at three different scattering angles. The uppermost graph shows the line of the incident radiation, of wavelength 0.71 Å. The abscissa is proportional to the wavelength and the ordinate is a measure of intensity. The peaks at left in the lower three graphs show that part of the scattered radiation is of the same wavelength as the incident radiation. The peaks at right show the Compton-scattered radiation, at a shifted frequency. The shift in the frequency increases with the angle of scattering in accordance with Compton's formula. (*Courtesy of The Physical Review.*)

† A. H. Compton and A. W. Simon, "Directed quanta of scattered X-rays," *Physical Review* **26**, 289 (1925). See also C. T. R. Wilson, "Investigations on X-Rays and β-rays by the Cloud Method," *Proceedings of the Royal Society (London)* **104**, 1 (1923).

consistency of our ideas, and among other things, we test special relativity.

There exists today overwhelming support for the generality of the relation $E = \hbar\omega$. It forms, we may say, an integral part of present-day physics. To study further the implications of this relation, let us consider two other phenomena: the emission of X-rays in an X-ray tube, and the annihilation of an electron-positron pair.

22 In an X-ray tube, shown schematically in Fig. 22A, electrons emitted from an incandescent cathode (heated by a filament) are accelerated in a potential drop, V_0, between the filament and the *anode*. As the electrons hit the anode, or *target*, they are stopped, and on the basis of classical electromagnetic theory we expect that this deceleration will be accompanied by the emission of electromagnetic radiation. The presence of this radiation was first detected by W. C. Röntgen, in 1895.† The emitted rays are known as X-rays, or Röntgen rays.

The true nature of the rays was at first a matter of some controversy, but in the beginning of this century it became increasingly clear that X-rays are in fact electromagnetic radiation. Through an ingenious double-scattering experiment in 1904 C. G. Barkla was able to show that the rays are transversely polarized. The most conclusive evidence came in 1912 when W. Friedrich and P. Knipping could show, following the suggestion by M. von Laue, that X-rays are diffracted in crystals, as already mentioned in Chap. 1.‡

23 After techniques had been developed making possible the spectroscopic study of X-rays, the intensity of the emitted rays as a function of the wavelength could be measured for a wide variety of experimental conditions. A typical plot of intensity versus wavelength for three different substances but the same voltage V_0, is shown in Fig. 23A. We see that superimposed on a continuous background there are several sharp "spikes," or intensity maxima. It has been found that the *locations* of these spikes is a property of

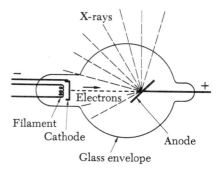

Fig. 22A Very schematic figure showing operation of an X-ray tube. Electrons emitted from a cathode heated by a filament are accelerated towards the anode. X-rays are emitted when the electrons hit the anode. Part of the radiation is characteristic radiation for the material in the anode, and part is bremsstrahlung.

---

† W. C. Röntgen, "Über eine neue Art von Strahlen," *Sitzungsberichte Med. Phys. Ges. Wurzburg,* 1895, p. 137; 1896, p. 11. These papers have been translated: W. C. Röntgen, "On a New Kind of Rays," *Science* **3,** 227 (1896); "A New Form of Radiation," *Science* **3,** 726 (1896).

‡ C. G. Barkla, "Polarized Röntgen Radiation," *Phil. Trans. Roy. Soc.* **204,** 467 (1905). C. G. Barkla, "Polarization in Secondary Röntgen Radiation," *Proc. Roy. Soc. (London)* **77,** 247 (1906). (The latter paper reports on the double-scattering experiments.)

W. Friedrich, P. Knipping and M. von Laue, *Annalen der Physik* **41,** 971 (1913).

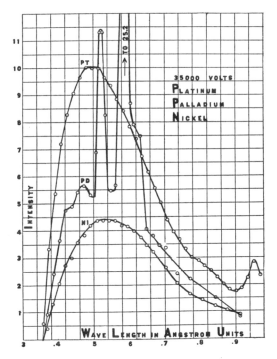

Fig. 23A Graph showing intensity of emitted X-rays versus wavelength, for three different substances, but the same accelerating potential $V_0 = 35,000$ volts. The sharp spikes correspond to the characteristic radiation of the substance. The continuous background is due to bremsstrahlung.

The figure is taken from C. T. Ulrey, "An Experimental Investigation of the Energy in the Continuous X-Ray Spectra of Certain Elements," *Physical Review* **11**, 401 (1918). *(Courtesy of The Physical Review.)*

the material in the target. The continuous background, on the other hand, is, for the same accelerating potential V_0, of the same shape for all materials. Examination of all the experimental material has led to the conclusion that there are two different mechanisms responsible for X-ray emission. The sharp spikes are analogous to the light emitted by atoms undergoing collisions: this radiation is known as the characteristic radiation of the substance, and it is an emission by an *atom* which has been excited in the collision with one of the energetic incident electrons. The continuous background, on the other hand, is an emission by the *electron* which is being decelerated in the target. It is known under the German name *bremsstrahlung*. This word has become incorporated in the English language by now: literally translated it means "braking radiation."

It has furthermore been found experimentally that for a given accelerating potential V_0 there is *no* radiation of shorter wavelength than a certain minimum wavelength, λ_{min}, which depends on the potential V_0, but not on the substance in the target: this we can see illustrated in Fig. 23A.

24 Let us see how we can understand the last mentioned circumstance theoretically.

First of all, we note that according to classical electromagnetic theory a uniformly moving electron cannot emit radiation. We can draw the same conclusion on the basis of the photon theory as follows. Consider the rest-frame of the electron *before* any possible emission: in this frame the total energy is mc^2. *If* an emission of one or more photons could take place these photons would carry away energy, and the total final energy after the emission would be larger than mc^2, which violates energy conservation. The emission should therefore not take place.

The situation is different, however, when the electron passes through the strong electric field of a nucleus in the target. It is then possible for the electron to transfer some energy and momentum to the nucleus, and the conservation equations for energy and momentum can be balanced. Let us see how this works out. The nucleus, of mass M, is originally at rest (in the laboratory frame), and the electron, of mass m and initial momentum $\mathbf{p}_i$, impinges upon it. After the collision the electron has momentum $\mathbf{p}_f$, and the nucleus has momentum $\mathbf{p}_n$. In addition, there emerges a photon of momentum $\mathbf{p}$ and frequency $\omega = pc/\hbar$. The conservation equations read

$$\mathbf{p}_i = \mathbf{p}_f + \mathbf{p}_n + \mathbf{p} \qquad \text{(momentum)} \qquad (24a)$$

$$E_i + Mc^2 = E_f + E_n + \hbar\omega \qquad \text{(energy)} \qquad (24b)$$

where E_i and E_f are the initial and final energies respectively of the electron, and where E_n is the final energy of the nucleus.

These equations taken together give us four conservation equations. There are, however, nine variables which characterize the final situation, namely the nine components of the three vectors $\mathbf{p}_f$, $\mathbf{p}_n$ and $\mathbf{p}$. The detailed investigation of the permissible range of these vectors is somewhat involved and we shall not attempt it. One can show, that for any given direction, the photon can emerge with an energy ranging from zero to some maximum. This maximum actually occurs when both the electron and the nucleus have the same velocity, say $\mathbf{v}$, after the collision: that this must be so is immediately obvious if we consider the problem in the center of mass system. Let us rewrite the conservation equations for the case when the final electron and nuclear velocities are indeed equal:

$$\mathbf{p}_i - \mathbf{p} = \frac{(M + m)\mathbf{v}}{\sqrt{1 - (v/c)^2}} \tag{24c}$$

$$E_i + Mc^2 - cp = \frac{(M + m)c^2}{\sqrt{1 - (v/c)^2}} \tag{24d}$$

Multiplying the first equation by c, and subtracting the square of the result from the square of the second equation gives us

$$\hbar\omega = pc = \frac{E_i - mc^2}{1 + (E_i - p_i c \cos\theta)/(Mc^2)} \tag{24e}$$

where θ is the angle between the emerging photon and the incident electron. The above formula thus gives the *maximum* photon energy at the angle θ. We note that is is approximately equal to $(E_i - mc^2)$, the kinetic energy of the incident electron, which in turn is equal to eV_0. The second term in the denominator in the right side of (24e) is very small for X-ray tubes since the constant $Mc^2 \sim 940\,A$ MeV, for a nucleus of mass number A, is large compared with E_i, which may range from 1 keV to 100 keV.

25   In the limit when the nucleus is assumed to be infinitely heavy, we thus obtain for the minimum wavelength $\lambda_{\min}$ the expression

$$\lambda_{\min} = \frac{2\pi c}{\omega} = \frac{ch}{eV_0} \tag{25a}$$

This we could, of course, have concluded directly: the energy of the emitted photon cannot exceed the kinetic energy of the incident electron, and for an infinitely heavy nucleus the maximum energy must occur when the electron is brought to complete rest in the collision.

The minimum wavelength $\lambda_{\min}$ is known as the *quantum limit*. Its existence is a manifestation of quantum phenomena: classical theory predicts that arbitrarily short wavelengths will be emitted.

Very precise measurements of the quantum limit as a function of V_0 have been carried out† and such measurements lead to precise values of the constant e/ch (and e/h).

26 Let us finally consider the annihilation of an electron-positron pair. Positrons were first observed by C. D. Anderson in 1932, in the cosmic radiation (see Fig. 26A). Positrons are known today to arise in the decays of many unstable particles, as for instance in the decay of the radioactive phosphorus isotope P^{30}. Positrons are also observed when very energetic gamma rays pass through matter, and as we mentioned in Chap. 1, our picture of this phenomenon is that in the electric field of a nucleus a gamma ray is able to create an electron-positron pair. This process is known as electromagnetic pair production.

When a positron collides, or interacts, with an electron, this pair of particles may *annihilate,* which means that the particles disappear and their energy is completely converted into electromagnetic radiation. This annihilation phenomenon is observed when positrons impinge upon matter in bulk. According to our present picture, the positron entering the material first loses most of its kinetic energy in collisions with the atoms of the material, although some of the positrons may annihilate in a direct collision with an electron before slowing down. The positrons which have slowed down diffuse around in the material, and eventually are captured by the electrons in the atoms. Under favorable circumstances a positron may actually form a hydrogen-like "atom" with a single electron, known as positronium. The slowed-down positrons interact with the electrons, and eventually the annihilation takes place.

To the best of our knowledge the mass of the positron is equal to the mass of the electron.

27 Let us now consider the annihilation process, which we represent by the reaction formula

$$e^+ + e^- = n\gamma$$

where the symbol γ stands for a photon (a gamma quantum). Suppose that the electron and the positron are practically at rest when the reaction takes place (in the laboratory frame), and suppose

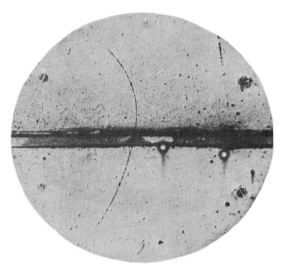

Fig. 26A The cloud chamber picture in which Anderson's discovery of the positron was presented to the world. [From C. D. Anderson, "The Positive Electron," *Physical Review* **43**, 491 (1933).] A positron of an energy of 63 MeV passes through the horizontal lead plate (of 6 mm thickness,) and emerges with an energy of 23 MeV. The tracks are curved because the chamber was placed in a magnetic field perpendicular to the plane of the picture. The quality of the picture is poor near the edge of the chamber, and the very faint portions of the tracks near the edge, which show that the positron did pass through the chamber, are therefore somewhat hard to see.

For some interesting questions concerning this picture, see Prob. 11 at the end of this chapter. (*Courtesy of The Physical Review.*)

† J. A. Bearden, F. T. Johnson and H. M. Watts, "A New Evaluation of h/e by X-rays," *The Physical Review* **81**, 70 (1951).

furthermore that the reaction takes place in free space, far away from all other particles.

First of all we note that there must be at least two gamma rays: $n \geq 2$, since we cannot otherwise conserve energy and momentum. (If the electron and positron are initially at rest the initial momentum is zero: if only *one* photon is emitted, the final momentum would *not* be zero.)   Let us therefore assume that *two* photons are emitted.   Since the initial momentum is zero it follows that the final (total) momentum must also be zero, and the momenta of the two photons are thus equal and opposite, which means that their energies, and hence their frequencies are also equal.   Let us denote the frequency by ω: the conservation of energy then implies that

$$2\hbar\omega = 2mc^2 \quad \text{or} \quad \lambda = \frac{2\pi c}{\omega} = \frac{h}{mc} \tag{27a}$$

The wavelength of the emitted photons therefore equals the Compton wavelength of the electron, $h/mc = 0.0243$ Å: to this wavelength corresponds the rest energy of the electron, $mc^2 = 0.511$ MeV.[†]

For the positrons which have been slowed down and captured in the material we can assume that the above prediction is valid: the presence of the other particles in the material may have an effect but this effect should be small since the atomic binding energies are very small compared to the rest energy of the electron.

We may therefore look for the two gamma rays formed in the annihilation.   They should emerge in opposite directions, and their wavelengths should be the Compton wavelength of the electron. It has been found experimentally that these predictions are correct in every detail: annihilation into two gamma rays indeed takes place.[‡]   In addition it has been found that annihilation into three gamma rays also takes place.

28   There is one point which we should clarify.   We argued that an electron-positron pair cannot annihilate into a *single* photon in empty space since energy and momentum cannot then be conserved.   It follows that the inverse process, in which a single photon suddenly changes into an electron-positron pair must also be impossible.   On the other hand we said that electron-positron pairs are created when very energetic photons traverse matter in bulk.   The

---

[†] Note that the quantity $\hbar/mc = 0.00386$ Å is also often called the Compton wavelength.

[‡] See, for instance, O. Klemperer, "On the Annihilation Radiation of the Positron," *Proceedings of the Cambridge Philosophical Society* **30**, 347 (1934).

resolution of this apparent contradiction is that this process *can* and does take place in the field of a nucleus. A certain amount of energy and momentum is transferred to the nucleus, and it is then possible to balance the conservation equations.

The inverse of the annihilation process which we have discussed is the process in which two colliding photons would produce a pair. This process has never actually been observed, and the reason for this is that we cannot produce sufficiently intense photon beams at high energy for the process to take place at an observable rate: we believe firmly that we would see this phenomenon if we could produce such photon beams. The inverse of the pair-production process in the field of the nucleus is the process in which an electron-positron pair annihilates in the field of the nucleus into *one* photon, with the nucleus taking up the energy and momentum necessary to balance the conservation equations. This process does take place, but it turns out that in general the two-photon annihilation process is more probable, and therefore dominates.

29 Since we are on the subject of positrons, we may speculate a bit about particles and anti-particles. Our present-day formulation of quantum electrodynamics is a theory in which the electron and the positron play a completely symmetric role. This is a general feature of our theories of the fundamental particles: for every particle we believe there also exists an anti-particle (some particles, such as the neutral pion, may be their own anti-particles), and the world is believed to be symmetric (in a sense) under the exchange of particles for anti-particles.[†] An anti-particle has the same mass, but the opposite charge as the particle. It has thus been found experimentally that both antiprotons and antineutrons exist,[‡] which we find pleasing from the theoretical standpoint: it enables us to persist with our symmetric theories of nature.

It is a fundamental property of particles and their anti-particles that they can annihilate each other, say, into photons. However, it is often the case that the annihilation is into other particles. For instance, the proton-antiproton system tends to annihilate into

---

[†] Very recently some experiments have been performed which seem to indicate that the *weak* interactions are not invariant under the exchange of particles for anti-particles. This implies that whereas the strong and electromagnetic interactions might obey the symmetry principle mentioned, the weak interactions do not. Since the strong and electromagnetic interactions are the dominant interactions in the world, it is perhaps fair to say that the symmetry principle is *almost* (but not quite) true.

[‡] For the discovery of the antiproton, see O. Chamberlain, E. Segrè, C. Wiegand and T. Ypsilantis, "Observation of Antiprotons," *The Physical Review* **100**, 947 (1955).

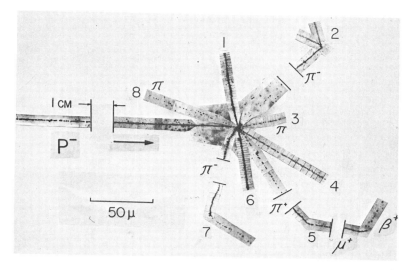

mesons, that process being more probable than annihilation into photons.

30 The reader may wonder why, if the particles and anti-particles really play an almost completely symmetric role in the world, anti-particles are not more evident. Why are positrons not "more commonly encountered," and why were they not discovered earlier? Our world as we know it does not seem to be in a symmetric *state* at all: our world consists of protons, neutrons, electrons and hydrogen atoms, etc., but *not* of antiprotons, antineutrons, positrons or anti-hydrogen atoms. The reason for this lack of symmetry is that the symmetric state would not be stable against annihilation: matter and antimatter cannot peacefully coexist in an intimate mixture. Granted that the earth exists, it must consist either of matter or else of antimatter. It could not possibly exist as a mixture.

It is an interesting question whether this lack of symmetry of the state of the world extends to the *entire* universe. It could possibly happen that there are galaxies consisting of antimatter: since the average separation between the galaxies is of the order of three million light years, the annihilation cannot take place readily. This question cannot be answered at present, although one might be inclined to believe that anti-galaxies do not exist. How the galaxies were formed is not known, but if we assume that they were formed through some kind of condensation process from "dust" it becomes hard to understand how a segregation of matter and antimatter

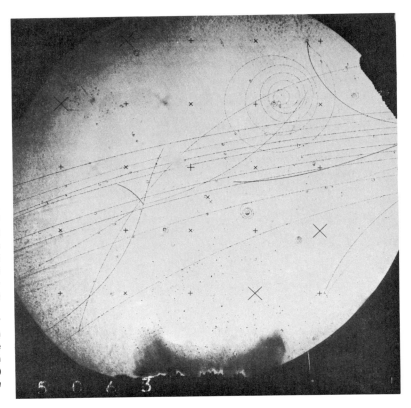

Bubble chamber photograph showing a charge-exchange scattering of an antiproton by a proton, followed by an annihilation reaction of a proton and the antineutron produced in the first event. See the schematic figures on the next page for identification of the tracks. The events take place in a liquid hydrogen bubble chamber, placed in a magnetic field perpendicular to the plane of the figure. (The reader should try to determine which way the magnetic field vector points.) Neutral particles do not leave visible tracks, but charged particles leave tracks which are curved because of the magnetic field. In the above situation positive particles turn in the clockwise direction, whereas negative particles turn in the opposite direction.

It so happens that this photograph shows another interesting event, namely the decay of a positive pion into a positive muon and a neutrino, followed by the decay of the muon into a positron, a neutrino and an antineutrino. The neutrinos (and the antineutrino) are neutral and leave no visible tracks. *(Photograph by courtesy of Dr. P. Schmidt, Berkeley.)*

could have taken place so that some galaxies consist of matter whereas others consist of antimatter. If we reject the idea of antigalaxies there then remains the mystery of why the state of the world is so unsymmetric in that one kind of matter dominates, in spite of the fact that the basic laws of physics seem to be almost completely symmetric.

Can Photons Be "Split?"

31   Our discussion so far suggests an interesting and very fundamental question: is it possible to split a photon of frequency ω into two parts, such that each part carries some fraction of the energy $\hbar\omega$, but such that each part is still of frequency ω?

We know that classical electromagnetic theory describes a wide variety of experiments with light very accurately. We have also said that the relation between energy and momentum of a "photon" is something which can be derived within classical electromagnetic theory. Is it not then permissible to say that a photon is simply a wave packet, or a wave train, of radiation which is governed by the laws of classical electromagnetic theory? We have clearly now asked a question of fundamental importance. *If* photons can be split in the above sense then the whole structure of ideas which we have tried to build in this chapter would totter.

To answer our question, we must turn to experiments. In order to see what kind of experiments we should perform, we will now adopt a completely classical point of view and then make some predictions which we can test in actual experiments to decide whether the classical picture is correct.

32 A classical wave train of electromagnetic radiation may arise as follows. As the source of radiation, we have an antenna and a transmitter which we can switch on and off at will. We let the transmitter, which operates at a frequency ω, be on for a certain length of time, and the antenna then emits a wave train of finite duration: this wave train is the classical almost monochromatic "photon." We may imagine that an excited atom acts like an antenna of this kind.

Let us again emphasize that we are trying to compare the behavior of a physical photon, such as it manifests itself in actual experiments, with the behavior of the classical wave train. This discussion involves the comparison of an object *actually occurring in nature*, namely the photon, with something which, as we shall see, *does not occur in nature at all*, namely an electromagnetic wave train which follows the laws of classical electrodynamics *exactly*. We are therefore comparing fact and fantasy: to avoid mixing fact and fantasy, we shall call the real thing the *photon* and the imaginary thing the *wave train*. To convince ourselves that the wave train is not the real thing, we must make some definite predictions based on the wave train picture which we can then test through experiments.

33 We consider the emission of light from a mercury atom which has been excited in a collision. The emitted light is blue, of frequency ω. That the emitted wave train should always be of the same frequency ω is not hard to understand: this frequency must correspond to some natural frequency of oscillation of the atom. It *is* hard, however, to understand, on the basis of classical theory, why the energy carried by the wave train should always be $\hbar\omega$. We would expect the collisions to differ in violence: sometimes

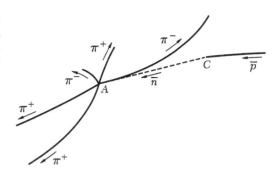

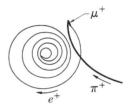

These two figures identify the tracks seen in the bubble chamber photograph on the preceding page. In the upper figure an incident antiproton collides with a proton at the point C. A neutron and an antineutron result from this reaction. The invisible track of the antineutron is shown dotted above. At the point A the antineutron undergoes an annihilation reaction with a proton. Five charged pions are produced in this reaction. The incident antiproton is one among a number of negative particles in a beam of particles traversing the chamber from right to left. These particles are probably all antiprotons.

The lower figure shows the tracks of the charged particles involved in the sequential decays of a positive pion. The spiral track is due to the positron. As it moves through the liquid hydrogen it looses energy, and the radius of curvature of the track accordingly decreases. Its ultimate fate is to disappear in an annihilation reaction with an electron in the liquid.

more, and sometimes less energy becomes available for light emission. What is even harder to understand is why two totally *different* atoms, say a sodium atom and a mercury atom, which emit light of different frequencies, ω_{Na} and ω_{Hg}, should emit wave trains of total energy $\hbar\omega_{\text{Na}}$ and $\hbar\omega_{\text{Hg}}$, respectively. From a classical standpoint the appearance of the *universal* constant of proportionality $\hbar$ is very mysterious.

If we think about the totality of all the many experimental facts which we discussed in Chapter 3 it is *very* clear that the phenomena cannot be understood on a classical basis. Let us, however, try to forget temporarily what we already know about the emission and absorption processes, and instead concentrate on the study of "isolated" photons. We consider wave trains which have been emitted from some source, and we study these wave trains with a photocell in the experimental region.

34 We study, in other words, the photoelectric effect. We set the retarding potential in the cell at a value V_0. If W is the work function of the photosensitive surface we will thus detect a wave train (the register attached to the photocell will "click") provided the energy carried by the wave train exceeds the value

$$E_{\min} = eV_0 + W \tag{34a}$$

We set V_0 such that

$$\hbar\omega > E_{\min} > \tfrac{2}{3}\hbar\omega \tag{34b}$$

where ω is the frequency of the light. (We arbitrarily pick $\tfrac{2}{3}$ as a number larger than $\tfrac{1}{2}$ but smaller than one.) Therefore, if we concentrate the entire energy of the emitted wave train into the photocell, then the register will click. However, if only half of this energy reaches the cell, the register will not click, because then the energy imparted to the electron could not possibly overcome the retarding potential.

35 The classical picture suggests an obvious experiment in which we split a wave train, for instance with the arrangement shown in Fig. 35A. Light from a source of very low intensity falls on a *beam splitter*, say a half-silvered mirror, or a suitable beam-splitting prism. We can arrange it so that the intensity of the transmitted beam equals the intensity of the reflected beam, and so that the intensity of each one of these beams is half the intensity of the beam emerging from the source through the slit. This is, in other words, a possible realistic experiment, and we do find that the *intensities* of the transmitted and reflected beams are as stated.

Robert Andrews Millikan. Born 1868 in Morrison, Illinois. Died in 1953. After studies in the United States and Germany, Millikan held a professorship at the University of Chicago, and later at California Institute of Technology, Pasadena. Millikan is known in particular for his determination of the charge of the electron, and for his work on the photoelectric effect. He received the Nobel prize in 1923. *(Photograph by courtesy of Professor L. B. Loeb, Berkeley.)*

Classically, we can understand these facts easily: each wave train arriving at the mirror is split into two parts.

Consider now what happens when a single wave train arrives at the mirror. On the classical model, we expect it to be split into two parts, and in such a way that the energy carried by the transmitted part of the wave train is one-half the energy of the incident wave train. Therefore, the photocell 2 should never click!

This prediction, based on classical theory, is in flat contradiction with experience. The light which goes through is still blue, of frequency ω, and the register of cell 2 does click as long as $\hbar\omega > E_{\min}$, which shows that the energy of the transmitted light comes in packets of $\hbar\omega$. What does happen when the mirror is inserted is that the *counting rate* is only half its value in the absence of the mirror.

36 How convincing is the evidence that photons cannot be split with the experimental arrangement shown in Fig. 35A, or with any other arrangement of a similar kind? The evidence is *exceedingly good*, and, in fact, photon-splitting experiments go on all the time. Every optical device in which we have a photocell, or a photographic plate, can be regarded as an instrument in which we try to split photons, but fail. The simplest observation of this kind is the observation of the photoelectric effect at different distances r from the light source. If an atom is like an antenna, then it should emit light in the form of a spherical wave train. The intensity of the emitted light is proportional to $1/r^2$, and on the classical picture, the amount of energy carried by a single wave train through a unit area at a distance r is proportional to $1/r^2$. Therefore, since the photocell has some definite cross section, it would seem that by merely placing the photocell at a sufficiently large distance, the fraction of the energy in a wave train which can possibly be active in the photocell can be made as small as we please, and for a fixed retarding potential, the photocell should cease entirely to register as soon as the distance exceeds a certain limit. This is most certainly not what we observe: all that happens is that the *counting rate* of the photocell diminishes as $1/r^2$. Perhaps the most striking example is the observation of the photoelectric effect for light from a very distant star. The wave train was emitted thousands of years ago, and should have spread out over a large portion of space. Only a minute fraction of the energy carried in the wave train should be able to reach the photocell through the telescope. Nevertheless, the amount of energy given to the electron in the photocell is found to be $\hbar\omega$, just as if the source were a lamp close to the photocell.

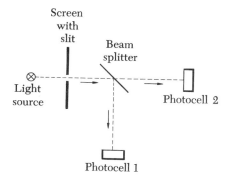

Fig. 35A Schematic figure to illustrate the discussion in Sec. 35. The light beam from the source is divided into two parts by the beam splitter, which may be a half-silvered mirror. Are the *individual photons* split?

37 Any attempted explanation of these facts along the lines that we might be seeing a cumulative effect in which very many "fractional photons" transfer a small amount of energy to the electron in the photosensitive surface and in such a way that the electron is finally ejected, when the accumulated energy is large enough, is wholly untenable. If this were the explanation, then this cumulative effect should *also* work when the retarding potential is set such that $E_{min} > 100 \; \hbar\omega$, which is certainly not what we observe: if the retarding potential is too large, the photocell *never* registers.

38 The experimental facts connected with the photoelectric effect therefore lead to the inescapable conclusion that almost monochromatic photons cannot be split into two photons of the same frequency which carry only a fraction of the energy of the original photon: photons *do not* behave like classical wave trains in this respect. This conclusion is further supported by the experimental results concerning the Compton effect, X-ray emission, pair production and pair annihilation which we have discussed earlier in this chapter. In our theoretical analysis of these phenomena we made the definite assumption that the relation $E = \hbar\omega$ always holds, i.e., that "fractional photons" do not exist, and on this basis we could indeed understand many of the experimental facts.

Something must therefore be wrong with the classical ideas, and we want to find out just what it is that has to be changed. We should be careful not to draw any hasty conclusions at this point. Let us instead consider some other experimental facts which also have a bearing on the question of whether photons can be "split." Our discussion so far has shown us that in *one* definite sense of the word, photons *cannot* be split. This does not exclude the possibility that photons can be "split" in some other sense.

39 Consider a two-slit diffraction experiment, as shown schematically in Fig. 39A. The opaque screen has two slits, U and L, perpendicular to the plane of the paper. The light source S illuminates the two slits with light (photons) of a well-defined frequency ω. For simplicity, we assume that the slits are identical in size, and with a width small compared to the wavelength. We furthermore assume that the separation, $2a$, between the slits is comparable to the wavelength $\lambda = 2\pi c/\omega$.

We carry out measurements of the intensity of the diffracted light as a function of the angle of observation θ, and at a distance r from the screen which is *large* compared with the slit separation $2a$. To measure this intensity, we can imagine that we employ a photo-

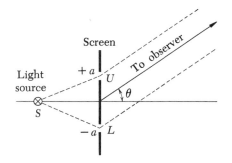

Fig. 39A Very schematic figure to illustrate our discussion of two-slit diffraction. Does an individual photon go through only one of the slits, or can it go through both as the classical wave train picture suggests? Does the two-slit interference pattern change as the intensity of the incident light is diminished?

cell: the intensity is then proportional to the counting rate observed with the cell.

40   Let us now see what classical electromagnetic theory says about the intensity distribution to the right of the screen. Our assumption that the width of the slits is small compared to the wavelength means that if either slit is covered, then the angular distribution of the diffracted radiation is a smooth function of the angle θ. Let A_0 denote the amplitude of the diffracted wave for the case when there is only *one* slit, either the upper or the lower slit shown in Fig. 39A. $A_0 = A_0(r,\theta)$ is, of course, a function of r and θ, and using the complex representation, we can write

$$A_0 = f(r,\theta)\, e^{-i\omega t} \tag{40a}$$

where $f(r,\theta)$ describes the *spatial* dependence of the amplitude.

With the experimental arrangement shown in Fig. 39A the diffracted wave observed at a large distance from the screen is the sum of the two waves from the two slits. They are of equal amplitude, but the wave from the lower slit is retarded in phase by the amount $(4\pi a/\lambda) \sin \theta$ relative to the wave from the upper slit. The amplitude of the combined wave is thus

$$A = f(r,\theta)e^{-i\omega t}\left[\exp\left(\frac{i\omega a}{c}\sin\theta\right) + \exp\left(-\frac{i\omega a}{c}\sin\theta\right)\right]$$

$$= 2A_0 \cos\left(\frac{2\pi a}{\lambda}\sin\theta\right) \tag{40b}$$

The *intensity* of the diffracted radiation is proportional to the absolute square of the amplitude, and we accordingly have

$$I(r,\theta) = |A|^2 = 4I_0(r,\theta)\cos^2\left(\frac{2\pi a}{\lambda}\sin\theta\right) \tag{40c}$$

where

$$I_0(r,\theta) = |A_0|^2 \tag{40d}$$

is the intensity observed with only a single slit. The intensity $I(r,\theta)$ in the two-slit experiment is thus equal to the product of the intensity in the single-slit experiment and the factor $4\cos^2[(2\pi a/\lambda)\sin\theta]$, which describes the effect of interference of the waves emerging from the two slits. We note that because of this interference we will observe zero intensity in certain directions, provided $4a/\lambda > 1$. In certain other directions the intensity will be *four* times as large as in the single-slit experiment. We are here specifically interested in these interference effects, as described by our relation (40c).   That the intensities I and I_0 should be related in this manner is the essence of our classical prediction.

41   In view of what we have learned about the impossibility of "splitting" photons we might be tempted to conclude that the classical prediction contained in Eq. (40c) must be wrong.   We could argue as follows.   Each photon must go through one of the slits since photons cannot be split.   Suppose a particular photon goes through the upper slit.   In that case the presence of the lower slit cannot affect the diffraction of this photon, and the intensity pattern of all the photons going through the upper slit must be given by $I_0(r,\theta)$.   Similarly for the photons which go through the lower slit, and we might conclude that the intensity I^* with both slits open is given by

$$I^*(r,\theta) \,=\, 2I_0(r,\theta) \tag{41a}$$

We have denoted the predicted intensity by I^* to indicate that we arrived at this prediction by rejecting the classical ideas leading to the prediction of Eq. (40c).   Now the reader should note that we have not said that we are *forced* to the conclusion (41a) by our earlier discussion of photon splitting: we merely wanted to explore this possibility.

42   The experimental evidence is conclusively in favor of the wave-theory prediction (40c).   We can regard our simple two-slit diffraction experiment as the prototype of a large class of interference experiments among which we note measurements with diffraction gratings, and X-ray diffraction experiments with crystals. The equation (41a) states that the waves diffracted by the two slits do not interfere with each other, and if this prediction were correct for the two-slit experiment it would clearly follow that we could not see any interference effects with diffraction gratings or crystals either.

Before we reject the prediction (41a) as absolutely wrong we should worry about the following: could it perhaps be the case that the interference phenomenon described by Eq. (40c) arises because of some kind of "interaction" between *several* photons?   With a sufficiently intense light source we will have several photons in transit at the same time, i.e., several photons passing through the slits simultaneously, and we might wonder whether the interference effects are possibly "many-photon phenomena."   This line of thinking leads us to inquire whether the prediction (41a) is correct for extremely feeble sources, such that only one photon is effectively in transit at a time, whereas the prediction (40c) is correct for sufficiently intense sources. In other words: Is it so that the nature of the diffraction pattern changes from that described by Eq. (40c) to that described by Eq. (41a) as the intensity of the source is decreased?

The answer to this question is *no:* there is not the slightest evidence that diffraction patterns would change their nature as the intensity of the radiation tends to zero. Our accumulated results of experiments on diffraction and interference support, beyond reasonable doubt, the ideas on which the prediction (40c) is based.

43 An experiment directly bearing on this question was carried out in 1909 by G. I. Taylor.[†] Taylor photographed the diffraction pattern in the shadow of a needle illuminated by an extremely feeble light source. In one of these experiments the exposure time was 2000 hours, or about 3 months. In this case the intensity is so low that only a very small number of photons could have been effectively present within the apparatus at any one instant. The diffraction pattern produced was, however, just as clear and sharp as the pattern produced with a strong light source. The *precise* theoretical analysis of Taylor's experiment is a bit tricky (among other things because his description of what he actually did is not sufficiently detailed) and we shall not attempt it here. We can, however, assert with confidence that the intensity of the light was indeed so low that if the nature of the diffraction pattern really changes as the number of photons in transit is reduced, then this effect should have shown up in the experiment. As we have said, there was not the slightest hint of any such effect.

We want to emphasize that our belief that diffraction patterns do not arise as the result of some kind of "interaction" between a large number of photons does not rest only on Taylor's somewhat barren experiment. It rests on a very large number of other interference experiments which we have found that we can describe correctly on the basis of the wave picture, irrespective of the intensity of the radiation present.

44 Let us now try to formulate a simple theory in terms of which we can account for the experimental results which we have discussed so far. Our theory goes as follows:

I. Almost monochromatic radiation, of the approximate frequency ω, emitted by a light source can be thought of as consisting of discrete "packets of radiation," which we call photons.

---

† G. I. Taylor, "Interference Fringes with Feeble Light," *Proc. Cambridge Phil. Soc.* **15,** 114 (1909).

II. The propagation of each photon in space is correctly described by Maxwell's equations of classical electromagnetic theory. Each photon can, for the purposes of this description, be regarded as a classical wave train defined by the two vector fields $\mathbf{E}(\mathbf{r},t)$ and $\mathbf{B}(\mathbf{r},t)$ which satisfy Maxwell's equations with the appropriate boundary conditions determined by the physical situation at hand. In particular, photons will be diffracted by obstacles, and the diffracted waves can be described within classical theory. A wave incident on a half-silvered mirror, or on a screen with two slits, will indeed be "split" into two waves, and these two waves can interfere with each other as predicted by classical theory.

III. It is not correct to interpret the sum of the squares of the amplitudes $\mathbf{E}$ and $\mathbf{B}$ as representing the energy density in space associated with the photon. This classical idea is wrong. Instead, every quantity depending *quadratically* on the wave amplitude is to be interpreted as being proportional to a *probability* for something to happen. For instance, the integral over the sum of the squares of the amplitudes $\mathbf{E}$ and $\mathbf{B}$ over some finite region in space does not equal the energy carried by the photon in this region, but is instead proportional to the probability that the photon will be observed in this region if we try to "catch" it with a photocell. Similarly the classically computed flux of radiation passing through a hole in a screen is to be reinterpreted as being proportional to the probability that the photon will be detected if we place a photocell immediately behind the hole.

IV. If a photon is detected (with a photocell) anywhere in space the energy delivered to the detector is always equal to $\hbar\omega$. Since the probability of detecting the photon is proportional to the sum of the squares of the amplitudes $\mathbf{E}$ and $\mathbf{B}$ we can conclude that the classical energy density integrated over a region equals the product of the energy carried by a photon with the probability of finding the photon in the region. Therefore, if the source of light is kept steady for a long time, such that a large number of photons are emitted, then the *average* energy which can be observed in a region is indeed equal to the classically computed energy in the region.

45 We have now departed from the ideas of classical electromagnetic theory. The new idea which we have introduced is the *probability* interpretation of all quantities which depend quadratically on the electromagnetic field amplitudes. We can continue to study the propagation of photons in space in terms of Maxwell's equations, but we have a new interpretation for the classically computed energy density and the classically computed flux of radiant energy. These quantities are to be interpreted as *average* quantities which we will observe with a very large number of photons.

It follows that in an experiment in which we only measure these averages but do not attempt to observe individual photons, the classical theory will appear to be correct. On the other hand, if we do observe individual photons, with a photocell, then the limitations of the classical theory immediately become apparent.

46 Let us now see how we can describe the observational facts in some concrete cases in terms of our new ideas. We consider the situation discussed in Sec. 36 in which we observe the photo-electric effect at various distances from a steady light source located at the origin. Suppose the source is approximately monochromatic, and suppose that it emits, on the average, N photons per second of frequency ω. The photocell is placed at some fixed distance from the source. It is coupled to a register so that we can count the number of photons detected by the cell.

Consider now a typical photon emitted by the source. It can be regarded as a wave train of finite duration, spreading out in all directions in space, and carrying a total energy $\hbar\omega$. We compute, *classically,* the total flux of energy E_c which this wave carries into the photocell. This energy is a certain fraction $q = E_c/\hbar\omega$ of the total energy emitted. However, according to our new interpretation of quantities which depend quadratically on the wave amplitude, q is in fact equal to the *probability* that the photon enters the photocell. (For simplicity we can assume that our photocell has 100 per cent counting efficiency, in which case q is equal to the probability that the counter will click when a photon is emitted by the source.)

For each photon emitted by the source we cannot predict whether the counter will actually click or not, but we can say that the probability for this to happen is q. *If* the counter clicks, then the amount of energy transmitted from the source to the photocell equals $\hbar\omega$. It follows that the *average* power transmitted from the source to the photocell, when the source is kept steady, equals $W_{av} = qN\hbar\omega = NE_c$. This average power is in accordance with the classical prediction.

The classically computed quantity E_c is naturally proportional to $1/r^2$, where r is the distance of the photocell from the source. It follows that $q = E_c/\hbar\omega$ is also proportional to $1/r^2$, and since the counting rate of the photocell equals qN we see that the counting rate is inversely proportional to the square of the distance, in accordance with what is actually observed.

47 Many people have felt that there is something paradoxical about the circumstances described above. They have argued as follows. Suppose the distance r is very large, say one light year.

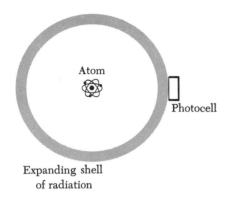

**Expanding shell
of radiation**

Fig. 47A The atom, shown in the center, emitted light one year ago. The spherical shell of radiation accordingly has a radius of one light year. It is about to reach the photocell at right. If the cell registers, the entire energy of the wave is suddenly concentrated in the cell. How is this possible? How can the energy from the far side of the shell reach the photocell in less time than two years?

The "paradox" evaporates if we give up the classical idea that the energy density is proportional to the square of the field amplitude. According to quantum mechanics the transfer of energy from the atom to the cell is governed by a probabilistic law, and the square of the field amplitude must be interpreted as a probability density.

After the photon has been emitted it spreads out like a spherical shell. By the time the wave arrives at the detector the energy carried by the wave has spread out over a very large region in space, say within some spherical shell of radius one light year. How is it then possible that the entire amount of this energy can suddenly become concentrated within the photocell in case the cell does register? It should take more than a year for the energy within the "far side" of the shell to reach the cell, otherwise we violate the principle that no signal can propagate faster than the velocity of light.

The fallacy in this line of reasoning is to believe in the classical expression for the energy density in terms of the electric and magnetic fields. We must remember that the whole purpose of introducing the concept of the electromagnetic field in physics is to describe the interactions between *charges*. In Vol. II of this series we learned that it is a *convenient* concept, and we also learned that it is sometimes convenient to *imagine* (in typical macroscopic situations) that energy is distributed in space with a density proportional to the square of the field amplitude. There was, however, no physical fact discussed in Vol. II which says that we must take this idea literally, and we now know that the classical expression for energy density refers to the *average* energy density which we will observe for a very large number of photons, but it does not describe the energy density associated with a *single* photon.

The real question is this: what laws govern the transfer of energy from an atom in the source to an electron in the detector? This is what we are studying, and we have now discovered some features of these laws.

48 Let us return to the diffraction experiment discussed in Secs. 39–42. Suppose that we observe the photons in some direction θ with a photocell. By observing the counting rate as a function of θ (while holding the source steady) we can observe the diffraction pattern. Suppose now that the counter has just clicked. *Question:* Through which slit did this photon come? *Answer:* It came through *both* slits: partly through slit U and partly through slit L.

This answer is in the spirit of our simple theory of Sec. 44. If the object under study were a billiard ball obeying the laws of classical mechanics then the answer would be shocking. Since, however, we deal with photons there is nothing surprising about this answer: it simply corresponds to what actually takes place.

Question: Can we arrange it so that we *know* through which slit the photon comes? *Answer:* Yes, very easily. We simply cover slit U, and then we know that all the photons detected must have come

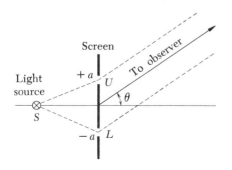

through slit *L*. If we do that we will, of course, not see the two-slit diffraction pattern, but only the one-slit pattern. That does not answer the real question, says the reader. We want to do this experiment with the help of some ingenious device without covering either slit. In other words: we want to preserve the two-slit diffraction pattern in *exactly* the form it has without the ingenious device, but nevertheless be able to tell through which slit each particular counted photon came. Can we do this?

Suppose it were possible. In that case we would simply first discard all the counts for which the photons came through slit *U*, and then plot a diffraction pattern on the basis of the remaining counts, for which the photons came through slit *L*. What would this pattern look like? It would have to look like the single-slit pattern, because we are assured that "nothing came through slit *U*," and in that case slit *U* might as well have been closed. Similarly the pattern obtained on the basis of all the counts associated with photons which came through slit *U* must be the single-slit pattern. *All* counts taken together would then lead to a pattern as predicted in Sec. 41, i.e., the pattern would *not* be the actually observed two-slit pattern. In an experiment in which we observe the two-slit pattern we cannot tell through which slit any particular photon came. The pattern can arise only if the photons go partly through both slits, and it is then *meaningless* to ask which one of the slits the photon came through.

49 We have now learned many interesting things about the behavior of photons. The simple theory which we formulated in Sec. 44 is the first step toward the formulation of a quantum-mechanical theory of electromagnetic radiation. Naturally our discussion in this section is not the whole story of quantum electrodynamics: there is much more to be learned. In particular there is much more to be said about processes involving many photons. Our aim in this chapter was, however, to arrive at a simple preliminary quantum-mechanical formulation in terms of which we can describe the most basic facts having to do with photons, and this we have done. The essence of our theory is that whereas the wave *amplitude* associated with a photon can be discussed as in classical electromagnetic theory, all quantities which depend quadratically on the amplitude must be interpreted in terms of probabilities. A photon can be "split" in the sense that the wave can be divided into two, or several, parts by half-silvered mirrors or other devices, just as in classical electromagnetic theory. However, an almost monochromatic photon cannot be "split" in the sense that we can detect, with a photocell, a "fractional photon" which carries only some fraction of the energy $\hbar\omega$, where ω is the frequency of the photon. These ideas constitute a clear departure from the ideas of classical electromagnetic theory. It would, however, be an exaggeration to say that the classical theory has been *completely* overturned: we have merely discovered limitations of the classical theory.

We want to emphasize that there is nothing paradoxical or mysterious at all about the experimental facts which we have discussed. Naturally we may occasionally feel perplexed about what we find in nature, and the reason for this is that we look at the facts with prejudiced minds. We have opinions about how things ought to be, and then we feel unhappy when our expectations are not fulfilled. We must learn, however, to accept things as they are, and try to find simple and consistent descriptions of the observed phenomena.

The reader should understand clearly that our theoretical ideas in this chapter derive from experimental facts. The results of one set of experiments never enable us to derive, by pure logic, what *must* happen in some different set of experiments. We might be led to *guesses,* but that is another thing. There is no reason in the world why things have to be the way we have described them in this chapter. It could just as well have been the case that "fractional photons" occur, or that a diffraction pattern does change its character as the light intensity is reduced.

50 We conclude this chapter by urging the reader to think about the remarkable theoretical value of an "Optics Kit" containing a few photocells, with associated electronic counting circuits, diffraction gratings, some monochromatic light sources, and a few other standard optical devices. With such a kit we can learn a lot about fundamental physics. In its high ratio of educational to monetary value the Optics Kit is quite unique among the apparatus of experimental physics.

References for Further Study

1) A set of reprints has been published by the American Institute of Physics, 335 East 45th Street, New York, N.Y., under the title: *Quantum and Statistical Aspects of Light.* As the title indicates, these papers deal with various properties of photons, and the reader may find some of them interesting. A short survey of the literature is included.

2) We refer again to the books *The World of the Atom,* edited by H. A. Boorse and L. Motz, Vols. I and II, (Basic Books, Inc., New York 1966), for translations and reprints (with editorial comments) of many early papers relevant to the subject matter of this chapter.

3) The following articles in the *Scientific American* can properly be read at this stage of our discussion:

a) G. E. Henry: "Radiation Pressure," June 1957, p. 99.

b) W. H. Jordan: "Radiation From a Reactor," Oct. 1951, p. 54 (discusses the Cerenkov radiation).

c) G. Burbidge and F. Hoyle: "Anti-Matter," April 1958, p. 34.

d) G. B. Collins: "Scintillation Counters," Nov., 1953, p. 36.

Problems

1 An atom, or nucleus, of mass M_i, decays by the emission of a photon. The final mass of the particle is M_f (after the emission of the photon). The emitted photon is observed in the inertial frame in which the atom was originally at rest: let the frequency of the photon be ω. Let us define ω_0 by $\omega_0 = (M_i - M_f)c^2/\hbar$.

(*a*) Show that

$$\omega = \frac{(M_i + M_f)}{2M_i} \omega_0 = \omega_0 \left[1 - \frac{\omega_0 \hbar}{2M_i c^2} \right]$$

(*b*) Compute $(\omega_0 - \omega)/\omega$ for the yellow line emitted by sodium. Similarly compute $(\omega_0 - \omega)/\omega$ for the 113 keV gamma ray emitted by the hafnium isotope $_{72}\text{Hf}^{177}$.

The formula above describes the *recoil-effect* in photon emission. As we see, the emitted photon is always of smaller frequency (in the rest frame of the emitter) than the frequency ω_0 which it would have for an infinite M_j. For optical photons emitted by atoms the effect is extremely small.

2 Consider the inverse of the process discussed in Prob. 1. An atom, or nucleus, of mass M_f, originally at rest in the laboratory, absorbs a photon of frequency ω. The final mass of the atom (or nucleus) is M_i. Let again $\omega_0 = (M_i - M_f)c^2/\hbar$. Derive a relation between ω, ω_0, M_i and M_f. Note that for small relative changes in mass the frequency ω is very close to ω_0.

3 On the basis of the data given in the graph in Fig. 23A, determine h/e to the accuracy permitted by the accuracy of the graph. (The velocity of light is regarded as known.)

4 Examine Compton's curves in Fig. 20A. The abscissa is roughly proportional to the wavelength. Using the data in the third graph, try to predict the displaced maxima in the second and fourth graphs, and compare with the actual curves.

5 Examine the graphs in Fig. 16A. We notice that the abscissa is expressed in two different ways: as a velocity and as a frequency. The energy of the gamma ray emitted from the excited Fe^{57} nucleus is 14.4 keV. Knowing this, can you relate the velocity to the frequency in the two scales shown on the graph?

6 If we think about the graphs in Fig. 16A we notice a remarkable feature: the recoil effect discussed in Prob. 2 is absent. This phenomenon is known as the Mössbauer effect after its discoverer.† Can you think of any explanation of this effect? After you have thought about it you can peek in the literature: it is an interesting phenomenon.

7 Gamma rays of wavelength 0.710 Å are scattered in a thin foil of aluminum. The scattered radiation is observed at an angle of 60° from the incident direction. What wavelengths would you expect to see?

8 Suppose an electron-positron pair annihilates into *three* gamma rays. If we observe one of the gamma rays in the rest frame of the electron-positron pair (we assume that the annihilation takes place when the electron and the positron are almost at rest), what are the possible energies of the photon?

9 Photons are incident normally on the plane boundary which separates a uniform dielectric, of refractive index n, from the vacuum. We assume that the photons, of frequency ω, enter from the vacuum.

(*a*) What is the frequency, and energy, of a photon inside the dielectric?

(*b*) Can one assign a momentum to the photon inside the dielectric? If so, give an expression for the momentum. How is the momentum related to the wavelength, and what is the wavelength inside the dielectric?

10 A charged particle traveling in empty space at a uniform velocity cannot radiate electromagnetic radiation (photons): conservation of energy and momentum forbids it. Investigate whether a charged particle traveling inside a dielectric at a uniform speed greater than the velocity of light in the medium can emit photons. It turns out that this is possible, and the radiation is known as Cerenkov radiation. (We are here only concerned with balancing energy and momentum, and not with the detailed "mechanism" responsible for the emission.) The emitted photons emerge at a certain angle with respect to the direction of the charged particle. Find this angle under the assumptions that the refractive index is 1.5, the particle is a pion of energy 5 BeV, and the photon is in the optical range. Detectors for charged particles

---

† R. L. Mössbauer, "Kernresonanzfluoreszenz von Gammastrahlung in Ir^{191}," *Zeitschrift für Physik* **151**, 124 (1958). A translation of this article, and reprints of other articles on the same subject can be found in *The Mössbauer Effect*, edited by H. Frauenfelder (W. A. Benjamin, Inc., New York, 1962).

in which the phenomenon of Cerenkov radiation is exploited are commonly employed in high-energy physics. The mass of a pion is 140 MeV.

11 (*a*) When a charged particle moves in a plane perpendicular to a uniform magnetic field its trajectory will be a circular arc. Assuming that the particle carries one elementary charge, show that the momentum of the particle is proportional to the quantity Br, where B is the magnitude of the field, and r is the radius of the trajectory. Find the conversion factor which gives the momentum in units of MeV/c if the quantity Br is given in gauss-cm. (c is here the velocity of light.)

(*b*) In analyzing his cloud chamber picture (see Fig. 26A) Anderson determined the energy of the positron from the known magnetic field and the observed curvature of the tracks. He expresses the momenta in the two portions of the trajectory as $Br = 2.1 \times 10^5$ gauss-cm, and $Br = 7.5 \times 10^4$ gauss-cm. Show that this corresponds to the energies 63 MeV and 23 MeV.

(*c*) Can the sign of the charge, and the direction of motion of the particle, be determined from a picture such as the faked cloud chamber photograph on this page? How did Anderson know (see Fig. 26A) that the particle was a positron rather than an electron going in the opposite direction?

(*d*) In Fig. 26A the magnetic field is perpendicular to the plane of the figure. Does it point *into* the paper, or *out of* the paper?

See Anderson's paper [*Phys. Rev.* **43**, 491 (1933)] for his arguments ruling out the possibility that his photograph shows the track of a *proton*.

12 We consider a refinement of the two-slit diffraction experiment discussed in Secs. 39–42. (See figure on this page.) We place polarization filters in front of the slits, in front of the light source, and in front of the observer. The problem is to find expressions for the intensity, analogous to Eq. (40c), for various combinations of filters. We assume that the light source itself emits unpolarized light, and that the slits are not sensitive to the state of polarization. Consider the following cases:

P_S	P_U	P_L	P_O
absent	absent	horizontal	absent
absent	horizontal	vertical	absent
circular	horizontal	vertical	circular
circular	horizontal	horizontal	circular
circular	horizontal	vertical	absent

In the table above "horizontal" refers to a filter which lets through only light polarized in the horizontal direction, "vertical" refers to a filter which lets through only light polarized vertically, and "circular" refers to a filter which lets through only left-circularly polarized light.

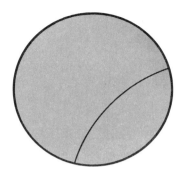

A faked cloud chamber picture, showing the track of a charged particle in a magnetic field which points out of the plane of the figure. (This figure refers to Prob. 11 on this page.)

Is this the track of a positron, and if so, in what direction is it going? Or is this the track of an electron going in the opposite direction?

How did Anderson know that *his* picture (see Fig. 26A) shows the track of a positron, rather than an electron?

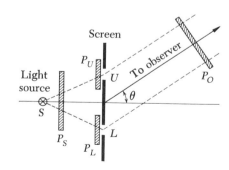

A refinement of Fig. 39A. Polarization filters have been placed as follows: P_S in front of the source, P_U and P_L in front of the upper and lower slits, and P_O in front of the observer.

This figure refers to Prob. 12 on this page. What kind of fringes will we see, for various choices of filters?

Chapter 5

Material Particles

Chapter 5   *Material Particles*

The de Broglie Waves

1 In this chapter we shall study the properties of *material* particles, by which we mean particles of non-zero rest mass, such as electrons, protons, neutrons, mesons, molecules, etc.

It is a simple experimental fact that material particles have wave properties. This fact is known today not only by people educated primarily in the physical sciences, but also by many others. We must remember, however, that the wave nature of an object such as the electron was once regarded as quite surprising. The reason for this surprise was simply that physicists had become accustomed to thinking about the electron as if it were a classical corpuscle. The early experiments concerning electrons seemed to imply such a model, and before 1927 nobody had performed any *clear-cut* experiments in which the wave nature manifested itself. In his high school physics course† the reader has very likely already learned about some of the experiments through which the wave nature of the electron was established, and we shall discuss these experiments further in this chapter.

For photons the wave properties were discovered first and the corpuscular properties later. For electrons the order was reversed. Because of this historical sequence of events‡ the general public tends to believe that light consists of waves, and that electrons are corpuscles. This is an incomplete picture. In the future it will undoubtedly become much more generally known that photons and electrons, and indeed *all* particles, are very similar in the sense that they all have *some* corpuscular properties and *some* wave properties.

2 It is of interest to retrace the prediction and discovery of matter waves because this was an important advance in our knowledge of physics. We shall follow a quasi-historical approach in the first part of this chapter, and the reader is asked to forget, temporarily, what he learned in high school about matter waves. We will imagine that we have gone back in time to the period around 1923. At that time quite a bit was known about the electron as a classical

---

† See, for instance, PSSC, *"Physics"* (D. C. Heath and Company, Boston, 1965), Part IV.

‡ The author believes that the historical order of discovery can be understood theoretically, on the basis of the smallness of the fine structure constant α.

particle, but nothing was known about its wave properties. That the photon has some corpuscular properties was, however, known.

Let us play a make-believe game in which we ask whether a material particle, say an electron, might possibly have some properties of a wave. To find out we must turn to experiments, but before we do that we should try some theoretical ideas first to see what we might expect.

3 To associate a wave with a particle may seem like a highly unmotivated act, and we certainly do not claim that we can *prove* logically that such a wave must exist. There are, however, analogies with optics to which we can point. Consider an optical instrument, and the passage of light through this instrument. We know that we could in principle describe the instrument by solving Maxwell's equations with the appropriate boundary conditions, and if we do this we can describe the propagation of the waves from the light source to the image of the source. There is, however, a simpler way of discussing optical instruments, namely by the method of ray optics. We can show, basing our discussion on the rigorous wave equations, that this method must yield an *approximate* solution. We trace the passage of a *light ray* through the instrument, and this ray can be regarded as the trajectory of a photon. How is the ray related to the waves? The ray is at each point perpendicular to the wave front: in each small region in space the wave appears approximately like a plane wave, and a light ray passing through this region is perpendicular to the planes of constant phase. We have here an association between a "particle" and a wave, and it is this optical analogy which we try to exploit to formulate a wave theory of material particles.

Ideas along these lines were first put forward by L. V. de Broglie around 1923.† We can well admire him for the intellectual courage he showed in presenting a new idea of this kind.

4 Let us now follow in de Broglie's footsteps and suppose, for the purpose of theoretical experimentation, that there is a wave associated with every moving particle. Let the particle be moving in the absence of any external forces, in which case the motion is uniform. Let the energy be E and the momentum be **p**, and let the mass of the particle be m.

If there is a wave associated with a particle moving in this man-

---

† L. V. de Broglie, "Ondes et quanta," *Comptes Rendus* **177**, 507 (1923); "A tentative theory of light quanta," *Philosophical Magazine* **47**, 446 (1924); "Recherches sur la théorie des quanta," *Annales de Physique* **3**, 22 (1925).

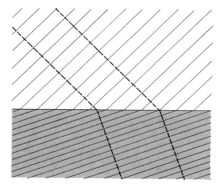

Fig. 3A The refraction of a plane wave at the plane interface of two homogeneous media of different refractive indices. The wave fronts, i.e., surfaces of constant phase, are here planes. They are represented by the thin whole lines in the figure. The rays, which are perpendicular to the wave fronts, are drawn dotted. We can think about them as representing trajectories of photons. To a given family of wave fronts corresponds a set of trajectories, of which the two trajectories shown are particular members.

Actually the wave is also partially reflected, although this is not shown in the figure for aesthetic reasons.

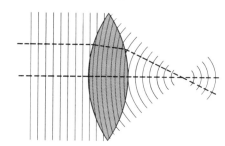

Fig. 3B This figure is analogous to Fig. 3A, and is intended as an illustration of the discussion in Sec. 3. Wave fronts are shown for the case of a plane wave incident on a lens from left. Two rays, or trajectories of photons are also drawn. Note that they intersect at the focal point. To the system of wave fronts corresponds again a set of trajectories.

Close inspection reveals some imperfections in this figure. These are not merely due to defective draftsmanship, but reflect the fact that a perfect lens does not exist. The figure is strictly correct only in the paraxial region, i.e., in the immediate neighborhood of the axis.

Reflections will, of course, take place at the various interfaces. This is not shown in the figure.

ner we would expect that this wave must be moving in the same direction as the particle. We represent the wave by the complex wave function

$$\psi(\mathbf{x},t) = A \exp (i\mathbf{x} \cdot \mathbf{k} - i\omega t) \tag{4a}$$

where A is the constant amplitude of the wave, $\mathbf{k}$ is the wave vector and ω is the frequency. Our problem is to try to guess the relationship between $\mathbf{k}$ and ω which characterize the wave, and the variables $\mathbf{p}$, E and m which characterize the particle.

The wave described by the wave function $\psi(\mathbf{x},t)$ is a plane wave: the planes of constant phase being given by $(\mathbf{x} \cdot \mathbf{k} - \omega t) = $ const. These planes, and thereby the wave, propagate with the *phase velocity*

$$\mathbf{v}_f = \frac{\omega \mathbf{k}}{k^2} \tag{4b}$$

We might at first be tempted to equate the phase velocity $\mathbf{v}_f$ with the velocity of the particle, $\mathbf{v} = \mathbf{p}c^2/E$, but on second thought it is the *group velocity* which we expect to be equal to the velocity of the particle. The group velocity is the velocity by which a signal, or an amount of energy, can be transmitted in space, and we may properly think about the particle as a "packet" of energy.

5 In Vol. III of this series† we derived an expression for the group velocity v of a wave packet, namely

$$\frac{1}{v} = \frac{dk}{d\omega} \quad \text{or} \quad v = \frac{d\omega}{dv}\frac{dv}{dk} \tag{5a}$$

We argued that the group velocity v must be the velocity of the particle. To go further we must make a guess on how the frequency ω depends on $\mathbf{p}$ and E. Let us guess that the relation $E = \hbar\omega$, which holds for photons, also holds for material particles. We then have

$$\hbar\omega = E = \frac{mc^2}{\sqrt{1 - (v/c)^2}} \tag{5b}$$

Inserting this into the second equation (5a) and rearranging, we obtain

$$\frac{dk}{dv} = \frac{1}{v}\frac{d\omega}{dv} = \left(\frac{m}{\hbar}\right)\left(1 - \frac{v^2}{c^2}\right)^{-3/2} \tag{5c}$$

---

† Berkeley Physics Course, Vol. III, *Waves*, Chap. 6.

Integrating this equation, assuming that $k = 0$ if $v = 0$, we obtain

$$\hbar k = \frac{mv}{\sqrt{1 - (v/c)^2}} = p \tag{5d}$$

or, in vector form,

$$\hbar \mathbf{k} = \mathbf{p} \tag{5e}$$

This was the relation proposed by de Broglie.

6 In arriving at the relation $\hbar \mathbf{k} = \mathbf{p}$ we made the somewhat dubious assumption expressed by the left side of equation (5b). We may ask whether we might not obtain the same result from a less drastic assumption supplemented by the requirement of relativistic invariance. Let us explore this possibility and at the same time convince ourselves that our equations (5b) and (5d) are consistent with the theory of special relativity.

First of all we must find out how $\mathbf{k}$ and ω transform under Lorentz transformations. Suppose that the wave is described by the wave function $\psi(\mathbf{x},t)$, as given by Eq. (4a) in the *unprimed frame*. The same wave is described in the *primed frame*, which moves with the velocity $\mathbf{v}$ with respect to the unprimed frame, by the wave function

$$\psi'(\mathbf{x}',t') = A' \exp (i\mathbf{x}' \cdot \mathbf{k}' - i\omega't') \tag{6a}$$

where A' is a constant amplitude which may or may not be equal to A.

Let us suppose that the primed frame is the *rest frame* of the particle. In this frame we thus have $\mathbf{k}' = 0$, $\mathbf{p}' = 0$, and $E' = mc^2$. Let us furthermore suppose that the relation (5b) holds in the rest frame (but perhaps not in any other frame): under this assumption we have $\omega' = mc^2/\hbar$.

Louis Victor de Broglie. Born 1892 in Dieppe, France. De Broglie first studied history, but later changed over to physics. He obtained his doctoral degree from the University of Paris in 1924. Since then he has held positions at the Sorbonne, at the Henri Poincaré Institute, and at the University of Paris. He received the Nobel prize in 1929.

The title of de Broglie's doctoral thesis was "Recherches sur la Théorie des Quanta," and it contained the essence of his ideas about matter waves. (*Photograph by courtesy of Physics Today.*)

7 The *phase* of the wave is given in any one frame by the expression $(\mathbf{x} \cdot \mathbf{k} - \omega t)$, and this quantity we assume to be an *invariant*: if the phase has a certain value at the point $\mathbf{x}'$, at the time t', in the primed frame, then the phase should have the same value at the corresponding point $\mathbf{x}$, at the corresponding time t, in the unprimed frame. We defend this assumption by pointing to the periodic nature of the wave. If the phases for two events in space-time differ by an integral multiple of 2π in *one* frame of reference, then the phases of the same wave must differ by the same integral multiple in *every* frame of reference. From this it follows that the phases in the primed and unprimed frames can differ by at most a constant, and this constant can be absorbed in the ratio A/A', in which case the phase is an *invariant*, as we have assumed. With this assumption, and with our choice of the primed frame as the rest frame of the particle, we have

$$\mathbf{x} \cdot \mathbf{k} - \omega t = -\omega' t' = -\left(\frac{mc^2}{\hbar}\right) t' \tag{7a}$$

The quantity t' may be expressed in terms of $\mathbf{x}$ and t, and the velocity $-\mathbf{v}$ by which the unprimed frame moves with respect to the primed frame, and the relation between these quantities is given by the Lorentz transformation discussed in Vol. I of this series,† namely

$$t' = \frac{t - (\mathbf{x} \cdot \mathbf{v})/c^2}{\sqrt{1 - (v/c)^2}} \tag{7b}$$

and if we insert this expression into (7a) we obtain

$$\mathbf{x} \cdot \mathbf{k} - \omega t = \frac{(mc^2/\hbar)\left((\mathbf{x} \cdot \mathbf{v})/c^2 - t\right)}{\sqrt{1 - (v/c)^2}} \tag{7c}$$

Since this relation must hold for *every* $\mathbf{x}$ and *every* t it follows that

$$\omega = \frac{(mc^2/\hbar)}{\sqrt{1 - (v/c)^2}} \tag{7d}$$

$$\mathbf{k} = \frac{(m\mathbf{v}/\hbar)}{\sqrt{1 - (v/c)^2}} \tag{7e}$$

On the other hand the velocity of the particle in the unprimed frame is just $\mathbf{v}$, since the particle was assumed to be at rest in the primed frame. The energy E and momentum $\mathbf{p}$ of the particle in the unprimed frame are therefore given by

$$E = \frac{mc^2}{\sqrt{1 - (v/c)^2}}, \qquad \mathbf{p} = \frac{m\mathbf{v}}{\sqrt{1 - (v/c)^2}} \tag{7f}$$

Taken together the equations (7d) to (7f) lead to

$$E = \hbar\omega, \qquad \mathbf{p} = \hbar\mathbf{k} \tag{7g}$$

We accordingly recover the result (5e) and furthermore we see that the equation (5b) which was introduced ad hoc in Sec. 5 indeed must be generally true if it holds in the rest frame. This line of reasoning thus shows us that the relations (7g) are consistent with

---

† Berkeley Physics Course, Vol. I, *Mechanics*, Chap. 11.

special relativity: we have now in fact derived these relations on the basis of relativistic invariance.

8   Following in de Broglie's footsteps we have thus been led to the hypothesis that there may be a wave associated with a moving particle, and that this wave is characterized by the wave vector **k** related to the momentum **p** of the particle by $\mathbf{p} = \hbar\mathbf{k}$. Differently stated the wavelength λ of the matter wave will be given by

$$\lambda = \frac{h}{p} = \frac{2\pi}{k} \tag{8a}$$

which is known as the de Broglie equation, the wavelength λ being known as the *de Broglie wavelength* of the particle. Notice that these relations also hold for *photons*.

In order to see how the de Broglie wavelength depends on the parameters of a moving particle, let us write the relation (8a) in several alternative forms. The form

$$\lambda = \left(\frac{h}{mc}\right)\frac{\sqrt{1 - (v/c)^2}}{(v/c)} \tag{8b}$$

shows us that λ decreases as the velocity v increases. For a fixed velocity v the wavelength λ is inversely proportional to the mass m.

9   If E denotes the *total* energy of the particle, as before, we can write

$$\lambda = \frac{hc}{\sqrt{E^2 - m^2c^4}} = \frac{(hc/E)}{\sqrt{1 - (mc^2/E)^2}} \tag{9a}$$

which shows us that for m fixed the wavelength λ decreases when E increases. For a fixed total energy E the wavelength λ increases with the mass m. A massless particle has the smallest de Broglie wavelength (for a given energy), and it is given by

$$\lambda = \frac{hc}{E} \tag{9b}$$

Since this expression is obtained from (9a) by setting $(mc^2)/E = 0$, we see that it also holds approximately in the extreme relativistic limit when the velocity v is very close to c, or, differently stated, when the total energy is very large compared with the rest energy.

Let T denote the *kinetic* energy of the particle, in which case

$$E = T + mc^2 \tag{9c}$$

We substitute this expression for E into (9a) and we obtain

$$\lambda = \frac{hc}{\sqrt{T(T + 2mc^2)}} = \frac{h}{\sqrt{2mT}} \; \frac{1}{\sqrt{1 + T/(2mc^2)}} \qquad (9d)$$

For a fixed rest mass m the wavelength λ decreases when the kinetic energy T is increased. For a fixed kinetic energy T the wavelength λ likewise decreases as m is increased.

In the limiting case when the velocity of the particle is very small compared to c, the ratio T/mc^2 becomes very small. Setting this ratio equal to zero in Eq. (9d) we thus obtain an expression for the wavelength λ in the non-relativistic approximation

$$\lambda \cong \frac{h}{\sqrt{2mT}} \cong \frac{h}{mv} \qquad (9e)$$

which we could, of course, also have obtained directly from (8a).

10 We now wish to see whether de Broglie's hypothesis about the matter waves is confirmed by experiments. First of all we should convince ourselves that the idea of the matter waves does not contradict our common-sense knowledge of macroscopic physics.

Consider a particle which is small for the macroscopic standpoint. Suppose, for instance, that the mass m of the particle is 10^{-5} gm, or 10 micrograms, and suppose that this particle moves with the velocity $v = 1$ cm/sec. Using the nonrelativistic expression (9e) for the de Broglie wavelength we obtain, for our case, $\lambda \cong 6.6 \times 10^{-22}$ cm, which is a ridiculously small wavelength. The smallness of this wavelength explains why the matter waves, if they exist, are not more apparent in macroscopic physics: the wave lengths are simply too small to be observed. We may refer to the optical analogy to understand this point clearly. The method of ray-optics is more accurate the smaller the wavelength is compared to all relevant dimensions of the optical instruments. To see the wave properties of light in an optical experiment we have to arrange it so that some geometric parameters of the instrument are comparable to the wavelength of the light: only then will we observe deviations from ray-optics in the form of interference and diffraction effects. To detect the presence of the matter waves, we must likewise arrange it so that the wavelength is comparable with some geometric parameter of the instrument. Specifically we should try to find a grating by which we can observe diffraction effects.

11 Inspection of the formula (8b) shows that if we wish to have a large wavelength we should try to perform the experiment with a particle with the smallest possible mass, namely the electron, and furthermore we should try to keep the velocity as small as is feasible. Since we want to consider the case when the velocity is very small we can employ the nonrelativistic approximate expression (9e) for the de Broglie wavelength. If we rewrite this expression specifically for electrons, of mass m, and *kinetic* energy T, we obtain

$$\lambda = \frac{h}{\sqrt{2mT}} = \sqrt{\frac{150.4 \text{ eV}}{T}} \text{ Ångströms} \qquad (11a)$$

The wavelength will thus be one Ångström unit $= 10^{-8}$ cm if the kinetic energy of the electron is 150.4 eV. This wavelength is of the same order of magnitude as the lattice constant of crystals, and, just as in the case of X-rays, we could try to employ a crystal lattice as a grating.

Experiments along these lines were first performed by C. J. Davisson in collaboration with L. H. Germer, and independently by G. P. Thomson, in 1927.† In the experiment by Davisson and Germer the *reflection* of electrons from the face of a crystal was studied, whereas in Thomson's experiment the *transmission* of electrons through a thin crystalline film was studied.

12 Let us consider the Davisson-Germer experiment in some detail: the experimental arrangement is shown schematically in Fig. 12A.

We shall let Davisson describe the history of his own experiment: the quotation is taken from his Nobel lecture, given in Stockholm in 1937. (Davisson and Thomson shared the 1937 Nobel prize for their discoveries.) This quotation is interesting because it reveals that the experimental issue was not as clear-cut in 1927 as it seems in retrospect. After a preliminary discussion of de Broglie's hypothesis Davisson continues:

> It was implicit in the theory that beams of electrons like beams of light would exhibit the properties of waves, that scattered by an appropriate grating they would exhibit diffraction, yet none of the chief theorists mentioned this interesting corollary. The first to draw attention to it was Elsasser, who pointed out in 1925 that a demonstration of diffraction would establish the physical existence of electron waves. The setting of the stage for the discovery of electron diffraction was now complete.

---

† C. J. Davisson and L. H. Germer, "Diffraction of electrons by a crystal of nickel," *Physical Review* **30**, 705 (1927).

G. P. Thomson, "Experiments on the diffraction of cathode rays," *Proceedings of the Royal Society (London)* **117A**, 600 (1928), and "The diffraction of cathode rays by thin films of platinum," *Nature* **120**, 802 (1927).

It would be pleasant to tell you that no sooner had Elsasser's suggestion appeared than the experiments were begun in New York which resulted in a demonstration of electron diffraction—pleasanter still to say that the work was begun the day after copies of de Broglie's thesis reached America. The true story contains less of perspicacity and more of chance. The work actually began in 1919 with the accidental discovery that the energy spectrum of secondary electron emission has, as its upper limit, the energy of the primary electrons, even for primaries accelerated through hundreds of volts; that there is, in fact, an elastic scattering of electrons by metals.

Out of this grew an investigation of the distribution-in-angle of these elastically scattered electrons. And then chance again intervened; it was discovered, purely by accident, that the intensity of elastic scattering varies with the orientations of the scattering crystals. Out of this grew, quite naturally, an investigation of elastic scattering by a single crystal of predetermined orientation. The initiation of this phase of the work occurred in 1925, the year following the publication of de Broglie's thesis, the year preceding the first great developments in the wave mechanics. Thus the New York experiment was not, at its inception, a test of the wave theory. Only in the summer of 1926, after I had discussed the investigation in England with Richardson, Born, Franck and others, did it take on this character.

The search for diffraction beams was begun in the autumn of 1926, but not until early in the following year were any found—first one and then twenty others in rapid succession. Nineteen of these could be used to check the relationship between wavelength and momentum and in every case the correctness of the de Broglie formula, $\lambda = h/p$ was verified to within the limit of accuracy of the measurements.

I will recall briefly the scheme of the experiment. A beam of electrons of predetermined speed was directed against a (111) face of a crystal of nickel as indicated schematically in [Fig. 12A]. A collector designed to accept only elastically scattered electrons and their near neighbors, could be moved on an arc about the crystal. The crystal itself could be revolved about the axis of the incident beam. It was possible thus to measure the intensity of elastic scattering in any direction in front of the crystal face with the exception of those directions lying within 10 to 15 degrees of the primary beam.

13 In the experiment the beam of electrons was produced in an electron gun, in which the electrons were accelerated to the desired energy, of the order of 50 eV. The crystal was, of course, placed in a vacuum. The electrons were incident perpendicularly on a particular crystal plane, technically known as the (111) plane. In this plane we can imagine a lattice of regularly spaced atoms on the surface of the crystal. To understand the principle involved we first consider a simple one-dimensional model, shown schematically in Fig. 13A. (We shall consider the general theory a bit later.)

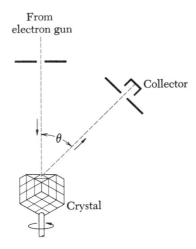

Fig. 12A Schematic figure showing the diffraction of electrons from the surface of a single crystal. For a fixed energy of the incident electrons the intensity of the *elastically* scattered beam is observed as a function of the angle θ.

The incident wave is diffracted by each atom in the row. In certain directions (in the plane of the figure) the diffracted waves from all the atoms will reinforce each other, whereas in other directions they will tend to cancel. The condition for constructive interference (i.e., mutual reinforcement of the diffracted waves) is that the differences in the distances from different atoms to the point of observation shall be integral multiples of the wavelength. If we imagine that the point of observation is very far away we easily see, by inspection of Fig. 13A, that the condition for constructive interference is

$$d \sin \theta = n\lambda \tag{13a}$$

where n is an integer. This relation simply says that the difference in path from two *neighboring* atoms to the point of observation is an integral multiple of the wavelength. We thus expect diffraction maxima in the directions for which the angle θ satisfies the condition (13a). We regard the lattice spacing d as known: it can be determined by other means, for instance through X-ray diffraction measurements.

Our simple theory obviously also applies to the case of a two-dimensional lattice if we imagine that each dot in Fig. 13A actually represents a row of atoms, in a direction perpendicular to the plane of the figure.

In a typical experiment the data were as follows: $d = 2.15 \times 10^{-8}$ cm, $E = 54$ eV, and a maximum was observed at $\theta = 50°$. For $n = 1$ the experimentally observed θ gives a wavelength of 1.65 Å, whereas the wavelength computed from Eq. (11a) is 1.67 Å, which is a satisfactory agreement. Davisson also observed higher order maxima, corresponding to $n > 1$, all in accordance with the theoretical predictions.

14 Thomson's method is analogous to the so-called Debye-Scherrer method in X-ray diffraction work. A unidirectional monochromatic incident beam of X-rays or electrons is scattered by a sample consisting of a very large number of small randomly oriented microcrystals. The theory predicts that the diffracted waves will emerge along the surfaces of circular cones, centered about the incident direction. (See Fig. 14A.) If the scattered radiation is recorded photographically on a plate perpendicular to the incident direction we thus obtain a set of concentric circles. The pattern of circles depends on the crystal structure in a characteristic way, and if the wavelength is known it is possible to determine completely the geometry of the crystal lattice.

Figs. 14B and 14C are two photographs obtained by this method,

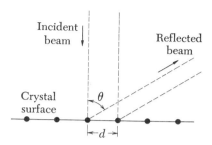

Fig. 13A To illustrate discussion in Sec. 13. Figure shows linear array of equally spaced atoms. We can also interpret a dot as a line of atoms perpendicular to the plane of the figure. Diffraction maxima occur in directions such that $d \sin \theta$ is an integral multiple of the wavelength.

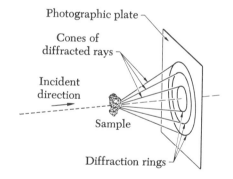

Fig. 14A Diffraction of X-rays or electrons from a sample which is an aggregate of small randomly oriented crystals. The diffracted rays lie along a set of circular cones, the pattern of which depends on the crystal structure and on the incident wavelengths.

The photographs 14B–C, 22A and 22C were obtained by this method. In electron diffraction work the sample must be inside the vacuum of the diffraction apparatus, because the electrons would be strongly scattered by air and by any intervening "window" in the tube. X-rays are scattered much less, and the sample can be kept in air, outside the X-ray tube.

Fig. 14B Photograph showing electron diffraction by white tin, by the method shown in Fig. 14A. Very small tin crystals (size about 300 Å) were deposited on a thin film of silicon monoxide. The film was placed as the sample in an electron microscope which was here used as an electron diffraction apparatus. The sample was illuminated by electrons of an energy of 100 keV. (This corresponds to a wavelength of about 0.04 Å.) The diffraction rings which can be seen correspond to the intersections of the cones in Fig. 14A with the photographic plate.

The purpose of this particular diffraction experiment was to check the crystal structure of the very small tin crystals formed by an evaporation process. *(Photograph by courtesy of Dr. W. Hines and Professor W. Knight, Berkeley.)*

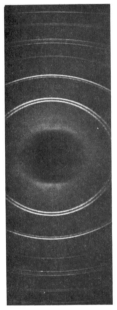

Fig. 14C Photograph showing X-ray diffraction by white tin, by the method shown in Fig. 14A. (This is actually not a flat photographic plate, but a strip of film which was curved as a circular arc during the exposure. This does not change the essence of the experiment.) The sample was in the form of a small amount of finely powdered tin, with an average crystal size of about one micron. The wavelength used was about 1.5 Å.

This picture should be carefully compared with Fig. 14B. The similarity is striking, and there can be no doubt that electrons and X-rays are diffracted by tin crystals in the same way. *(Photograph by courtesy of Mr. George Gordon, Berkeley.)*

the first with electrons and the second with X-rays. The samples were in both cases aggregates of small crystals of white tin. The similarity in the pattern of circles is very striking. Even if we know nothing about the detailed theory of diffraction of waves in lattices, a glance at these two photographs will at once convince us that X-rays and electrons are diffracted in the same way.

15 The experiments of Davisson and Germer and the related experiments of Thomson thus show beyond any reasonable doubt that matter waves do exist, and that their wavelength (at least for electrons) is given by the de Broglie relation. In 1929 Estermann and Stern† showed that helium atoms and hydrogen molecules are also diffracted in accordance with the de Broglie theory. Their experiments greatly strengthen our belief in the universality of the matter waves since they involve two new particles very different from the electron. Apart from the difference in mass the helium atoms and hydrogen molecules differ from the electron in that they are clearly composite systems whereas the electron is (perhaps) an elementary particle. The experiments thus show that the atom as a whole and the molecule as a whole are waves, and we are perhaps now willing to believe that under suitable experimental conditions a grand piano will also behave like a wave.

Later the diffraction of very slow neutrons in crystal lattices was shown to take place, and out of these observations there have developed techniques which are used today in a routine manner to

† I. Estermann and O. Stern, "Beugung von Molekularstrahlen," *Zeitschrift für Physik* **61**, 95 (1930).

investigate the structure of crystals and molecules as a complement to X-ray and electron diffraction methods.†

Theory of Diffraction in a Periodic Structure‡

16 Let us consider the diffraction by a one-, two-, or three-dimensional lattice in some detail. A lattice is a periodic structure which we can imagine built up of replicas of a *unit cell*. Figs. 16A–C illustrate this idea. For a one-dimensional lattice the unit cell is simply a line segment; for a two-dimensional lattice a parallelogram, and for a three-dimensional lattice a parallelepiped. Let us imagine, for simplicity, that an atom (of a given kind) is located at each corner of the unit cell. The positions of all the atoms in the lattice are then given by

$$\mathbf{x} = n_1 \mathbf{e}_1 \tag{16a}$$

for the linear lattice and by

$$\mathbf{x} = n_1 \mathbf{e}_1 + n_2 \mathbf{e}_2 \tag{16b}$$

for the plane lattice and by

$$\mathbf{x} = n_1 \mathbf{e}_1 + n_2 \mathbf{e}_2 + n_3 \mathbf{e}_3 \tag{16c}$$

for the three-dimensional lattice. The numbers n_1, n_2, and n_3 are here *integers*, and the vectors $\mathbf{e}_1$, $\mathbf{e}_2$, and $\mathbf{e}_3$ define the unit cells, as shown in Figs. 16A–C.

In the following we shall imagine that the lattice contains a finite but very large number of atoms. To prevent misunderstandings, let us also state explicitly that we are considering one-, two-, and three-dimensional arrays embedded in *three*-dimensional space, and not, say, a two-dimensional lattice in a two-dimensional world.

17 We consider the situation illustrated schematically in Fig. 17A. A wave is emitted by a source located at the point $\mathbf{x}_i$. It is diffracted by an array of identical atoms, and the diffracted, or *scattered*, wave is observed at the point $\mathbf{x}_0$. We shall assume that the center of the array (occupied by one of the atoms) is at the origin, and that the distances $x_i = |\mathbf{x}_i|$ and $x_0 = |\mathbf{x}_0|$ are very large compared to the linear dimensions of the array. We first consider the case of a one-dimensional array. Very similar considerations

---

† D. P. Mitchell and P. N. Powers, "Bragg reflection of slow neutrons," *The Physical Review* **50**, 486 (1936). See also E. O. Wollan and C. G. Shull, "Neutron diffraction and associated studies," *Nucleonics* **3**, 8 (1948).

‡ Sections 16–22 can be omitted in a first reading, but do not fail to look at the photographs in Sec. 22.

Fig. 16A   Linear array of equally spaced atoms.

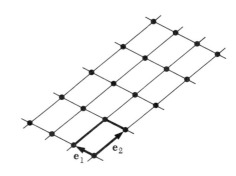

Fig. 16B   Two-dimensional lattice. The unit cell is defined by the two vectors $\mathbf{e}_1$ and $\mathbf{e}_2$. Its edges are drawn heavier in the figure. The entire lattice is built up of replicas of the unit cell.

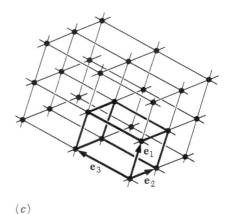

(c)

Fig. 16C   Three-dimensional lattice. The edges of the unit cell are again drawn heavier. The position vector of any lattice point is some linear combination with integral coefficients of the vectors $\mathbf{e}_1$, $\mathbf{e}_2$, and $\mathbf{e}_3$. (These vectors are not necessarily perpendicular.)

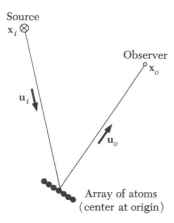

Fig. 17A Diffraction by a linear array. (To illustrate discussion in Sec. 17.) In the text it is assumed that the distances from the array to the source and to the observer are large compared with the size of the array. The array itself consists of a finite but very large number of atoms.

The unit vector $\mathbf{u}_i$ points in the incident direction, and the unit vector $\mathbf{u}_o$ in the direction of the scattered beam.

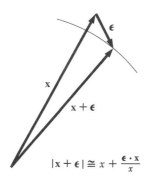

Fig. 17B To illustrate an important approximation which is frequently made in discussions of physics. If the length of the vector $\boldsymbol{\varepsilon}$ is very small compared with the vector $\mathbf{x}$, this vector is almost parallel with the vector $\mathbf{x} + \boldsymbol{\varepsilon}$. The length of the latter vector is approximately equal to the length of $\mathbf{x}$ plus the projection of $\boldsymbol{\varepsilon}$ on the direction of $\mathbf{x}$.

apply to a two-dimensional or to a three-dimensional scattering array.

The length of path from the source to the observer via the origin is $s_0 = x_i + x_0$. Let $s(n_1)$ be the length of path from the source to the observer via the atom whose position is described by the integer n_1, according to the formula (16a). We then have

$$s(n_1) = |\mathbf{x}_i - n_1\mathbf{e}_1| + |\mathbf{x}_0 - n_1\mathbf{e}_1| \tag{17a}$$

The waves arriving at the observer from the different atoms interfere with each other, and the resultant wave amplitude is the sum of the amplitudes from each atom. For a diffraction maximum all the waves must arrive in phase with each other, otherwise the waves from the different atoms will cancel. The condition for this is that for each atom, i.e., for each integer n_1, the path difference $s(n_1) - s_0$ must be an integral multiple of the wavelength λ.

Since we assumed that the size of the array is very small compared to the distances to the source and to the observer, the vector $n_1\mathbf{e}_1$ will be very small compared to the vectors $\mathbf{x}_i$ and $\mathbf{x}_0$. We can then write an approximate expression for the two distances in the right side of Eq. (17a), namely

$$|\mathbf{x}_i - n_1\mathbf{e}_1| \cong x_i - n_1\frac{(\mathbf{x}_i \cdot \mathbf{e}_1)}{x_i} \tag{17b}$$

$$|\mathbf{x}_0 - n_1\mathbf{e}_1| \cong x_0 - n_1\frac{(\mathbf{x}_0 \cdot \mathbf{e}_1)}{x_0} \tag{17c}$$

The geometric significance of this approximation is at once apparent from Fig. 17B.

We then have for the difference in path

$$s(n_1) - s_0 \cong -n_1\mathbf{e}_1 \cdot \left(\frac{\mathbf{x}_i}{x_i} + \frac{\mathbf{x}_0}{x_0}\right) \tag{17d}$$

18 Let $\mathbf{u}_i$ be the unit vector in the direction of the incident beam, and let $\mathbf{u}_0$ be the unit vector pointing in the direction of the diffracted beam. We then have

$$\mathbf{u}_i = -\frac{\mathbf{x}_i}{x_i}, \qquad \mathbf{u}_0 = \frac{\mathbf{x}_0}{x_0} \tag{18a}$$

If we now let x_i and x_0 in (17d) tend to infinity we obtain

$$s(n_1) - s_0 = n_1\mathbf{e}_1 \cdot (\mathbf{u}_i - \mathbf{u}_0) \tag{18b}$$

and the condition for a diffraction maximum reads

$$\frac{n_1\mathbf{e}_1 \cdot (\mathbf{u}_i - \mathbf{u}_0)}{\lambda} = n_1' \tag{18c}$$

where n_1' must be an integer for every choice of the integer n_1. This will obviously be the case if and only if

$$\frac{\mathbf{e}_1 \cdot (\mathbf{u}_i - \mathbf{u}_0)}{\lambda} = m_1 \qquad (18\text{d})$$

where m_1 is an integer. This we might have concluded immediately. The waves from any pair of atoms arrive in phase with each other if and only if the waves from two *neighboring* atoms arrive in phase with each other, and this is just what the condition (18d) says.

Making use of the de Broglie relation we can rewrite (18d) in a physically interesting form, as follows. Let $\mathbf{p}_i$ be the incident momentum, and let $\mathbf{p}_0$ be the momentum in the scattered beam. We then have

$$\frac{\mathbf{u}_i}{\lambda} = \frac{\mathbf{p}_i}{h}, \qquad \frac{\mathbf{u}_0}{\lambda} = \frac{\mathbf{p}_0}{h} \qquad (18\text{e})$$

and the condition (18d) can be written

$$\mathbf{e}_1 \cdot (\mathbf{p}_i - \mathbf{p}_0) = \mathbf{e}_1 \cdot \mathbf{q} = m_1 h \qquad (18\text{f})$$

where $\mathbf{q} = \mathbf{p}_i - \mathbf{p}_0$ is the momentum transfer to the array. For a one-dimensional array the condition for a diffraction maximum is thus that the scalar product of the momentum transfer $\mathbf{q}$ with the vector $\mathbf{e}_1$ must be an integral multiple of h: the momentum transfer in the direction of the array is "quantized."

19 It has been tacitly assumed in our discussion that the scattering process is *elastic*, which means that the energy (or frequency) of the scattered particle is the same as the incident energy (or frequency). This implies another condition: the magnitude of the incident momentum is the same as the magnitude of the scattered momentum. The locations of the diffraction maxima are thus determined by the two conditions

$$\mathbf{e}_1 \cdot (\mathbf{p}_i - \mathbf{p}_0) = \mathbf{e}_1 \cdot \mathbf{q} = m_1 h \qquad (19\text{a})$$

and

$$|\mathbf{p}_i| = |\mathbf{p}_0| \qquad (19\text{b})$$

where m_1 is any integer.

For an *infinite* array the scattered momentum must satisfy the conditions (19a) and (19b) exactly. For a *finite* array we also observe some scattering outside the directions defined by the above conditions. The sharpness of the diffraction maxima (as functions of the angle) depends on the number of atoms in the array. We assume that the number is large and the scattered particles therefore emerge in very well-defined directions, as given by (19a) and (19b). These equations define a set of cones, one for each integer m_1. These integers are, of course, subject to the constraint

$$|m_1| \leqq 2|\mathbf{e}_1||\mathbf{p}_i|/h \qquad (19c)$$

since the momentum transfer cannot exceed twice the incident momentum.

20 We can easily find the corresponding conditions for diffraction maxima for a two-dimensional array. The condition (19a) must hold in each lattice direction, i.e., for each line containing more than one atom. In particular it must hold for the edges of the unit cell, and we thus have the conditions

$$\mathbf{e}_1 \cdot (\mathbf{p}_i - \mathbf{p}_0) = m_1 h, \qquad \mathbf{e}_2 \cdot (\mathbf{p}_i - \mathbf{p}_0) = m_2 h \qquad (20a)$$

$$|\mathbf{p}_i| = |\mathbf{p}_0| \qquad (20b)$$

where m_1 and m_2 are any integers. We can again say that the momentum transfer in the plane of the lattice is "quantized." To show this more clearly, let us define two vectors $\mathbf{q}_1$ and $\mathbf{q}_2$ in the $(\mathbf{e}_1, \mathbf{e}_2)$-plane by the conditions

$$\begin{aligned} \mathbf{e}_1 \cdot \mathbf{q}_1 = h, \qquad \mathbf{e}_2 \cdot \mathbf{q}_1 = 0 \\ \mathbf{e}_1 \cdot \mathbf{q}_2 = 0, \qquad \mathbf{e}_2 \cdot \mathbf{q}_2 = h \end{aligned} \qquad (20c)$$

These equations always have a unique solution. Notice that the vectors $\mathbf{q}_1$ and $\mathbf{q}_2$ will in general not have the same directions as the vectors $\mathbf{e}_1$ and $\mathbf{e}_2$, unless the lattice is rectangular.

The conditions (20a) then read

$$\mathbf{q} = \mathbf{p}_i - \mathbf{p}_0 = m_1\mathbf{q}_1 + m_2\mathbf{q}_2 + \mathbf{q}^* \qquad (20d)$$

where m_1 and m_2 are any integers, and where the vector $\mathbf{q}^*$ is an arbitrary vector perpendicular to the lattice plane. The momentum transfer *in* the lattice plane is quantized, but the orthogonal component is not. Its magnitude is determined by the condition (20b) which says that the scattering is elastic. We can therefore find several solutions to the equations (20a) and (20b) provided the incident momentum is not too small (i.e., provided the wavelength is not too large). The diffracted rays in this case emerge in a number of sharply defined discrete directions, and not along cones, as was the case for the one-dimensional array.

In the experiment of Davisson and Germer the low-energy electrons do not significantly penetrate the crystal. The diffraction is

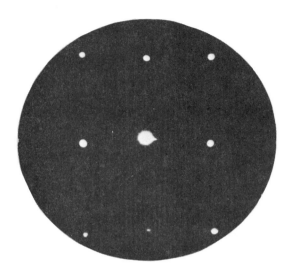

Fig. 20A Above is shown the diffraction pattern of electrons scattered backward from the face of a nickel crystal. The electrons were incident perpendicularly on the crystal surface, and their energy was 76 eV. This is a typical case for which our theory of diffraction by a two-dimensional lattice applies.

Fig. 20B Below is shown the planar symmetry of this particular face of the crystal. We can imagine that the small circles represent the nickel atoms in the surface layer. The diffraction pattern exhibits a similar rectangular symmetry. Question for the reader: are the two figures correctly oriented with respect to each other, or should the lower figure have been turned 90°?

(Electron diffraction photograph by courtesy of Dr. A. U. MacRae, Bell Telephone Laboratories, New Jersey.)

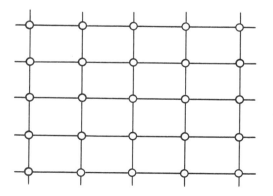

by the atoms in the surface layer, and the theory for the two-dimensional lattice therefore applies.

21   For the three-dimensional array we have

$$\mathbf{e}_1 \cdot (\mathbf{p}_i - \mathbf{p}_0) = m_1 h,$$
$$\mathbf{e}_2 \cdot (\mathbf{p}_i - \mathbf{p}_0) = m_2 h, \qquad (21a)$$
$$\mathbf{e}_3 \cdot (\mathbf{p}_i - \mathbf{p}_0) = m_3 h$$

$$|\mathbf{p}_i| = |\mathbf{p}_0| \qquad (21b)$$

where m_1, m_2, and m_3 are any integers.   In analogy with what we did in the preceding section, let us define three vectors $\mathbf{q}_1$, $\mathbf{q}_2$, and $\mathbf{q}_3$ by the conditions

$$\mathbf{e}_1 \cdot \mathbf{q}_1 = h, \qquad \mathbf{e}_2 \cdot \mathbf{q}_1 = 0, \qquad \mathbf{e}_3 \cdot \mathbf{q}_1 = 0$$
$$\mathbf{e}_1 \cdot \mathbf{q}_2 = 0, \qquad \mathbf{e}_2 \cdot \mathbf{q}_2 = h, \qquad \mathbf{e}_3 \cdot \mathbf{q}_2 = 0 \qquad (21c)$$
$$\mathbf{e}_1 \cdot \mathbf{q}_3 = 0, \qquad \mathbf{e}_2 \cdot \mathbf{q}_3 = 0, \qquad \mathbf{e}_3 \cdot \mathbf{q}_3 = h$$

These equations always have a unique solution.   We can then write the conditions (21a) in the form

$$\mathbf{q} = \mathbf{p}_i - \mathbf{p}_0 = m_1 \mathbf{q}_1 + m_2 \mathbf{q}_2 + m_3 \mathbf{q}_3 \qquad (21d)$$

The momentum transfer $\mathbf{q}$ is "quantized" in such a way that it must be a linear combination, with integral coefficients, of three vectors $\mathbf{q}_1$, $\mathbf{q}_2$, and $\mathbf{q}_3$ determined by the geometric structure of the lattice.   If we think about Eq. (21d) we notice that the possible values of the momentum transfer form a lattice, in momentum space.   This lattice is known as the *reciprocal lattice* of the crystal.

For an arbitrary incident momentum it is in general not possible to satisfy *both* the equations (21d) and (21b).   The equations (21a) and (21b) taken together constitute four equations to determine the three components of the final momentum $\mathbf{p}_0$.   A solution will exist only if the crystal happens to be oriented in the right way.

22   Suppose now that we do a diffraction experiment with a sample which consists of a very large number of randomly oriented microcrystals.   There will then always be some microcrystals so oriented in the sample that the conditions (21a) and (21b) can be satisfied (at least approximately).   For such a sample we thus have two conditions for diffraction maxima, namely

$$|\mathbf{p}_i - \mathbf{p}_0| = |m_1 \mathbf{q}_1 + m_2 \mathbf{q}_2 + m_3 \mathbf{q}_3| \qquad (22a)$$

$$|\mathbf{p}_i| = |\mathbf{p}_0| \qquad (22b)$$

where m_1, m_2, and m_3 are any integers, and where $\mathbf{q}_1$, $\mathbf{q}_2$, and $\mathbf{q}_3$

Figs. 22A-B The upper photograph shows electron diffraction rings obtained with the method illustrated schematically in Fig. 14A. The sample consisted, as for the photograph 14B, of small white tin crystals deposited on a thin film of silicon monoxide. The lower photograph shows the appearance of the sample in the electron microscope (8 mm corresponds to 1000 Å). The dark spots are the images of the crystals (their darkness depends on their orientation). The lightest spots are pits in the SiO once occupied by crystals which have disappeared in the preparation of the sample. The average size of the crystals is about 600 Å.

In obtaining the diffraction photograph the electron beam was limited to a comparatively small area of the sample. In view of the theory in Sec. 22 we expect to see individual dots, as the photograph indeed shows, rather than well-developed rings.

(Photographs by courtesy of Dr. W. Hines and Professor W. Knight, Berkeley.)

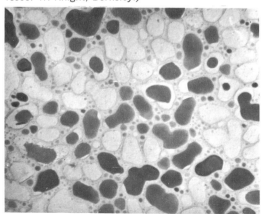

are the vectors discussed in the preceding section, for some particular orientation of the crystal lattice. The above equations do have solutions, and we see that the diffracted rays emerge along a set of cones centered about the incident direction.

Fig. 14A shows schematically how a diffraction experiment based on the above theory is performed. In X-ray work the sample is often a small amount of a fine powder consisting of many small microcrystals. This is the way the photograph 14C was obtained. The lines on the film strip are the intersections of the cones [defined by the conditions (22a) and (22b)] with the film strip.

We can easily understand that if the sample is too small, in the sense that it does not contain sufficiently many crystals, then the distribution of the diffracted rays along the cones will be very non-uniform. We will not see continuous circles on the photographic plate, but rather individual dots. This effect is shown very beautifully in the photographs 22A and 22C. These photographs, which should be compared with the photograph 14B, show the diffraction of 100 keV electrons by tin crystals. In this case the electron waves completely penetrate the small crystals. An electron microscope was used as a diffraction apparatus. Photographs 22B and 22D, taken with the same electron microscope, show the appearance of the sample.

There Is But One Planck's Constant

23 The subheading above perhaps surprises the reader. Of course there is only one Planck's constant, by definition. What deep conclusion does the author intend to draw from this trivial fact?

The entirely nontrivial point is that we do not *need* more than one "Planck-type constant" in physics. Consider the de Broglie relation written in the form

$$h = \lambda p \tag{23a}$$

where p is the momentum of the particle, and λ its de Broglie wave length. Both p and λ are independently measurable quantities, and by measuring a pair of corresponding variables (p, λ) we can determine Planck's constant h. It is a remarkable *empirical* fact that we always get the same value for h, irrespective of which *kind* of particle we are observing, and that this is so is not a triviality.

The reader is perhaps not impressed by this. We could, after all, *derive* this relation on the basis of some very simple ideas. Let us, however, examine the premises of our derivation.

24 In our discussion in Secs. 3–5 we assumed that there is a wave associated with every material particle, and in such a way that the group velocity of the wave equals the velocity of the particle. Furthermore we assumed that the particle-wave description satisfies the principle of special relativity, which means that the relationship between the wave vector and frequency of the wave, and the momentum and energy of the particle, must be the same in *every* inertial frame.   On this basis we drew the conclusion that

$$E = \hbar\omega, \qquad \mathbf{p} = \hbar\mathbf{k} \tag{24a}$$

where E is the energy, $\mathbf{p}$ the momentum, ω the frequency and $\mathbf{k}$ the wave vector, and where $\hbar$ is a constant defined by

$$E_0 = mc^2 = \hbar\omega_0 \tag{24b}$$

in terms of the rest energy E_0 and the "rest frequency" ω_0.

How did we conclude that the constant $\hbar$ is actually Planck's constant?   By *guessing*.   The relation $E = \hbar\omega$ holds for photons, and it is tempting to guess that it will also hold for material particles. But that is just the crucial point: Does the first relation (24a) *really* hold for all material particles?

Therefore, what we really derived in Secs. 3–5 was that the connection between energy, momentum, frequency and wave vector is

$$E = C\omega, \qquad \mathbf{p} = C\mathbf{k} \tag{24c}$$

where C is a constant characteristic of the particle, and this constant is defined, for instance, by

$$C = \frac{E_0}{\omega_0} \tag{24d}$$

There is, however, no reason why C has to be the *same* constant for *all* particles.   Our world might have been different, and we might have found, experimentally, that $C = \hbar$ for photons, $C = 7\hbar$ for electrons, $C = 17\hbar$ for protons, and, to top it all, we might have found that whereas electrons and protons are associated with de Broglie waves, there are *no* matter waves associated with neutrons!

25 Fortunately the available experimental material seems to rule out the horrible possibility that the "Planck-type constants" C are different for different kinds of particles.   We say "fortunately" because our aesthetically pleasing present-day formulation of quantum mechanics depends in an essential way on the assumption that $C = \hbar$ is a universal constant, independent of the kind of particle.

Figs. 22C-D These two photographs were obtained in the same way as the photographs 22A–B.   The sample here consists of smaller crystals (average size about 200 Å), and the diffraction pattern arises from a much larger number of crystals.   The rings are better developed, although individual dots can be seen.   The photographs 22A and 22C should be compared with the photograph 14B, in which individual dots can no longer be seen.   The latter photograph was taken with an electron beam traversing a much larger portion of the film.   We accordingly expect a well-developed ring pattern since all orientations of the crystals are well represented in the sample.

The electron energy was 100 keV for the pictures 14B, 22A and 22C.   This corresponds to a wavelength of about 0.04 Å.

(Photographs by courtesy of Dr. W. Hines and Professor W. Knight, Berkeley.)

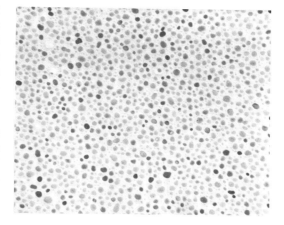

If this were *not* the case the theory of the elementary particles and their interactions would have to be very different indeed.

How well is the hypothesis that $C = \hbar$ for *every* kind of particle verified experimentally? Direct experiments, analogous to the experiments of Davisson and Germer, or Thomson, have been performed only for a few kinds of particles. Such experiments are very readily interpreted as tests of the relation $h = \lambda p$, but they are naturally of limited precision. They support our belief in the universality of the relations (24a), but the real basis of our faith in these relations is the general success of quantum mechanics. There exists an enormous amount of experimental evidence which supports the relations (24a) indirectly. The interpretation of this evidence is not always as clear-cut and simple as in the case of electron diffraction in crystals, but in its totality it is very convincing. Our belief that the relations (24a) are *exactly* true is somewhat analogous to our belief that the relation $E_0 = mc^2$ is also *exactly* true. The direct evidence for this latter relation is quite strong, but the totality of the indirect evidence, concerning the general validity of the ideas of special relativity, is what really convinces us. There is not the slightest hint in our experimental material that the relations (24a) or the relation $E_0 = mc^2$, might be only approximately true. We assume that they are true exactly, and we regard them as cornerstones of physical theory.

Let us recall our discussion in Sec. 12, Chap. 2. We argued that because of the fundamental role played by the constants c and $\hbar$ in relativistic quantum physics one might well select a system of units in which $\hbar = c = 1$. Such a system of units would clearly not make much sense if there were a different Planck-type constant C for every particle. Since we believe that there is only *one* such constant it means that, for instance, mass, energy and frequency are always related in the same way, and we can regard the words "mass," "energy," and "frequency" as different names for the same thing.†

The reader may wonder whether sound waves also obey the relations (24a). Indeed they do, and we call the "particles" of sound waves *phonons*. The energy of a sound wave, say in a solid, comes in packets of magnitude $\hbar\omega$, where ω is the frequency.

We do not regard phonons as *elementary* particles, because they can be completely explained in terms of the "real" particles in the solid. Elastic waves are collective motions of electrons and nuclei. It is, however, often useful to think about phonons in the same way as we think about other particles, and conversely it is sometimes useful to think about the "real" particles as if they were "sound waves in the ether."

26 In view of the relations (24a) we can restate the conservation laws for energy and momentum which hold for collision processes.

Understood in a general sense a collision event can be described as follows. There is present, at some early time, a number of particles moving in such a way that they are well separated from each

----

† The author would prefer to use the term "mass" only to denote the "rest-mass" of an isolated system (i.e., the rest energy divided by c^2). With this usage "the mass of a particle," whether moving or not, means the rest-mass of the particle. Other writers often talk about the "mass" of a particle when they mean the total energy divided by c^2.

other. Let their momenta be $\mathbf{p}'_1, \mathbf{p}'_2, \ldots, \mathbf{p}'_i$, and let their energies be $E'_1, E'_2, \ldots, E'_i$. When we say that the particles are initially "well separated" from each other we mean that the particles initially move in such a way that all interactions between the particles are effectively absent at the early time. This idea makes sense if we assume that the inter-particle forces tend rapidly to zero as the distance between the particles increases. At first each particle therefore moves as if the other particles were absent. As time goes on the particles may converge in a "collision region," and the inter-particle forces come into play. An interaction takes place, and in this process the particles are deflected. Furthermore some of the particles may be destroyed, and new particles may be created.

If we wait for a sufficiently long time the particles involved in the collision event will again be dispersed, and the interactions between the particles will effectively cease for the simple reason that they are no longer close together. At some late time each particle will move as if the other particles were absent. Let the momenta of the particles after the collision event be $\mathbf{p}''_1, \mathbf{p}''_2, \ldots, \mathbf{p}''_j$, and let their energies be $E''_1, E''_2, \ldots, E''_j$.

The conservation laws then read

$$\sum_{r=1}^{i} E'_r = \sum_{s=1}^{j} E''_s, \qquad \sum_{r=1}^{i} \mathbf{p}'_r = \sum_{s=1}^{j} \mathbf{p}''_s \qquad \text{(26a)}$$

The total initial energy equals the total final energy, and the total initial momentum equals the total final momentum. The condition that the particles effectively do not interact with each other at the "early time" or at the "late time" is essential; because otherwise the total energy would not be equal to the sum of the energies of the individual particles. If the particles do interact with each other we have to include an "interaction energy" in the expression for the total energy.

The reader should note that the particles need not be elementary particles; they can just as well be composite particles, like atoms or nuclei. When we discuss collision events we mean by "particle" any object which is reasonably stable so that it can be assigned a momentum, an energy and a (rest) mass as soon as it is well separated from other similar objects. As an example we can consider the collision between a neutral helium atom and an electron. Suppose that the helium atom is ionized in the collision. There are then *two* initial particles, namely the electron and the neutral helium atom. There are *three* final particles, namely two electrons and one singly-charged helium ion. (This is, of course, not the only possible outcome of the collision event. The helium atom might lose both its electrons in the event, or it might lose none. Furthermore the event can lead to the emission of one, or several, photons.)

27 If we now recognize that because of the relations (24a) there is associated a frequency and a wave vector with each one of the initial and final particles, we can rewrite the conservation laws (26a) in the form

$$\sum_{r=1}^{i} \omega'_r = \sum_{s=1}^{j} \omega''_s, \qquad \sum_{r=1}^{i} \mathbf{k}'_r = \sum_{s=1}^{j} \mathbf{k}'_s \qquad (27a)$$

The sum of the initial frequencies equals the sum of the final frequencies, and the sum of the initial wave vectors equals the sum of the final wave vectors. These conservation laws are completely equivalent to the conservation laws (26a). Each set of laws implies the other. This is so because there is only one Planck's constant.†

Can Matter Waves Be Split?

28 In the preceding chapter we discussed the sense in which photons can, and cannot, be "split." We should now undertake a similar discussion for matter waves. We can be quite brief, because the behavior of matter waves is quite analogous to the behavior of photons. In this respect nature is simple.

To be specific we will now discuss electrons, but our results are completely general and apply equally well to any other particle.

In the preceding chapter we concluded that a monochromatic photon, of frequency ω, cannot be split in the sense that we could detect, with a photocell, a "fractional photon," which carries only some fraction of the energy $\hbar\omega$. In an analogous sense an electron cannot be split either, because nobody has ever detected a "fractional electron."

---

† To the reader with an advanced knowledge of quantum mechanics: it may seem as if the relations (27a) could be derived independently, on the basis of the homogeneity of physical space. Such a derivation can indeed be given, *provided* we accept certain ideas characteristic of quantum mechanics. On the other hand it is clear that no purely logical argument could tell us that there is a de Broglie wave associated with protons, given that electrons have wave properties. Similarly pure logic cannot tell us that the constant C must be the same for all particles. Momentum and wave vector have independent operational definitions, and they need not necessarily be connected by the de Broglie relation.

29 Let us consider an electron-diffraction experiment, shown schematically in Fig. 29A. The beam of electrons incident on the crystal surface has a well-defined momentum. The reflected electrons are detected by four counters, C_1 to C_4, and we imagine that counters C_1 and C_4 are located at different diffraction maxima, whereas C_2 and C_3 are located at diffraction minima.

The first thing to note is that, experimentally, the counting rate of each counter remains proportional to the incident flux of electrons as this flux tends to zero. This circumstance rules out any explanation of the observed diffraction phenomena as collective effects involving a large number of electrons: it is really each *individual* electron which shows the wave behavior. To make the issues clear we may assume that the counting rates in counters C_1 and C_4 are equal, and that the counting rates in counters C_2 and C_3 are zero.

Suppose now that we regard an electron as a *classical* wave packet. We then expect that this wave is "split" in the reflection at the crystal: part of the wave is reflected in the direction of counter C_1, and part in the direction of counter C_4, but nothing in the direction of counters C_2 and C_3. Since the original incident packet is "split" in this manner we would expect that this "splitting" manifests itself in some way. For instance: that the energy carried by the "part" reflected toward counter C_1 is some fraction of the energy of the incident electron. This is, however, *not* what is found experimentally, as we may recall from Davisson's own account: the reflected electrons carry the *full* energy of the incident electrons. If an electron is detected by a counter at all, it is the *whole* electron which is detected, with the full electronic charge and electronic mass. As we have said nobody has ever seen a third of an electron. Electrons have wave properties but they are most certainly not classical waves: the electron-wave packet cannot be split like a classical wave packet.

30 Now it may happen that the reader does not have any firm opinions about the properties of a "classical wave," and therefore the statement that an electron is *not* a classical wave might appear a bit colorless. What we have in mind here is that for a classical wave the absolute square of the wave amplitude at a given instant of time, and at a given point in space, represents a physical quantity such as a charge density, or an energy density. This idea is analogous to the idea in classical electromagnetic theory that the squares of the electric and magnetic fields represent energy densities.

Suppose, for instance, that the square of the wave amplitude is proportional to the charge density. We could then compute the flux of charge into one of the counters, and since the wave is

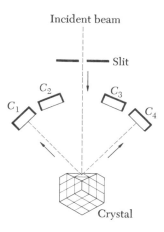

Fig. 29A Schematic figure (to illustrate discussion in Sec. 29) showing set-up to observe electron diffraction in various directions from the surface of a crystal. Since the incident wave is "split" by the crystal, will we find only half an electron with counter C_1?

"divided" between the counters C_1 and C_4 we would expect to find only half the electronic charge in counter C_1. This might be true *on the average:* if we perform the diffraction experiment with a very large number of electrons the flow of charge into counter C_1 may indeed be one-half of the total incident flow of charge.† However, each individual electron is either detected by counter C_1 or by counter C_4: the charge of an individual electron is *not* split.

In the spirit of quantum mechanics we describe what happens as follows. The incident electron wave is divided into two parts by the crystal. One part of the wave propagates in the direction of counter C_1, and the other part in the direction of counter C_4. The *intensity* of the wave in a given direction is proportional to the absolute square of the wave amplitude. In quantum mechanics the intensity has a *probabilistic* interpretation: a quantity which depends quadratically on the amplitude always represents the *probability* for something to happen. The classically computed flux into one of the counters is proportional to the probability that the counter will click.

This probability interpretation of intensities is a distinguishing feature of quantum mechanics, and it is clearly contrary to the spirit of a classical wave theory.

31 In analogy with our discussion in Sec. 47, Chap. 4, the reader should consider an imaginary experiment in which the arrangements are as in Fig. 29A, but with the counters at a very large distance, say one light year, from the crystal. Suppose that an electron is detected by counter C_1. On the basis of the classical wave theory it would again be hard to understand how the physical quantities, such as charge, energy and mass, carried by the wave could suddenly be concentrated in the counter C_1 after having first been spread out over a large region in space. With the quantum-mechanical probability interpretation this difficulty evaporates: we can describe what happens in a consistent way.

32 We have said that the wave is divided into two (or several) "parts" in the diffraction experiment shown in Fig. 29A. The reader may then ask: Can the wave traveling in the direction of counter C_1 be made to interfere with the wave traveling in the direction of counter C_4? If an electromagnetic wave is divided by a half-silvered mirror, then the two "parts" can certainly interfere with each other, and we expect the same behavior from the de Broglie waves. In other words, if we somehow deflect the wave

---

†This may not be true in practice, but, for the sake of argument, we can assume that every incident electron goes into either counter C_1 or counter C_4.

traveling in the direction of counter C_4, and "mix" it with the wave traveling in the direction of counter C_1, will we then see interference effects?

The answer is that we certainly expect to see interference effects. On the other hand it must be admitted that it would be very difficult in practice to perform this experiment exactly as described, with electrons. Fortunately we do not have to do this experiment, because the very fact that we can observe electron diffraction at all with the crystal is conclusive proof of the reality of the interference effects. Each atom in the crystal surface gives rise to a diffracted wave when "illuminated" by the incident wave, and all these diffracted waves combine to produce the overall interference pattern which we observe with the crystal. What does it mean that the waves diffracted by the individual atoms "combine?" How do we describe the "combination?" We describe it by adding the *amplitudes* of all the separate waves to obtain the total amplitude of the wave emerging from the crystal. The square of this resultant amplitude is an intensity variable, to be understood quantum mechanically, which describes the response of a detector.

33   In Secs. 39–42, Chap. 4, we discussed a two-slit diffraction experiment with photons. Suppose we perform the same experiment with electrons. The arrangement is shown schematically in Fig. 33A. This figure is identical, except for the text, with Fig. 39A, of Chap. 4. The analysis is also the same, and the intensity $I(r,\theta)$, observed at a distance large compared with the separation between the slits, is given by

$$I(r,\theta) = 4I_0(r,\theta) \cos^2 \left(\frac{2\pi a}{\lambda} \sin \theta \right) \qquad (33a)$$

where $I_0(r,\theta)$ is the intensity which we would observe with only *one* slit open.

The dependence of the intensity on the angle θ can be determined by counters, and the intensity is then simply proportional to the counting rate when the experiment is performed with a *beam* of electrons.

Experiments entirely analogous to this oversimplified imaginary experiment have been performed, and we can conclude from the results that the prediction of Eq. (33a) is correct.†

† G. Möllenstedt and H. Düker, "Beobachtungen und Messungen an Biprisma-Interferenzen mit Elektronenwellen," *Zeitschrift für Physik* **145**, 377 (1956). See also R. G. Chambers, "Shift of an electron interference pattern by enclosed magnetic flux," *Physical Review Letters* **5**, 3 (1960). This latter paper reports on a very interesting effect, which we shall not discuss in this book, but which the reader may want to study on his own.

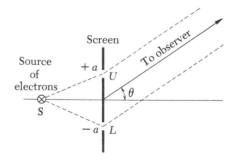

Fig. 33A   An imaginary experiment on two-slit diffraction of electrons. This figure is the same as Fig. 39A in Chapter 4, except that the light source S is replaced by a source of electrons.

34 If we are to see the interference effect *both* slits must be open and each electron must thus pass through both slits. If we want to make sure that the electron passes through only *one* of the slits we must close the other slit, but then we will naturally not see the two-slit diffraction pattern. If we try to find out through which slit the electron went by placing counters immediately behind the slits we will also destroy the interference pattern. The counting rates observed in the two counters will be the same. For each electron incident on the screen one, but only one of the counters may click, and the electron so detected will carry the full charge and the full energy of the incident electron. We cannot say beforehand *which one* of the counters will click, but we can compute and predict the probability for registration by finding the intensity of the wave penetrating the slit.

The reader should return to our discussion, in Sec. 48, Chap. 4, in which we prove that the two-slit pattern is incompatible with a knowledge of which slit the photon went through. The same argument applies to electrons. There is no clever arrangement whereby we can determine through which slit the electron went without disturbing the two-slit pattern.

35 Let us tighten up our language a bit. When we discussed the discovery of the de Broglie waves we talked about "waves associated with a particle." This is bad language because it sounds as if we would have a classical corpuscle traveling together with a wave in some fashion. Some people like to call the de Broglie waves "guide waves," or "pilot waves," but this terminology is also bad. The de Broglie waves are not waves traveling together with, and "guiding," a classical corpuscle. The de Broglie wave and the particle are the *same thing;* there is nothing else. The real particle, found in nature, has wave properties and that is a fact. If we want to emphasize this fact we might talk about the de Broglie wave of an electron, but this term is really a synonym for "electron." Our excuse for our previous bad language is that our discussion was at first tentative, as well as historical, and the cautious term "wave associated with a particle" was therefore defensible. The time has now come for us to be more precise and definite, and we should reject a terminology which might lead our thinking astray.

Consider again the two-slit experiment. There is nothing in this experiment which suggests to us that there might be a classical corpuscle passing through one of the slits, "guided" by a wave which passes through both slits. Better stated: Our description of what takes place in no way improves if we try to introduce this idea. It is quite enough to discuss the wave only, with the quantum-mechanical interpretation of intensities as probabilities. Any talk about the "hidden" corpuscle is metaphysics unless the assumption that the corpuscle exists has some definite experimental consequences which cannot be predicted on the basis of the quantum-mechanical wave theory alone. No such experimental circumstances are known, and in view of this we must firmly reject all mental images of classical corpuscles guided by waves.

The Wave Equation and the Superposition Principle

36 We now want to present arguments in favor of a differential equation, known as the Klein-Gordon equation, in terms of which we can describe the propagation of matter waves in *empty* space.

Our most important assumption is that the wave equation which describes a *single* particle, of mass m, shall be a *linear* differential equation. This means that the solutions of the equation satisfy a *principle of superposition:* Any linear combination of two solutions of the equation is again a solution of the equation. Furthermore we assume that every solution of the equation which satisfies certain mild conditions represents a possible physical situation, at least in principle. The physical implications of these assumptions are far reaching. Amplitudes of matter waves can be added just like amplitudes of electromagnetic waves. (Maxwell's equations are also *linear* differential equations.)

The reader should note that in our discussion of diffraction of matter waves by atoms on a crystal surface, or by a double slit, we have tacitly assumed linearity. We thus have added, for instance, the amplitudes of the waves emerging from the two slits to obtain a resultant amplitude. We here elevate this procedure to a general principle of physics.

37 Let us now find a differential equation satisfied by *all* matter waves describing a particle of mass m. The procedure is as follows. We first find a differential equation satisfied by all *plane* waves, of the form

$$\psi(\mathbf{x},t;\mathbf{p}) = \exp\,(i\mathbf{x} \cdot \mathbf{p} - i\omega t) \tag{37a}$$

We employ units such that $\hbar = c = 1$, and we denote the momentum (= wave vector) by $\mathbf{p}$, and the energy (= frequency) by ω. Each such plane wave is determined (except for a constant factor which fixes the amplitude of the wave) by the momentum $\mathbf{p}$. We try to write a *linear* differential equation in which $\mathbf{p}$ does not occur explicitly, and which is satisfied by *every* plane wave. Since it is linear this differential equation will then be satisfied by every linear combination of plane waves, and therefore, we shall argue, by *every* de Broglie wave describing the particle of mass m.

The energy ω and the momentum $\mathbf{p}$ are related by

$$\omega^2 - \mathbf{p}^2 = m^2 \tag{37b}$$

since the mass of the particle is m.

If we differentiate the wave function ψ twice with respect to the time t we obtain

$$\frac{\partial^2}{\partial t^2}\, \psi(\mathbf{x},t;\mathbf{p}) = -\omega^2\psi(\mathbf{x},t;\mathbf{p}) \tag{37c}$$

If we differentiate the wave function twice with respect to the coordinate x_1 we obtain

$$\frac{\partial^2}{\partial x_1{}^2}\, \psi(\mathbf{x},t;\mathbf{p}) = -p_1{}^2\psi(\mathbf{x},t;\mathbf{p}) \tag{37d}$$

and similarly for the second derivatives with respect to the other two space coordinates x_2 and x_3.

We thus obtain, taking into account the relation (37b),

$$\frac{\partial^2}{\partial t^2}\, \psi(\mathbf{x},t;\mathbf{p}) - \boldsymbol{\nabla}^2\psi(\mathbf{x},t;\mathbf{p}) = -m^2\psi(\mathbf{x},t;\mathbf{p}) \tag{37e}$$

where $\boldsymbol{\nabla}^2$ stands for the *Laplacian operator*, defined by

$$\boldsymbol{\nabla}^2 \equiv \frac{\partial^2}{\partial x_1{}^2} + \frac{\partial^2}{\partial x_2{}^2} + \frac{\partial^2}{\partial x_3{}^2} \tag{37f}$$

The equation (37e) is the desired wave equation. As we see, this equation is satisfied by *all* plane waves of the form (37a) i.e., for *all* momenta $\mathbf{p}$, and hence is also satisfied by any de Broglie wave which is a superposition of plane waves.

38 The wave equation (37e) is known as the Klein-Gordon equation. It is, in a sense, the simplest differential equation satisfied by the de Broglie waves. Note that this equation is also satisfied by electromagnetic waves in empty space, with the photon mass $m = 0$. The reader can easily convince himself that there cannot be any differential equation of *first* order, i.e., involving only first derivatives with respect to the independent variables, satisfied by *all* de Broglie waves. The equation must be at least of second order, and the reason is that the relation (37b) between energy and momentum is a quadratic algebraic relation.

We must repeat again, because this is an important point, that the equation (37e) can describe the propagation of a particle only in *empty* regions in space-time, i.e., far away from all other particles. Similarly the *homogeneous* Maxwell's equations, i.e., with the current density and charge density both equal to zero, describe the propagation of electromagnetic waves only in regions free of charges and currents, i.e., in regions free of other particles.

39   A superposition of two plane waves, i.e., a wave of the form

$$\psi(\mathbf{x},t) = A' \exp{(i\mathbf{x} \cdot \mathbf{p}' - i\omega't)} + A'' \exp{(i\mathbf{x} \cdot \mathbf{p}'' - i\omega''t)} \quad \text{(39a)}$$

where A' and A'' are two arbitrary complex constants, also satisfies the differential equation (37e).   In other words,

$$\frac{\partial^2}{\partial t^2} \psi(\mathbf{x},t) - \boldsymbol{\nabla}^2 \psi(\mathbf{x},t) = -m^2 \psi(\mathbf{x},t) \quad \text{(39b)}$$

Let us consider a more general (continuous) superposition of plane waves, of the form

$$\psi(\mathbf{x},t) = \int_{(\infty)} d^3(\mathbf{p}) \, A(\mathbf{p}) \exp{(i\mathbf{x} \cdot \mathbf{p} - i\omega t)} \quad \text{(39c)}$$

Here $A(\mathbf{p})$ is a complex-valued function of the vector $\mathbf{p}$.   The integral is over the entire three-dimensional $\mathbf{p}$-space.   The quantity ω is a function of $\mathbf{p}$ such that $\omega > 0$, and such that the equation (37b) is satisfied.   In other words,

$$\omega = \omega(\mathbf{p}) = \sqrt{\mathbf{p}^2 + m^2} \quad \text{(39d)}$$

The wave function $\psi(\mathbf{x},t)$ defined by the integral in Eq. (39c) also satisfies the differential equation (39b).   This is a very general de Broglie wave, in fact the most general such wave.   We assume, of course, that the function $A(\mathbf{p})$ is a reasonably well-behaved function of $\mathbf{p}$ so that the integral in Eq. (39c) makes good sense.

40   In the theory of the Fourier integral the following theorem can be proved: If $\psi(\mathbf{x},0)$ is any reasonably well-behaved function of $\mathbf{x}$, and if we define a function $A(\mathbf{p})$ by the integral

$$A(\mathbf{p}) = (2\pi)^{-3} \int_{(\infty)} d^3(\mathbf{x}) \, \psi(\mathbf{x},0) \exp{(-i\mathbf{x} \cdot \mathbf{p})} \quad \text{(40a)}$$

then it follows that

$$\psi(\mathbf{x},0) = \int_{(\infty)} d^3(\mathbf{p}) \, A(\mathbf{p}) \exp{(i\mathbf{x} \cdot \mathbf{p})} \quad \text{(40b)}$$

This is a theorem.   Its precise statement, and its proof, hinges on an appropriate definition of "a reasonably well-behaved function."

We shall not prove this theorem here, and we shall not really depend on the theory of the Fourier integral in our discussion in this book. In due time the reader will learn how to state this theorem precisely, and how to prove it, in her calculus course. Our aim here is to discuss the physical implications of the theorem, and thereby provide the reader with a strong "physical" motivation for learning about the Fourier integral. It is of central importance in physics.

41　Let us now see what the theorem implies. Suppose that $\psi(\mathbf{x},0)$ is a de Broglie wave function at the time $t = 0$. We can then associate with this wave function an amplitude $A(\mathbf{p})$ in momentum space, through the integral in Eq. (40a). In terms of the momentum space amplitude $A(\mathbf{p})$ we can then define a new wave function $\psi_1(\mathbf{x},t)$ by

$$\psi_1(\mathbf{x},t) = \int_{(\infty)} d^3(\mathbf{p})\, A(\mathbf{p}) \exp\left(i\mathbf{x} \cdot \mathbf{p} - i\omega t\right) \tag{41a}$$

If we set $t = 0$ in the above expression and compare with the formula (40b) we see that $\psi_1(\mathbf{x},0) = \psi(\mathbf{x},0)$. The new wave function $\psi_1(\mathbf{x},t)$, which satisfies the Klein-Gordon equation (39b), is thus identical with the wave function $\psi(\mathbf{x},0)$ at the "initial time" $t = 0$. This means that we have a procedure for solving the Klein-Gordon equation subject to the *initial condition* that the solution shall agree with a given function (of $\mathbf{x}$) at time $t = 0$.

42　Let us consider the question of the uniqueness of the solution of the Klein-Gordon equation which we find in this manner. What is true is that our procedure, according to which we construct the functions $A(\mathbf{p})$ and $\psi_1(\mathbf{x},t)$ from a given function $\psi(\mathbf{x},0)$, is a definite procedure which leads to a unique function $\psi_1(\mathbf{x},t)$ satisfying the equation (39b). The question arises whether there might not be other solutions of the differential equation (39b) which also agree with $\psi(\mathbf{x},0)$ at time $t = 0$. The answer is yes. The differential equation (39b) is also satisfied by wave functions of the form

$$\psi'(\mathbf{x},t) = \exp\left(i\mathbf{x} \cdot \mathbf{p} + i\omega t\right), \qquad \omega = \sqrt{\mathbf{p}^2 + m^2}$$

We call these solutions "negative frequency solutions," in contrast to the "positive frequency solutions" of the form (37a).

We exclude the negative frequency solutions on physical grounds. They do not represent particles of positive energy (= positive frequency). Now it is clear that for every positive frequency solution of the equation (39b) there also exists a negative frequency solution of the same momentum $\mathbf{p}$, and the Klein-Gordon equation therefore has twice as many solutions as we want. The reason for this is that the equation (37b) has *two* solutions ω for every **p**; one positive and one negative. Only the positive solution has physical meaning: the energy of a particle is a positive quantity.

The Klein-Gordon equation (39b) is therefore not the whole story of the de Broglie waves. We have to add the condition that *all negative frequency (= negative energy) solutions are to be excluded.* With this provision it can be proved that every permissible solution of the equation (39b) is uniquely determined by its values at $t = 0$, and that answers the question which we asked. We shall not prove this theorem here.

43 The important idea which emerges from our discussion is that every physically acceptable de Broglie wave function $\psi(\mathbf{x},t)$ can be represented in the form (41a) where $A(\mathbf{p})$ is uniquely determined, through Eq. (40a), by the wave function at some specific time, say time $t = 0$. Every matter wave can thus be regarded as a super-position of plane matter waves. If we like we can regard this as our basic assumption, and thus down-grade the importance of the Klein-Gordon equation. It is merely a nice differential equation satisfied by the physically acceptable wave functions.

44 By a suitable choice of the momentum space amplitude $A(\mathbf{p})$ in the Fourier integral (39c) [or (41a)] we can construct wave packets which are approximately localized in some region in space at a given time. Such a wave has the property that it is appreciable only in some limited region in space, but decreases rapidly to zero as $|\mathbf{x}|$ tends to infinity. A wave packet of this kind represents a particle which is approximately confined to a finite region in space. It is clear that all particles studied experimentally must be described by such wave functions. We assume, of course, that the particle is most likely to be found (when we look for it with a counter) in those regions in space where the wave function is large. This is in accordance with our quantum-mechanical interpretation of the absolute square of the amplitude: it has something to do with the probability for something to happen. For the time being it suffices for us to assume that "the particle is most likely to be found where the wave function is large." Later we will discuss a particular kind of wave function, for which we will state a precise prescription for how to compute the probability of finding the particle in a region.

We can conclude that a single plane wave cannot represent a particle in an actual experiment. For such a wave the absolute square of the amplitude is a constant, independent of **x** (and t), and the probability of finding the particle in any region of unit volume is independent of the location of the region. Since space can be built up of an infinite number of such regions it follows that the probability of finding the particle in any specific one of these must be zero. The probability of finding the particle within *any* finite region is also zero, which does not make physical sense.

Strictly monochromatic plane waves therefore do not exist. It is possible, however, for a wave to look like a plane wave of constant amplitude over an arbitrarily large region of space, although the amplitude does tend to zero outside this region. If this region includes the region in which the physical phenomena under study take place, we can think about the wave as an idealized plane wave. It is common practice in physics to talk about plane waves, and it is then understood that such a wave is approximately plane; it looks like a plane wave over a very large region in space.

45 The Klein-Gordon equation (39b) is satisfied by every wave function describing the state (of motion) of a particle of mass m. If we set $m = 0$ we obtain an equation satisfied by both the electric and magnetic vector fields of electromagnetic theory. The Klein-Gordon equation is, however, not identical with Maxwell's equations, and the reader may be bothered by this fact. Is it possible that Maxwell's equations say more than the Klein-Gordon equation? The answer is yes. Maxwell's equations also describe the *polarization* of the photon. The state of motion of a photon is not fully specified by giving its momentum and energy; we also have to give its polarization. For every momentum there are *two* linearly independent states of polarization, say the state of left-circular polarization and the state of right-circular polarization.

The question arises: can material particles also exhibit different states of polarization? The answer is that some particles do and some do not. The pion and the alpha-particle are examples of particles which cannot be polarized. The electron, the proton and the neutron are examples of particles which can be polarized. These latter particles all have an intrinsic angular momentum, or spin, and to the different orientations of the spin correspond different states of polarization. The pion and the alpha-particle, on the other hand, have no spin: there is nothing that defines a direction in the rest frame of these particles. They are spherically symmetric.

To describe the state of polarization of the particles which have a non-zero spin we must introduce, besides the variables x and t, a variable which describes the spin. The wave equation which describes the full story of electrons, protons and neutrons, is therefore more complicated than the Klein-Gordon equation (39b) but the wave function nevertheless *also* satisfies the Klein-Gordon equation. This equation describes, so to speak, the behavior of the particle in space-time without regard to the spin. We shall not discuss the quantum-mechanical description of polarization here. It is somewhat analogous to the description of the polarization of an electromagnetic wave.

46 In concluding this part of the chapter, let us restate the wave equation (39b) for the case that we employ the cgs- (or MKS-) system of units. The constants $\hbar$ and c are easily restored, and we have

$$\frac{1}{c^2}\frac{\partial^2}{\partial t^2}\,\psi(\mathbf{x},t) - \boldsymbol{\nabla}^2\,\psi(\mathbf{x},t) = -\left(\frac{mc}{\hbar}\right)^2\psi(\mathbf{x},t) \qquad (46a)$$

The reader should convince himself, through a dimensional argument, that this equation is correct. Note that each term has the dimension (wave function)/(length)^2. The restoration of the constants $\hbar$ and c is unique.

Advanced Topic: The Vector Space of Physical States†

47 Let us state concisely the Principle of Superposition which we assume holds for the matter waves.

Let $\mathcal{H}'$ be the set of all wave functions ψ which do not vanish identically, and which represent possible physical states of a particle of mass m. To this set of wave functions we add the wave function which is identically zero everywhere in space, at all times. The resulting set is denoted $\mathcal{H}$. This set has the following properties:

(a) If ψ_1 and ψ_2 are two wave functions in the set $\mathcal{H}$, then the sum $(\psi_1 + \psi_2)$ is also in $\mathcal{H}$.

(b) If ψ is in $\mathcal{H}$, and if c is any complex number, then the function $c\psi$ is also in $\mathcal{H}$.

The principle of superposition of wave functions specifically says that if ψ_1 and ψ_2 are two physically meaningful wave functions, and if c_1 and c_2 are any two complex numbers, then the function

$$\psi = c_1\psi_1 + c_2\psi_2 \qquad (47a)$$

is also a physically meaningful wave function provided that it does not vanish identically.

---

† Can be omitted in a first reading.

48 The properties of the set $\mathcal{K}$ are characteristic properties of an abstract mathematical object, namely an *abstract complex vector space*. Let us state the postulates in terms of which this object is defined.

A linear complex vector space $\mathcal{K}$ is a set of elements, called vectors, such that:

I. For any two vectors ψ_1 and ψ_2 in $\mathcal{K}$ there exists a unique vector ψ in $\mathcal{K}$, called the sum of ψ_1 and ψ_2, and denoted $\psi = \psi_1 + \psi_2$. The operation of forming the sum of two vectors satisfies the conditions

(a) $\psi_1 + \psi_2 = \psi_2 + \psi_1$, for any two ψ_1, ψ_2 in $\mathcal{K}$.

(b) $\psi_1 + (\psi_2 + \psi_3) = (\psi_1 + \psi_2) + \psi_3$, for any three ψ_1, ψ_2, ψ_3 in $\mathcal{K}$.

(c) There exists a unique vector 0 in $\mathcal{K}$, called the null vector, such that $\psi + 0 = \psi$ for every ψ in $\mathcal{K}$.

II. Given any vector ψ in $\mathcal{K}$, and any complex number c, there exists a unique vector in $\mathcal{K}$, denoted $c\psi$, called the product of the vector ψ by the scalar c. The operation of multiplying a vector by a scalar ($=$ a complex number) satisfies the conditions

(a) $(c_1 c_2)\psi = c_1(c_2\psi)$ for any vector ψ, and any two scalars c_1 and c_2.

(b) $(c_1 + c_2)\psi = c_1\psi + c_2\psi$ for any vector ψ, and any two scalars c_1 and c_2.

(c) $c(\psi_1 + \psi_2) = c\psi_1 + c\psi_2$ for any two vectors ψ_1 and ψ_2, and any scalar c.

(d) For the particular scalar 1 we have $1\,\psi = \psi$.

These are the postulates which define an abstract linear vector space over the field of complex numbers. The latter phrase means that the scalars by which the vectors may be multiplied are the complex numbers. If the scalars are restricted to be real numbers we speak of a linear vector space over the field of real numbers. For brevity we simply say "complex vector space," respectively "real vector space." The reader has already encountered an example of a real vector space, namely the three-dimensional Euclidean "physical space."

49 The postulate Ia is the commutative law for addition; the postulate Ib is the associative law for addition; postulate Ic concerns the existence and uniqueness of the null vector. The postulate IIa is the associative law for scalar multiplication, and the postulates IIb and IIc are distributive laws for scalar multiplication. Postulate IId says that the identity times a vector equals the vector.

From these postulates we may prove many almost self-evident

facts, such as

$$0\psi = 0, \quad (-1)\psi + \psi = 0, \quad (-c)\psi = -(c\psi), \quad \text{etc.}$$

We shall not list all the trivial theorems here, as the author is confident that the reader's intuition will not lead him astray.

What is the virtue of introducing the concept of an abstract complex vector space? The answer is that in our study of mathematical theories we encounter over and over again sets of elements which, among any other properties which these sets may have, have the particular property of satisfying all the axioms of an abstract complex vector space. When we encounter such a set we need not list the properties of an abstract vector space anew but we can simply say that the set is a complex vector space, and then everyone who knows the axioms for a vector space immediately knows quite a bit about the set.

50 We can now state that the set $\mathcal{3C}$ of all physically acceptable wave functions, with the identically vanishing wave function adjoined, is a complex vector space. It is a *concrete* complex vector space because the vectors are actually "tangible" complex functions of space and time. If we compare the postulates given in Sec. 48 with the properties of the set of all wave functions which we mentioned explicitly in Sec. 47 we notice that the list in Sec. 48 is longer. However, many of the postulates for an abstract vector space are trivially satisfied by the set of concrete wave functions, and there was no need to state these trivialities explicitly.

51 Let us note that in the definition of the abstract complex vector space there is no mention of the *dimensionality* of the vector space: it may be finite-dimensional or infinite-dimensional. Let us elaborate on this question a bit.

A set of N vectors $\psi_1, \psi_2, \ldots, \psi_N$ in a complex vector space $\mathcal{3C}$ are said to be *linearly independent* if the equation

$$\sum_{n=1}^{N} c_n \psi_n = 0 \tag{51a}$$

implies that $c_1 = c_2 = \ldots = c_N = 0$: otherwise they are said to be *linearly dependent*.

A complex vector space is of dimensionality N if it is possible to find a set of N linearly independent vectors in the space, but impossible to find a set of more than N linearly independent vectors. The vector space is infinite-dimensional if it is possible to find a

Paul Adrien Maurice Dirac. Born 1902 in Bristol, England. Dirac first studied electrical engineering, but later switched to theoretical physics. In 1932 he was appointed Lucasian Professor of Mathematics at Cambridge. He received the Nobel prize in 1933.

Dirac has made many important contributions to the development of quantum mechanics and quantum electrodynamics. His celebrated relativistic theory of the hydrogen atom led him to a theory of anti-particles which received its spectacular confirmation with Anderson's discovery of the positron.

In the early days of quantum mechanics Dirac did much to develop the algebraic approach to the theory. His ideas are presented in his book *The Principles of Quantum Mechanics*, 4th ed. (Oxford University Press, 1958). (Photograph by courtesy of *Physics Today*.)

set of N linearly independent vectors in the vector space for *every* integer N.

The vector space $\mathcal{3C}$ of all physically meaningful de Broglie wave functions is clearly *infinite*-dimensional; there is an infinite number of linearly independent wave functions.

52 We have been concerned with the solutions of the Klein-Gordon equation, but we can now conclude that if we consider the totality of the solutions of any *linear* differential equation, then this set is always a (complex) vector space. Many different kinds of linear differential equations have been proposed to describe, within quantum mechanics, the particles occurring in nature. The sets of all physically acceptable solutions of these equations always form vector spaces.

We can state this as follows: To describe a given kind of particle one may introduce a complex vector space, and associate a vector in this vector space with every possible state (of motion) of the particle.

This idea, which is the essence of the mathematical theory of quantum physics, is a great idea. At first it might not seem so; to say that a state (of motion) of a particle is described by a vector in a complex vector space might seem to be nothing more than a restatement of the principle of superposition satisfied by the solutions of the wave equation, and perhaps a restatement of questionable value. As we penetrate further into quantum physics we discover, however, the great merits of this idea. For instance: By noticing that the wave functions form a vector space we can actually achieve a considerable simplification in many practical computational problems. The computational techniques appropriate to vector spaces are in a sense *algebraic* in character, and we are accordingly led to consider the algebraic aspects of the solutions of differential equations. It turns out that in many problems the algebraic methods are vastly superior (in computational economy from the human standpoint) to the direct solution of the differential equations, especially in problems characterized by special symmetries. In this book we will not be able to demonstrate that this simplification takes place. The author felt that it was nevertheless worthwhile to state this fact: the seemingly abstract theory of vector spaces leads to great simplifications in the solution of practical problems. One minor aspect of this simplification is a certain simplification in notation. (Questions of notation are, by the way, not always so minor. A bad notation retards progress, whereas a good notation advances it.)

53　Heisenberg's Matrix Mechanics is a particular formulation of quantum mechanics in which the vector space aspect of the theory is emphasized, whereas the wave equations play a secondary role. At first Heisenberg's theory appears to be very different from the wave theories, such as Schrödinger's wave mechanics, but the different kinds of theories are in fact completely equivalent, and lead to the same physical predictions. They have a common abstract skeleton, and this skeleton is the abstract vector space theory. Since we cannot assume that the reader has already learned about matrices in his mathematics courses we will have to omit the discussion of Heisenberg's theory in this book. The theory is not particularly difficult, but since there are so many other things for the reader to learn we do not want to load the discussion with a presentation of matrix theory.

Werner Heisenberg's first paper on the subject appeared in 1925.[†] In this paper matrix theory is not mentioned explicitly because Heisenberg had not realized that his mathematical operations had a matrix theory interpretation. The connection with matrix theory was soon thereafter clarified in an important paper by Max Born and Pascual Jordan.[‡]

54　The reader should note that historically matrix mechanics was invented and developed *before* Schrödinger had invented his wave mechanics. We have said that it is a natural idea to regard the set of all solutions of a linear differential equation as a vector space, and thereby be led to consider algebraic aspects of the equation. There is no doubt that had Schrödinger's wave mechanics been invented first, matrix mechanics would soon have been discovered as a reformulation of the wave theory. This was, however, not the way things actually happened. The historical sequence of events is almost incredible to this author, and he regards the invention of matrix mechanics as one of the most astounding accomplishments in physical theory.

That matrix mechanics and wave mechanics are physically equivalent was proved by Schrödinger in 1926.[§]

Werner Karl Heisenberg. Born 1901, in Würzburg, Germany. Heisenberg studied under Sommerfeld at the University of Munich, and received his PhD degree in 1923. After a fruitful period as Born's assistant at the University of Göttingen, Heisenberg spent three years at Bohr's institute at Copenhagen. He later held positions at the University of Leipzig, and at the Max Planck Institute for Physics in Berlin. Since 1946 he has been director of the Max Planck Institute for Physics in Göttingen. He was awarded the Nobel prize in 1932.

Among Heisenberg's many important contributions to theoretical physics his discovery of matrix mechanics stands out as a most remarkable intellectual achievement. *(Photograph by courtesy of Physics Today.)*

[†] W. Heisenberg, "Über quantentheoretische Umdeutung kinematischer und mechanischer Beziehungen," *Zeitschrift für Physik* **33**, 879 (1925).

[‡] M. Born and P. Jordan, "Zur Quantenmechanik," *Zeitschrift für Physik* **34**, 858 (1925). The principles of quantum mechanics were further developed by these authors and by Heisenberg in M. Born, W. Heisenberg, and P. Jordan, "Zur Quantenmechanik II," *Zeitschrift für Physik* **35**, 557 (1926).

[§] E. Schrödinger, "Über das Verhältnis der Heisenberg-Born-Jordanschen Quantenmechanik zu der meinen," *Annalen der Physik* **79**, 734 (1926).

References for Further Study

1) For the history of the topics discussed in this chapter we again refer the reader to the books mentioned at the end of Chap. 1. (Items 3 and 5.)

2) There exists a vast literature on the mathematical theory of linear partial differential equations. The author does not expect the reader to go deeply into this theory at this time, but he wants to mention one treatise which has played an important role in physics, namely R. Courant and D. Hilbert: *Methoden der mathematischen Physik*, vols. I and II (Verlag von Julius Springer, Berlin, 1931 and 1937.) This work has been translated under the title: *Methods of Mathematical Physics*, vols. I and II (Interscience Publishers, Inc., New York, 1953 and 1962).

The second volume discusses partial differential equations. The first discusses a variety of subjects of interest in physics, such as Fourier analysis, the theory of matrices and vector spaces, the calculus of variations, and the theory of certain ordinary linear differential equations which appear in many physical problems.

It so happened that important developments in mathematics, which later were found to be "made to order" for quantum mechanics, took place about the time quantum mechanics was discovered. David Hilbert at the University of Göttingen played a central role in these developments, and the infinite-dimensional vector space in terms of which quantum mechanics is formulated today is called *Hilbert space*, after him. Hilbert did not originally develop his theory of linear spaces for physical applications, but the discovery of quantum mechanics naturally stimulated mathematical investigations of the questions raised by the physical applications. The period was one of considerable interaction between mathematicians and physicists.

The theory of quantum mechanics from a mathematician's standpoint is presented in J. von Neumann: *Mathematische Grundlagen der Quantenmechanik.* (Verlag von Julius Springer, Berlin, 1932. Reprinted by Dover Publications, New York, 1943.) An English translation has appeared under the title: *Mathematical Foundations of Quantum Mechanics* (Princeton University Press, 1955).

3) Matrix mechanics is discussed in most advanced textbooks on quantum mechanics. For an introductory account of quantum physics in which the algebraic approach is discussed and used we refer to R. P. Feynman, R. B. Leighton, and M. Sands: *The Feynman Lectures on Physics*, vol. III (Addison-Wesley Publishing Co., Inc., 1965). This is the final volume in a series of three books on basic physics. The presentation in these books is magnificent, and the reader is strongly advised to become acquainted with them.

4) We say very little about solid state physics in this volume. Among introductory books on this subject we mention C. Kittel: *Introduction to Solid State Physics* 3rd ed. (John Wiley and Sons, Inc., New York, 1966). Among many other topics the reader will find discussions of crystal structure, diffraction theory, and the theory of phonons.

Concerning crystals the reader should note the long article on the subject in the *Encyclopaedia Britannica* under the heading "crystallography."

5) The reader may find the following articles in the *Scientific American* interesting:

   a) K. K. Darrow: "The quantum theory," March 1952, p. 47

   b) K. K. Darrow: "Davisson and Germer," May 1948, p. 50

   c) E. Schrödinger: "What is matter?" Sept. 1953, p. 52

d) P. and E. Morrison: "The neutron," Oct. 1951, p. 44

e) G. Gamow: "The principle of uncertainty," Jan. 1958, p. 51

Problems

1 The *resolving power* of a microscope expresses the limit of our ability to see, with the microscope, fine details in the object under study. We can express the resolving power by the smallest distance between two points in the object such that the two points are clearly seen as two separate points. In an optical microscope the maximum possible resolution is clearly limited by the finite wavelength of the light used for illumination: we cannot expect to see features of the object which are much smaller than this wavelength. To overcome this limitation of the optical microscope, electron microscopes have been constructed. Instead of lenses made of glass an electron microscope employs suitably shaped electric and magnetic fields. Let us consider a typical electron microscope, in which the illumination is provided by a source of electrons of an energy of 50 keV. Compare the maximum possible resolving power of such an electron microscope with that of an optical microscope.

We must state that the actual resolving power achieved in a microscope (whether optical, or electron) is also dependent on certain design characteristics of the device, and these have to do with the magnitude of the angle within which "light" from the object is accepted by the microscope. For technical reasons this angle is much smaller in an electron microscope than in an optical microscope, and therefore the actual practical resolving power of the electron microscope is considerably smaller than the maximum theoretical resolving power. In spite of this the resolving power of an electron microscope is vastly superior to the resolving power of an optical microscope.

2 Consider helium gas at room temperature. This gas is monatomic. The average energy of a helium atom in a gas of temperature T is given by $E_{kin} = 3kT/2$, and from this expression we may find the average velocity (and momentum) of the helium atoms.

(*a*) Compute the average velocity (in cm/sec) of the helium atoms.

(*b*) Compute the de Broglie wavelength corresponding to this average velocity, in cm. Compare this wavelength with the mean spacing between the atoms in the gas. We assume that the pressure is one atmosphere, and the mean spacing may then be found from the known density.

One may suspect that quantum effects *could* play a role if the de Broglie wavelength is *larger* than the mean spacing, whereas a classical description should be adequate when the de Broglie wavelength is much *smaller* than the mean spacing. In the classical picture the gas is a collection of billiard balls, incessantly colliding with each other, whereas in the quantum-mechanical description the gas is a collection of interacting waves. It is therefore of great interest to carry out the above comparison for an actual gas.

(*c*) The density of liquid helium is about 0.15 gm cm^{-3}. Under atmospheric pressure this substance stays liquid at the lowest attainable temperatures. In analogy with the study in (*b*) compare the de Broglie wavelength with the mean separation at the very low temperature of 0.01°K.

3 Carry out the same comparison between the de Broglie wavelength and the mean spacing for a "gas" of electrons in a chunk of copper. There are models of a metal in which the electrons are regarded as forming a "gas," just like helium atoms in a container. Assume that there is one electron which can move freely in the lattice per copper atom. The spacing between the atoms is then the mean spacing between the electrons.

4 Consider a three-dimensional problem in which a particle is incident obliquely on the plane interface between two regions, R_1 and R_2. We assume that the potential energy of the particle has the constant value V_1 in most of R_1, and the constant value V_2 in most of R_2, except for the immediate vicinity of the interface in which the potential changes rapidly from V_1 to V_2. Inside the regions R_1 and R_2 the particle will therefore experience no forces, but it will experience a strong force in the neighborhood of the interface in the direction of the normal of the plane. Let us assume that the total energy of the particle is E, and that $E > V_1$ and $E > V_2$. The particle will then be refracted in the interface, and we wish to study the refraction both classically and quantum mechanically.

(*a*) Derive the law of refraction on the basis of classical mechanics. In this case there will be a change in the normal component of the momentum of the particle as it passes the interface, but no change in the lateral component. The energy principle gives the momentum in region R_2 if we know the momentum in region R_1, and we may thus derive the law of refraction.

(*b*) Derive the law of refraction on the basis of wave mechanics, and show that you can obtain the same result as in the classical case. In considering this problem within quantum mechanics you will again have to study the connection between the energy E, the momentum $\mathbf{p}$, the frequency ω, and the wave vector $\mathbf{k}$ of the particle. Our earlier discussion applied to the case of a region in which the potential is zero, and it might therefore not be valid in the present case. You are invited to present your ideas about how the theory is to be created. Questions to worry about include the following: Is the frequency the same on both sides of the interface? Must the tangential component of the wave vector be continuous at the interface? Is the relation $\mathbf{p} = \hbar \mathbf{k}$ always true? How about $E = \hbar \omega$?

In this particular problem you really know the answer: The law of refraction must be correctly given by the classical discussion in part (*a*). This helps in your search for good ideas: You know that your quantum mechanical theory *must* in this case yield a known result.

(*c*) According to classical dynamics the particle will not be *reflected* at the interface, but only *refracted*. Light incident on the interface between two different dielectrics is *both* reflected and refracted. Express your opinions about what the situation ought to be in the case of a quantum mechanical particle, i.e., a real particle.

5 We consider diffraction with a *ruled grating*, shown schematically in the adjoining figure. Such a grating consists of a large number of very fine parallel scratches of equal separation on a plane surface (made of glass, metal or plastic). For simplicity we shall regard this as a two-dimensional problem, which is permissible if we assume that the incident wave propagates in a direction contained in a plane perpendicular to the lines. The incident direction thus lies in the plane of the figure.

Let the incident wave be a plane wave of frequency (= energy) ω and wave vector (= momentum) $\mathbf{p}_i$. Find the possible directions of the diffracted waves, and show that they can be described as follows. A particle of momentum $\mathbf{p}_i$ collides with the grating. It emerges after the collision with momentum $\mathbf{p}_0$. The energy of the particle does not change in the collision, but an amount of momentum $\mathbf{q} = \mathbf{p}_i - \mathbf{p}_0$ is transferred to the grating. Show that the possible directions of the diffracted waves are determined by the simple rule that the component of the *momentum transfer* $\mathbf{q}$ along the grating, i.e., the vertical component in the figure, must be an integral multiple of $2\pi/a$, where a is the distance between the lines. The vertical component of the momentum transfer is "quantized."

6 (*a*) We consider the diffraction of visible light in a grating such as the one discussed in the preceding problem. Let the grating constant a be equal to twice the wavelength of the light, and let the angle of incidence be $45°$. Find all the angles at which the diffracted rays can emerge. Draw a figure.

(*b*) Let us modify the arrangement such that the grating is sandwiched between a plate of crown glass (refractive index 1.51) and a plate of flint glass (refractive index 1.74). Both plates are of a uniform thickness of 5 mm, and the crown glass plate is toward the side from which the light is incident. The wavelength, the grating constant and the angle of incidence are as in the first part of the problem. Find the directions at which the diffracted rays can emerge from the double-slab, and compare with the first part of the problem.

7 In a Davisson-Germer type experiment electrons of the energy 88 eV are incident normally on the surface of a metal crystal in which the atoms are arranged in a quadratic lattice of side length $a = 2.9$ Å. Draw a figure showing the points of intersection of the diffracted rays with a plane parallel

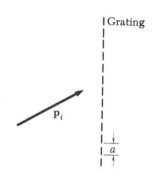

The figure above, which refers to Prob. 5, shows a diffraction grating schematically. The incident momentum $\mathbf{p}_i$ is represented by the vector. The spacing between adjacent lines of the grating is the grating constant a.

The figure below shows how the directions of the diffracted rays can be found by a simple geometric construction. The final momenta are determined by the intersections of a circle, corresponding to an unchanged magnitude of momentum, and a set of parallel lines corresponding to the allowed values of the vertical component of the momentum transfer to the grating. Ten possible final momenta, including the incident momentum, are shown by the vectors in the figure.

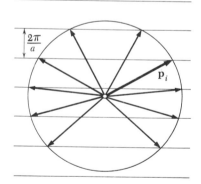

to the crystal surface, at a distance of 5 cm from the surface. This figure must be drawn in the correct scale, and all diffracted rays must be shown.

8 Once upon a time there was a physicist who performed experiments, such as described above, with a number of different metals. In his report of his results he said: "With metal A I observed a diffraction pattern of three-fold symmetry; with metal B a four-fold symmetry; with metal C a five-fold symmetry; and with metal D a six-fold symmetry." (The pattern is said to be of n-fold symmetry if it remains invariant under a rotation by a angle $2\pi/n$.) Evaluate this report in detail.

9 Neutrons from a reactor are made to pass through a column of (polycrystalline) beryllium. This material is chosen because it does not absorb neutrons appreciably. It is found that the neutrons which emerge at the other end are "cold," they have kinetic energies corresponding to temperatures below 50°K. The "warmer" neutrons, with kinetic energies corresponding to room temperature, are found to be strongly scattered out of the beam by the beryllium. Can you give an explanation of these phenomena?

10 Suppose that the wave function $\psi(\mathbf{x},t)$ is a positive-frequency solution of the Klein-Gordon equation (with mass m). We assume that this wave function represents a particle (wave packet) reasonably well concentrated in space, and moving in some more or less well-defined direction. Consider the function $\psi_R(\mathbf{x},t)$, defined by

$$\psi_R(\mathbf{x},t) = \psi(-\mathbf{x},t)$$

(a) Show that $\psi_R(\mathbf{x},t)$ is also a positive-frequency solution of the Klein-Gordon equation.

(b) The wave function $\psi_R(\mathbf{x},t)$ accordingly represents another state of motion of the particle. Describe *physically* how the state of motion described by $\psi_R(\mathbf{x},t)$ is related to the state of motion described by $\psi(\mathbf{x},t)$. (A nice and simple statement can be made. To guide yourself you can first think about the "average" trajectories in the two cases.)

11 The following problem is analogous to the Problem 10, but it is probably more difficult. Consider the function $\psi_T(\mathbf{x},t)$, defined by

$$\psi_T(\mathbf{x},t) = \psi^*(\mathbf{x}, -t)$$

where the star indicates the complex conjugate.

(a) Show that $\psi_T(\mathbf{x},t)$ is also a positive-frequency solution of the Klein-Gordon equation.

(b) Describe *physically* how the state of motion described by $\psi_T(\mathbf{x},t)$ is related to the state of motion described by $\psi(\mathbf{x},t)$.

Chapter 6

The Uncertainty Principle and the

Theory of Measurements

Chapter 6

The Uncertainty Principle and the Theory of Measurements

Heisenberg's Uncertainty Relations

1 In the preceding two chapters we have learned that the particles which occur in nature have wave properties. A moving particle with a well-defined momentum p can behave like a wave of wavelength $\lambda = h/p$, and this relation between wavelength and momentum is *universal*, i.e., valid for all real particles. We emphasized that one should not think about the wave properties in terms of a "guiding wave" associated in some manner with a classical corpuscle. A real physical particle is a single irreducible entity, and its wave properties and its corpuscular properties are manifestations of different aspects of its intrinsic nature.

2 We have learned that the state of motion of a particle can be described by a complex wave function $\psi(\mathbf{x},t)$. For an isolated particle this wave function satisfies the Klein-Gordon equation, subject to the additional condition that only positive frequencies will occur in the Fourier resolution of the wave function. As we have explained, it is possible to solve the Klein-Gordon equation subject to this condition, given an initial wave function $\psi(\mathbf{x},0)$ at time $t = 0$ (or at any other fixed time). The initial wave function is quite arbitrary, and we can therefore have a very wide variety of different waves corresponding to different states of motion of the particle. It is important to understand that a wave in quantum mechanics need not look like a sine wave: that is a very special case. The Klein-Gordon equation determines the time dependence of the wave function, but it imposes no restriction on the "appearance" of the wave at some *single* instant of time. It will, however, restrict the appearances of the wave at two different times. The wave function $\psi(\mathbf{x},t_1)$ at time $t = t_1$ uniquely determines the wave function at all other times, and thereby uniquely determines the state of motion of the particle. In this sense quantum mechanics is a deterministic theory.

3 Consider now a state of motion of a particle described by the initial wave function $\psi(\mathbf{x},0)$. What can we say about the position and momentum of the particle at time $t = 0$?

We have said that the wave amplitude must be given a probabilistic interpretation. The particle is most likely to be found in those regions in space in which the amplitude is large. More precisely

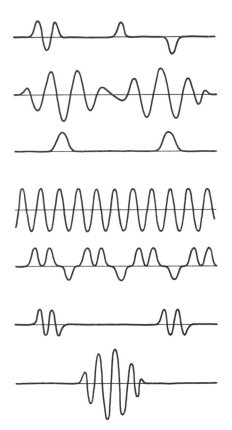

Fig. 2A An assortment of waves, shown to remind the reader that a wave in quantum mechanics need not look like a sine wave (at a given instant of time). An *arbitrary* wave can be an almost arbitrary function of position, and in its appearance it need not be as harmonious as the above waves. We have here plotted the real part of the (in general) complex wave function.

the absolute square of the wave amplitude at a point is a measure of the probability that the particle will be detected if we look for it with a (small) detector in a neighborhood of the point. If the initial wave function is such that the amplitude is zero except in a very small region, then we can say that the particle is *in* the region (at time $t = 0$): its position is accurately known. On the other hand, if the initial wave function is very much spread out such that its amplitude is approximately constant over a very large region we cannot assign a precise position to the particle: the position at time $t = 0$ is subject to a large uncertainty.

The idea that a precise position cannot in general be assigned to a particle (at a given time) follows naturally, as we have seen, from the wave picture. The precision with which the position is known depends on the state of motion of the particle. There is nothing that forbids a wave function (a state of motion) for which the position is known with extreme accuracy, and there is nothing that forbids another kind of wave function for which the position is not better known than to within a light year.

4 Analogous considerations apply to the momentum variable. Since momentum and wavelength are related by the de Broglie equation it is intuitively clear that the momentum cannot be well defined unless the wavelength is well defined. For the wavelength to be well defined it is necessary that the wave function exhibit some pattern of periodicity. A long sine wave has a well-defined wavelength, but for an arbitrary irregular curve the whole concept of wavelength has no precise meaning. We can therefore understand that the precision to which momentum is defined depends on the state of motion of the particle: it can be very well defined, or it can be very poorly defined.

It was realized by Heisenberg that whereas there are no limits to the accuracy to which *either* the momentum *or* the position can be defined, there is a fundamental limit to the accuracy to which position and momentum can be defined *at the same time* (i.e., for the *same* wave function). This insight finds its expression in the celebrated *uncertainty relations*, formulated by Heisenberg in 1927.† We shall now derive these relations through simple intuitive arguments.

5 Let us first consider de Broglie waves in a one-dimensional world. For simplicity we employ units such that $\hbar = 1$. Wave-

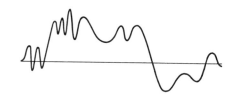

Fig. 4A An example of a wave train for which the concept of wavelength has very little meaning. For such a wave the momentum is very poorly defined. See also Fig. 2A: the momentum is poorly defined for all the waves shown except for the one in the middle.

---

† W. Heisenberg, "Über den anschaulichen Inhalt der quantentheoretischen Kinematik und Mechanik," *Zeitschrift für Physik* **43**, 172 (1927).

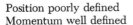

Position poorly defined
Momentum well defined

(A)

Position better defined
Momentum less well defined

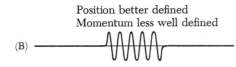

(B)

Position well defined
Momentum very poorly defined

(C)

Position very well defined
Momentum very poorly defined

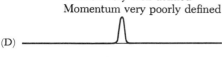

(D)

Figs. 5ABCD To illustrate our discussion of the position-momentum uncertainty relation. A well-defined position requires a short wave train. A well-defined momentum requires many well-developed sinusoidal cycles. The two requirements are in conflict with each other.

length and momentum are then related by $\lambda = 2\pi/p$, and we do not have to distinguish between wave vector and momentum.

We shall argue in terms of pictorial representations of waves, and for this purpose we have plotted four particular wave trains of finite length in Figs. 5ABCD. (The x-coordinate is the abscissa in these figures.) Now the reader should note that the wave function $\psi(x,0)$ is in general a complex-valued function, which fact creates problems when we want to represent it graphically. We can, however, plot the real part and the imaginary part of the function separately, and the reader can interpret the Figs. 5ABCD as showing either the real or imaginary parts of $\psi(x,0)$.

The graphs show "interrupted sine waves," described by the function $\sin(px)$ in the region in which the wave function does not vanish. The wave is, however, not *really* a pure sine wave, because it is "cut off" at both ends. For this reason the wavelength (and the momentum) are not precisely defined: these quantities can be precisely defined *only* if the wave is a pure sine wave.

Looking at the figures 5ABCD we can see very clearly that the better the position is defined, the more poorly is the momentum defined. Let us denote the uncertainty in the position x by Δx. As a rough measure of the uncertainty in position we may take the length of the wave train: if the wave train consists of n full waves we have

$$\Delta x \sim n\lambda = \frac{2\pi n}{p} \qquad (5a)$$

where λ is the wavelength. Now it is clear that the wavelength must be better defined the larger the number of full oscillations in the wave train. As a rough measure of the *fractional* uncertainty in the wave length we may take the quantity

$$\frac{1}{n} \sim \frac{\Delta\lambda}{\lambda} = \frac{\Delta p}{p} \qquad (5b)$$

where Δp is the uncertainty in the momentum. (Since $\lambda = 2\pi/p$ it follows that $\Delta\lambda/\lambda = \Delta p/p$.)

Combining (5a) with (5b) we obtain the order of magnitude relation

$$\Delta x \, \Delta p \sim 1 \qquad (5c)$$

We have here dropped the factor 2π because we are only interested in an order of magnitude estimate. Our definitions of Δx and Δp are not precise, but only qualitative, and because of this our result is also just a qualitative result.

6 The relation (5c) is the form the uncertainty relation takes for the particular kinds of waves shown in Figs. 5ABCD. The general uncertainty relation, which holds for *all* waves, is in the form of an *inequality*. To convince the reader of this fact we show another kind of wave in Fig. 6A. It is clear that for this wave the uncertainty in the position is about the same as in Fig. 5A. The uncertainty in the momentum (or wavelength) must, however, be much larger for the wave in Fig. 6A than for the wave in Fig. 5A. The correct position-momentum uncertainty relation must therefore be of the form

$$\Delta x \, \Delta p \gtrsim 1 \tag{6a}$$

The reader will recognize that this is the uncertainty relation which we discussed very briefly in Chap. 1.

Fig. 6A For the above wave train the position is as poorly defined as in Fig. 5A. The momentum is, however, *also* very poorly defined, and certainly much more poorly defined than in Fig. 5A. The correct uncertainty relation must be an inequality: it is possible to imagine wave trains for which the uncertainties in both momentum and position are arbitrarily large.

7 Let us next consider a wave in three-dimensional space. First of all we note that our discussion of the one-dimensional wave applies to each coordinate direction separately. Therefore, if x_α and p_α $(\alpha = 1, 2, 3)$ are the Cartesian position and momentum coordinates of the particle, we have

$$\Delta x_\alpha \, \Delta p_\alpha \gtrsim 1, \qquad \alpha = 1, 2, 3 \tag{7a}$$

On the other hand it is perfectly possible for the wave to be very well localized in *space* in, say, the x_1-direction, whereas the *momentum* of the wave is very well defined in the 2-direction. To see this the reader should think about a wave packet which is confined to a very small region around the 2-axis, but which has a large extent in the direction of this axis. The x_1-coordinate of the particle (= wave packet) is then very well known. In the direction of the x_2-axis we can, on the other hand, have a strictly periodic wave over a large distance along this axis, which means that the momentum p_2 can be very well defined. The precision with which the x_1-coordinate of the particle is known therefore places *no* limitation on the precision with which the momentum component p_2 can be known, and we have the general relations

$$\Delta x_\alpha \, \Delta p_\beta \gtreqless 0, \qquad \text{for } \alpha \neq \beta \tag{7b}$$

The inequalities (7a) and (7b) are the uncertainty relations for waves (= particles) in three-dimensional space.

8 To gain further insight, let us again consider the representation of an arbitrary wave as a superposition of plane waves:

$$\psi(\mathbf{x},0) \,=\, \int_{(\infty)} d^3(\mathbf{p}) A(\mathbf{p}) \exp\,(i\mathbf{x}\cdot\mathbf{p}) \tag{8a}$$

where

$$A(\mathbf{p}) \,=\, (2\pi)^{-3} \int_{(\infty)} d^3(\mathbf{x})\,\psi(\mathbf{x},0) \exp\,(-i\mathbf{x}\cdot\mathbf{p}) \tag{8b}$$

We discussed this representation in Secs. 39–44, Chapter 5, and as we said there, it is a fact that either one of these equations implies the other.

Suppose now that the function $A(\mathbf{p})$ is very well localized in momentum space. This means that $A(\mathbf{p})$ is large only in the immediate neighborhood of some point $\mathbf{p} = \mathbf{p}_0$, but small elsewhere. For simplicity we can assume that $A(\mathbf{p})$ *vanishes* outside some very small neighborhood of $\mathbf{p}_0$. If we look at the integral defining $\psi(\mathbf{x},0)$ we expect, intuitively, that the wave function $\psi(\mathbf{x},0)$ will then *not* be well localized. The wave function $\psi(\mathbf{x},0)$ will look approximately like a plane wave, of momentum $\mathbf{p}_0$. To see this the reader should think about the extreme case when the region in which $A(\mathbf{p})$ is different from zero shrinks to a point. (In going to this limit we must, of course, increase the amplitude $A(\mathbf{p})$ at the same time, otherwise the integral giving $\psi(\mathbf{x},0)$ will tend to zero.)

The author hopes that the reader can "see" that the more concentrated the function $A(\mathbf{p})$ is, the more spread out is the wave function $\psi(\mathbf{x},0)$. There is, however, a remarkable symmetry between the equations (8a) and (8b), and we can also conclude that the more concentrated the function $\psi(\mathbf{x},0)$ is, the more spread out is the function $A(\mathbf{p})$. If the function $\psi(\mathbf{x},0)$ is very concentrated, i.e., large only in some small region around a point $\mathbf{x}_0$, it means that the position of the particle is well defined. The momentum is, however, then poorly defined because a large range of momenta will contribute in the expansion (8a).

9 These ideas can be given a precise form, and we can relate the concentration of the function $A(\mathbf{p})$ to the concentration of the function $\psi(\mathbf{x},0)$. The result is an *uncertainty relation:* the precision with which the position is defined is inversely related to the precision with which the momentum is defined. Since we have promised the reader not to depend on the theory of the Fourier integral in this book we shall not present a rigorous derivation of the uncertainty relations.[†] The important thing for us is to understand

---

† For a standard derivation of these relations we refer the reader to L. I. Schiff, *Quantum Mechanics* (McGraw-Hill Book Company, New York, 1968), 3rd edition, p. 60.

qualitatively how the uncertainty relations arise. As we have seen, the idea is extremely simple. If the position of the particle is to be very well defined, the wave train must be very short. But this condition is incompatible with the condition for a well-defined momentum, namely that the wave train must resemble a sine wave over a range which includes a large number of complete periods. If we accept the wave description of particles we must therefore conclude that the position and the momentum of the particle cannot be known *simultaneously* to unlimited precision.

We remind the reader of our brief discussion in Secs. 20–26, Chap. 1, of the physical significance of the uncertainty relations. It should now be perfectly clear that these relations do not merely describe some unfortunate and unavoidable "perturbations" of our measuring instruments on the orderly classical motion of a classical corpuscle. They state instead the limit beyond which classical ideas cannot be pushed. It simply does not make any sense to talk about the simultaneous *precise* position and *precise* momentum of the quantum-mechanical particle (= wave packet).

10   What are the conditions which have to be satisfied so that we may think about an electron as a classical corpuscle; as a "charged billiard ball"? These conditions are analogous to the conditions for the validity of ray-optics. The linear dimensions of the apparatus through which the particle passes must be large compared to the wavelength, otherwise we will see the diffraction effects characteristic of waves. Let d signify some linear dimension of the instrument; d may be the diameter of a lens, or the width of a slit. Let λ be the de Broglie wavelength of the particle. For the classical corpuscle description to be sufficiently accurate we must have $d \gg \lambda$. Since $\lambda = 2\pi/p$ we can write our criterion in the form

$$dp \gg 1 \qquad\qquad (10a)$$

With the cgs-system of units this criterion reads $dp \gg \hbar$, and this is the same criterion which we discussed in Secs. 20–26 of Chap. 1.

11   To illustrate the implications of the uncertainty relations, let us study to what precision a classical trajectory can be assigned to an electron in a particular case. The situation is illustrated in Figs. 11ABCD. A beam of electrons, each one described by a plane wave, is incident from the left on the screen at left. The screen has a slit of width d. We wish to select d in such a way that the spot produced on the screen at right by the beam passing through

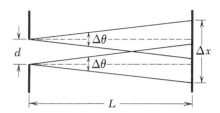

Fig. 11A We attempt to produce a narrow beam of electrons by limiting a broad beam incident from the left by the slit in the screen at left. The beam is diffracted at the slit, and the uncertainty $\Delta\theta$ in the angle at which the electrons leave the slit is inversely proportional to the width d of the slit. The size of the spot on the screen at right is given by $\Delta x \sim d + L\,\Delta\theta$.

the slit is as narrow as possible. The distance between the two screens is L.

We assume that the electrons all have the same incident momentum p. If an electron passes through the screen at left the uncertainty in its lateral position will be d. The uncertainty Δp in its lateral momentum is then given by

$$\Delta p \sim \frac{1}{d} \tag{11a}$$

If we assume that Δp is small compared to p we can restate (11a) in terms of the uncertainty $\Delta\theta$ in the angle θ (with respect to the incident direction) with which the electrons emerge, and we have

$$\Delta\theta \sim \frac{\Delta p}{p} \sim \frac{1}{pd} \tag{11b}$$

Let Δx measure the size of the spot produced on the screen at right. The magnitude of Δx is determined by two things: the size of the opening in the screen at left, and the spreading of the wave by diffraction at the slit. (See Fig. 11A.) We can therefore write

$$\Delta x \sim d + L\,\Delta\theta \sim d + \frac{L}{pd} \tag{11c}$$

Since the wavelength λ is given by $\lambda = 2\pi/p$, we can rewrite (11c) in the form

$$\Delta x \sim d + \frac{\lambda L}{d} \tag{11d}$$

where we have dropped a factor 2π in the last term. We are here only concerned with an order of magnitude estimate, and since our final result will be neater without the factor 2π we leave it out. We see that if we make d too small the second term in (11d) will become large because of the diffraction effects, whereas the first term is large if the width d of the slit is large. It is a simple problem in calculus to determine the optimum value d_0 of d for which the estimate (11d) for Δx assumes its minimum value $\Delta x_{\min}$, and we find

$$d_0 = \sqrt{\lambda L}, \qquad \Delta x_{\min} = 2d_0 = 2\sqrt{\lambda L} \tag{11e}$$

In the optimum case the spot on the right screen is twice as large as the slit in the left screen. (The factor 2 should not be taken too literally: remember that we are making only an order of magnitude estimate, and that we have set $2\pi \sim 1$.) Suppose $L = 1$ meter, and suppose that the energy of the electrons is 150 eV. Their wavelength is then 1 Å, and the estimate (11e) tells us that the spot on the screen at right can in principle be as small as 0.02 mm. The "tracks" of the electrons between the two screens are thus narrow and well defined from a macroscopic point of view.

12   The detailed investigation of the conditions under which a physical system appears to obey the laws of classical physics is an interesting, but non-trivial problem.   Some people describe what has to be done as follows.   First solve the problem quantum mechanically, and then set $\hbar = 0$ to obtain the classical limit.   This idea is not correct.   We cannot set $\hbar = 0$ because we know that $\hbar$ is really (with the right choice of units) equal to one.   The *real* problem is to show how it comes about that a system which obeys the laws of quantum mechanics, as *all* physical systems do, *appears* to obey the laws of classical physics, i.e., obeys these laws to a considerable accuracy.   In the study of this problem it is a good idea to employ units such that $\hbar = 1$, as we did in our example, because then one is forced to face the real issue.

The problem of how the classical limit is attained has many aspects, and it is quite impossible to give an exhaustive explanation in a single statement.   *One* condition, if we understand by the "classical limit" the classical theory of particle dynamics, is that the arrangement must be such that diffraction effects are not readily observable.   We have discussed this point in the preceding section. If a wave packet is to stay well localized, and is to have a well-defined trajectory which can be interpreted as the trajectory of a corpuscle, then the linear dimensions of the slits which define the trajectory must be large compared to the de Broglie wavelength. Classical dynamics is, however, not the only "classical limit."   It is also interesting to find out under what conditions classical electromagnetic theory appears to be valid.   In this case the condition is *not* that diffraction effects must be unobservable, but rather that the individual photons must not manifest themselves as *particles*.

We shall not pursue the question of the classical limit further. For the time being our rough qualitative understanding of this limit is sufficient.   The reader should think about these problems on his own.   As our discussion shows, what we mean by the "classical limit" depends on the system under consideration, and this is an important point to realize.

13   As a further example of arguments involving the uncertainty relation, let us try to estimate the binding energy of the hydrogen atom on the basis of this relation, as we promised to do in Sec. 26, Chap. 2.   In this discussion we want to employ the cgs-system of units, in which the uncertainty relation (6a) takes the form

$$\Delta x \, \Delta p \gtrsim \hbar \qquad (13a)$$

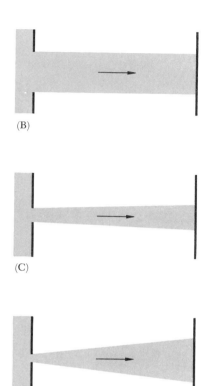

(B)

(C)

(D)

Figs. 11BCD   These three very schematic figures illustrate how the width of the beam depends on the width d of the entrance slit. (Note that the wavelength of the electrons is shorter in the above figures than in Fig. 11A.)   In B the size of the spot on the screen at right is large because the entrance slit is large.   If we make the width of the slit very small, as in D, the spot on the screen at right will be large because of diffraction effects.   We obtain the smallest spot by selecting $d \sim \sqrt{\lambda L}$, in which case the size of the spot is of the same order of magnitude.   Fig. C is intended to symbolize this optimum choice.

To make the estimate we assume that the classical expression

$$E = \frac{p^2}{2m} - \frac{e^2}{r} \qquad (13b)$$

for the total energy of an electron in the electrostatic field of the proton continues to have a meaning in quantum mechanics. The variable p then refers to the momentum of the electron wave, and the variable r is some kind of "position coordinate" for the wave.

The first term in the expression for E is manifestly positive, whereas the second term is negative. The ground state energy is the lowest possible energy of the system, and we know that it must be negative, otherwise there is no binding. *Classically* we can make the binding energy as large as we please simply by selecting an orbit for the electron which has a very small radius. For such a state of motion the uncertainty in *position* would be small, and if we now try to work in the spirit of quantum mechanics as well, we conclude from the uncertainty relation that the uncertainty in the momentum must be large, which means that the quantity $p^2/2m$ must be large. In other words, if we try to make the potential energy large (and negative) by making r small, then the kinetic energy will be large, which may lead to a large total energy if the kinetic energy term "wins." On the other hand, if we try to make the kinetic energy small by making p small, then r will have to be large, in which case the negative potential energy is small. We can well imagine that there will be some optimal radius for which the *total* energy assumes its smallest value.

14 To see how this "balance" between the kinetic and potential energies leads to binding we shall make a *rough* approximation as follows. Let us replace the uncertainty in position by r, and let us replace the uncertainty in momentum by p, and let us rewrite the uncertainty relation in the form

$$r p \sim \hbar \qquad (14a)$$

or, for definiteness, let us assume that

$$r p = \hbar \qquad (14b)$$

Let us now employ the relation (14b) to eliminate r from the expression (13b) for the total energy E. We obtain

$$E = \frac{p^2}{2m} - \frac{e^2 p}{\hbar} \qquad (14c)$$

As a function of p the energy E will have a minimum at a point $p = p_0$, which we determine by setting the derivative of E with respect to p equal to zero, i.e.,

$$\left(\frac{\partial E}{\partial p}\right)_{p=p_0} = \frac{p_0}{m} - \frac{e^2}{\hbar} = 0 \tag{14d}$$

Solving for p_0, and defining $r_0 = \hbar/p_0$, we thus obtain

$$p_0 = \frac{e^2 m}{\hbar}, \qquad r_0 = \frac{\hbar^2}{e^2 m} \tag{14e}$$

and

$$E = \frac{p_0{}^2}{2m} - \frac{e^2 p_0}{\hbar} = -\frac{e^4 m}{2\hbar^2} = -R_\infty \tag{14f}$$

Comparing these results with the results obtained in Sec. 23, Chap. 2, we find that the energy E given in Eq. (14f) is exactly right. The "radius" r_0, given by Eq. (14e) is also "right," it is the Bohr radius, $r_0 = a_0 = 0.53 \times 10^{-8}$ cm.

15   It is certainly an "accident" that our crude argument should give the correct binding energy. Whether we obtain the precise energy or not is, however, not the important point. The important thing is that we obtain the correct order of magnitude, both for the binding energy and size of the atom, *and that we can understand, on the basis of the wave theory, why atoms do not collapse.* The structure of an atom results from a compromise. The ground state energy is the lowest possible energy with which the atom can exist, and this energy is the sum of two terms of opposite sign. If we try to make the negative term, i.e., the potential energy, large by confining the electron waves to a very small region around the nucleus, then the kinetic energy term becomes large because the waves will carry a large momentum. On the other hand we must not spread out the waves too much, because then the potential energy term becomes insignificant. The ground state corresponds to the "best" possible compromise. These considerations are illustrated schematically in Figs. 15AB.

Our discussion also shows us that the idea of classical orbits in an atom is utterly incompatible with the wave picture. In the preceding section we found that the uncertainty in the position of the electron in the hydrogen atom must be of the order of the Bohr radius a_0. This estimate clearly applies to the position coordinate in each direction, and under these circumstances it makes no sense to talk about a circular orbit of radius a_0.

16   Let us next employ the uncertainty relation to obtain a crude estimate of the strength of the nuclear force. We consider a

Fig. 15A   If the electron is confined to a very small region around the nucleus the uncertainty in its position will be small. The uncertainty in its momentum must then be large, and this means that its kinetic energy must also be large. Its potential energy is, of course, negative, and large in magnitude.

Fig. 15B   If we want the kinetic energy to be very small we must allow enough space for the electron: the uncertainty in its position must be large. Its mean distance from the nucleus is then large, and the magnitude of its potential energy is small.

The ground state results from a compromise in which the *total* energy has the smallest possible value consistent with the uncertainty principle.

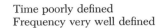

Time poorly defined
Frequency very well defined

(A)

Time better defined
Frequency less well defined

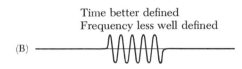

(B)

Time well defined
Frequency very poorly defined

(C)

Time very well defined
Frequency very poorly defined

(D)

Figs. 18ABCD To illustrate the time-frequency uncertainty relation. These figures are identical with Figs. 5ABCD, except for the legends.

nucleon, confined in a nucleus within a sphere of roughly the radius $r_0 = 1.2 \times 10^{-13}$ cm. The uncertainty relation then tells us that the momentum must be at least of order $p \sim \hbar/r_0$, and hence the kinetic energy of the nucleon must be of order

$$E_{\text{kin}} \sim \frac{1}{2M_p}\left(\frac{\hbar}{r_0}\right)^2 \sim 10 \text{ MeV} \tag{16a}$$

Since the nucleon is bound in the nucleus the average of the potential energy, denoted $\langle U \rangle$, must be negative and larger in magnitude than the kinetic energy, and we may conclude that

$$-\langle U \rangle \gtrsim 10 \text{ MeV} \tag{16b}$$

This estimate is *very rough*, but it does give an idea of the order of magnitude involved.

17 We note that the same kind of argument can be employed to refute the idea that nuclei consist of protons and *electrons*. If we look at Eq. (16a) we note that the kinetic energy is inversely proportional to the mass of the particle, and we would thus be led to the conclusion that the average potential energy of an electron would be about 2000 times larger than the estimate in Eq. (16b), which is totally incompatible with our experimental evidence that the dominant interactions of electrons are electromagnetic in nature.

18 We can state a time-frequency uncertainty relation which is completely analogous to the position-momentum uncertainty relation. Let $f(t)$ be the (complex) amplitude for some physical process. For instance, $f(t)$ may be the amplitude as a function of the time t of an electromagnetic wave at some fixed position in space. If the wave was emitted by an atom it will be a wave train of finite length, and the amplitude will tend to zero as t tends to $+\infty$ or $-\infty$. Such a wave can be regarded as a superposition of *monochromatic* waves, and the resolution of the wave into its monochromatic components is expressed by the Fourier integral

$$f(t) = \int_{-\infty}^{\infty} d\omega \, g(\omega) \, e^{-i\omega t} \tag{18a}$$

where the function $g(\omega)$ is given by

$$g(\omega) = (2\pi)^{-1} \int_{-\infty}^{\infty} dt \, f(t) \, e^{i\omega t} \tag{18b}$$

As we stated in Chap. 5 it is a theorem that either one of these integrals implies the other, for a large class of "well-behaved" functions $f(t)$ or $g(\omega)$. This theorem thus permits us to analyze an arbitrary time-dependent process in terms of its harmonic components.

If the function $g(\omega)$ in (18a) is large only in the immediate neighborhood of the point $\omega = \omega_0$ we can say that the frequency is very well defined: the amplitude $f(t)$ represents an almost monochromatic process. For a long time interval the amplitude $f(t)$ will be approximately of the form $f(t) = A\,e^{-i\omega_0 t}$. If, on the other hand, the amplitude $f(t)$ is large only in some small time interval about the time $t = t_0$, corresponding to the case when $f(t)$ represents a sharp pulse, then the frequency will be very poorly defined. The function $g(\omega)$, given by (18b), will be appreciable over a large interval of frequency. The frequency associated with the process, and the time at which the process takes place, cannot *both* be defined to arbitrary precision. The uncertainty $\Delta\omega$ in the frequency, and the uncertainty Δt in the time at which the process takes place, are subject to the uncertainty relation

$$\Delta\omega \, \Delta t \gtrsim 1 \tag{18c}$$

The reasoning which leads to this uncertainty relation is obviously entirely analogous to the reasoning leading to the position-momentum uncertainty relation. We have illustrated our ideas in Figs. 18ABCD.

19   The reader will recall that we discussed, in Secs. 20–23, Chap. 3, the relationship between the mean-life τ of an excited state, and the finite width ΔE of the corresponding energy level. We concluded that the width is inversely proportional to the mean-life. Let us now consider this relationship in the light of the time-frequency uncertainty relation.

Suppose that the system decays from the excited state to the ground state by the emission of a photon. The uncertainty in the frequency of the photon will then be $\Delta\omega = \Delta E/\hbar$, if ΔE is the width of the excited level. The duration of the emission process is of the order of the mean-life τ, and the uncertainty in the time at which the emission takes place is therefore of order τ. In view of (18c) we can then write

$$\tau\,\Delta\omega \sim 1 \quad \text{or} \quad \tau\,\Delta E \sim \hbar \tag{19a}$$

We have written an approximate equality rather than an inequality. We here deal with an exponentially damped harmonic oscillation, as shown schematically in Fig. 19A. The amplitude for this process is clearly more akin to the amplitude shown in Fig. 18A, for which the uncertainty product assumes its lower limit, than to

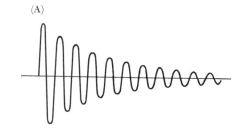

(A)

Figs. 19AB   Above is shown an exponentially damped harmonic oscillation. It is intuitively clear that the frequency is much better defined for such a process than for the "irregular" process illustrated by the curve in Fig. 19B below. For the upper curve it is reasonable to guess that the inequality in the general uncertainty relation is an approximate *equality*.

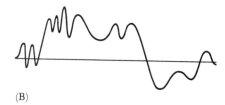

(B)

an amplitude such as the one shown in Fig. 18B, for which *both* the time and the frequency are very poorly defined.

The relations (19a) are just the relations we derived in Secs. 20–23, Chap. 3, by a seemingly different line of reasoning. If the reader thinks more about this matter he will notice that the basic ideas of our two derivations are not really so different. Our discussion in Chap. 3 could be characterized as "Fourier analysis in disguise."

Measurements and Statistical Ensembles

20 In the remainder of this chapter we shall discuss the process of measurement in physics. We shall do this by analyzing some simple physical situations in terms of what we now know. Our aim here is to find patterns of quantum-mechanical thinking, rather than to try to formulate some kind of complete theory of measurements. Physical measurements are of a most varied nature, and no short discussion could possibly reflect this variety. When we want to understand the implications of a theory it is natural that we consider highly idealized experimental situations in which the particular feature in which we are interested stands out as clearly as possible. We temporarily disregard all the *technical* experimental difficulties with which we are faced in the real world. Our *theoretical* discussion of measurements is therefore far from being a faithful account of what actually transpires in the laboratory.

21 It is often convenient to think about the process of measurement as if two stages were involved: the *preparation* of the system under study, followed by the actual *measurement*. This is certainly a schematic description because there is often no sharp distinction between the preparation and the measurement: some of the measuring process can just as well be regarded as being part of the preparation, and vice versa.

The analysis in terms of two stages is particularly appropriate when we consider scattering experiments. We study the interaction of a particle in a beam with a particle in a target. The preparation stage involves the setting up of the target and the production of the beam in an accelerator. The measurement stage involves the observation of the particles which emerge from the region in which the interaction takes place. Experiments with a beam of light belong to this class. The preparation of the photons takes place in the source, which might be some kind of "lamp" in conjunction with a system of lenses, polarizers, prisms, slits, etc.

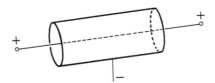

Very schematic picture of a Geiger-Müller counter. The device consists of two electrodes inside an envelope containing a suitable gas. In the figure the positive electrode is a thin wire, and the negative electrode is a cylinder centered about the wire. The electrodes are kept at a potential difference of about 1000 volts. A fast charged particle traversing the space between the electrodes will ionize the gas molecules along its path. The ions and electrons so produced will be accelerated toward the electrodes, and if the potential difference is sufficiently high, secondary ionization will occur to the extent that an avalanche of electrons is produced. The resulting current pulse can be amplified and registered, and the device can thus count individual charged particles. It is, of course, necessary to arrange it so that the discharge is "extinguished" after each pulse. This can be achieved either electronically through an auxiliary circuit which momentarily lowers the potential difference after each pulse, or else by using a filling gas such that the discharge will extinguish itself. Tubes of the latter kind are said to be self-quenching.

The measurements are carried out in some region of observation which is physically separated from the source; the measuring instrument might be a photomultiplier tube in association with other optical devices.

22 It is a characteristic feature of measurements in microphysics that we repeat the measurement a large number of times, always preparing the system in the same way. The statements of the results are typically statistical in character: we say that of N incident photons an average of N' photons were registered by a particular photomultiplier. The *single experiment*, or *single measurement*, involves only *one* photon, but our final report involves statistical averages of a large number of identical single experiments.

Now it is clear that two single experiments are not in principle identical since they are performed at two different times. We believe however, that the Laws of Nature are invariant under displacements in time, and the time at which the experiment was performed is therefore immaterial. For this reason a sequence of *repeated* single experiments can be regarded as a set of identical experiments in the sense that the preparation of the system was the same in each elementary experiment.

23 A beam consists of very many particles, but if the intensity of the beam is sufficiently low each individual scattering event will involve only *one* particle from the beam. This situation always prevails in scattering experiments with material particles, as well as in most experiments with photon beams. We can say that the beam is a *one-particle beam*. Doing experiments with a beam is the practical way of repeating the single elementary experiment (which involves only one particle at a time) a large number of times.

The target in a scattering experiment might be a thin slab or a thin foil of solid material, or it might be a container filled with a gas or a liquid. If the intensity in the beam is reasonably high it can well happen that two or more interactions take place at the same time in the target. This does not invalidate our description of the beam as a one-particle beam, because the two (or several) simultaneous events in the target are completely independent of each other. They correspond to two independent *elementary* experiments which just happened to take place at the same time.

In principle we can carry out our measurements with a beam of *very* low intensity, say one particle per minute, and we can then be sure that only one particle interacts with the entire target at a time. Since it is conceptually simple to think about beam experiments in terms of a succession of one-particle experiments, we shall assume

Two examples of commercially available Geiger-Müller counters. An important design consideration is to provide for the entry of the particles to be counted into the active space of the counter. The counters shown are provided with very thin mica windows for this purpose. The upper counter is a self-quenching counter suitable for counting alpha-particles, beta-particles and gamma quanta. The length of the counter is about 5 in., and the diameter about 3/4 in. The mica window can be seen at the lower end. The detection efficiency for very fast beta-particles is about 85 per cent. The lower counter is designed to have as large an entrance window as possible. (The diameter is about 1 3/4 in.) The metal envelope is one electrode, and the other electrode can be seen through the mica window. (*Photographs by courtesy of EON Corporation, Brooklyn, N.Y.*)

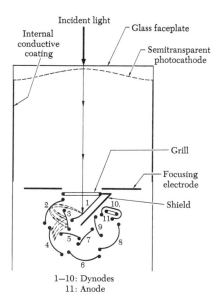

Incident light

Internal conductive coating

Glass faceplate

Semitransparent photocathode

Grill

Focusing electrode

Shield

2 1 10.
3 11
9
5 7
4 8
6

1–10: Dynodes
11: Anode

Fig. 24A The photomultiplier tube is extensively used as a photon detector. The figure shows such a tube schematically. Photons enter through the glass faceplate of the tube, and release photoelectrons from a very thin alkali-metal film on the inside of the faceplate. The electrons are accelerated and focused on the first dynode (denoted by the number 1 in the figure). Each electron which hits the first dynode gives rise to several secondary electrons, which are accelerated and focused on the second dynode where they give rise to more secondary electrons. These are accelerated and focused on the next dynode, and so on. For each detected photon an avalanche of electrons reaches the anode which is coupled to an external amplifier. The device is thus in effect a photo-electric cell with an amplifier in the same glass envelope. A current amplification of the order of 10^8 can readily be achieved.

in the following that our beams are of such a low intensity that only one particle is in transit at a time. In actual practice we would not deliberately limit the intensity of the beam, but, on the contrary, we would try in general to work with the highest intensity we could achieve.

24 To illustrate our ideas, let us consider an experiment with a beam of light. We shall analyze the *single experiment*, i.e., the sequence of events which takes place when a photon from the source arrives. Suppose that the detection system is some optical device fitted with photon counters (say photomultipliers). After the photon has arrived we find that some of the counters have "clicked" whereas some have not: we list in our record of the experiment the counters which have clicked. We imagine that all the detectors are restored to their original condition before the next photon arrives. When it does arrive some other counters may click, and these are not necessarily the same ones which clicked in the preceding measurement. We again record the facts, reset the detectors and wait for the next photon. We continue in this manner until we have repeated the elementary measurement a very large number of times, say after N photons have arrived from the source.

A *single measurement on the system* thus involves the observation of all the counters, and the basic data recorded are whether a particular counter clicked or not. After a sequence of N single measurements we might say:

(*a*) The counter 1 clicked an average number of p_1 times per incident photon. This average is defined experimentally by

$$p_1 = \frac{N_1}{N} \tag{24a}$$

where N_1 is the number of times counter 1 clicked in the sequence of N single experiments.

(*b*) The event that the counters 1 and 2 *both* clicked in a single experiment occurred an average number of p_{12} times per incident photon. This average is defined experimentally by

$$p_{12} = \frac{N_{12}}{N} \tag{24b}$$

where N_{12} is the number of elementary experiments in the sequence in which *both* counters 1 and 2 clicked.

(*c*) The counter 1 clicked an average number of $p(1;2)$ times per click of counter 2. This number is defined by

$$p(1;2) = \frac{N_{12}}{N_2} \tag{24c}$$

where N_2 is the number of times counter 2 clicked, and N_{12} the number of times *both* counters 1 and 2 clicked.

25   If we state our results in the above form we merely state what we have directly observed: the above numbers are our primary data.   We may, however, perform an abstraction and report the results of the measurements as follows:

   (*a*) The *probability* that counter 1 clicks under our experimental arrangements is p_1.

   (*b*) The *probability* that *both* counters 1 and 2 click (in a single experiment) is p_{12}.

   (*c*) The *probability* that counter 1 clicks given that counter 2 clicked is $p(1;2)$.

If we state our results in this manner we clearly make an assumption, and this assumption is that if we would continue the run of the experiment indefinitely then the numbers N_1/N, N_{12}/N and N_{12}/N_2 would tend to definite limits.   These hypothetical limits are what we try to determine: we designate the limits as the probabilities p_1, p_{12} and $p(1;2)$.   Since N is necessarily finite in any actual sequence of experiments the assumption that the limits exist and can be determined to any desired accuracy and with any desired degree of confidence provided we choose N to be sufficiently large is thus a pious expectation.   The nature of this expectation has given rise to much philosophical speculation.   We should regard it as an empirical fact that Nature possesses this kind of orderliness.

The results of a sequence of N single experiments may thus be stated in terms of probabilities, of which the numbers p_1, p_{12} and $p(1;2)$ are particular examples.   The number p_1 is a simple probability for the event that counter 1 clicks, p_{12} is a probability for the simultaneous occurrence of two events, and $p(1;2)$ is a conditional probability for the occurrence of an event given that another event occurred.   We may consider many other probabilities of a similar nature, say the probability that counter 1 clicked, given that counters 2 and 3 but no other counter clicked, etc.

26   We might describe our measurements as a sequence of experiments performed on a large number of photons, all prepared in the *same way* in the source.   Let us, however, think a bit about what it means to prepare a set of photons "in the same way."   Sup-

Fig. 24B An example of a commercially available photomultiplier tube.   The arrangement of the dynodes, which can be seen in the middle of the tube, is roughly as in the schematic figure 24A.   The light-sensitive cathode is inside the front end of the tube.   This particular kind of tube is intended for use with scintillation counters.   It is characterized by a very high quantum efficiency (see Fig. 24C).   *(Photograph by courtesy of Radio Corporation of America, Harrison, N.J.)*

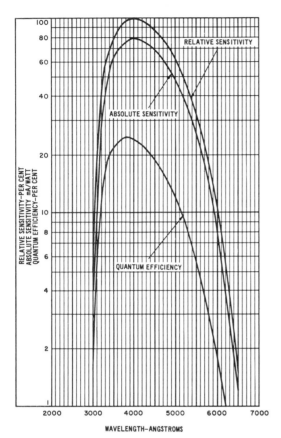

Fig. 24C Graphs showing detection efficiency of the photomultiplier tube in Fig. 24B. Note the curve labeled "quantum efficiency." It represents the probability for the detection of a photon, as a function of the wavelength. The maximum probability is about 25 percent, which is a very high efficiency for a phototube. The graph is taken from the manufacturer's booklet describing the tube. *(Illustration by courtesy of Radio Corporation of America, Harrison, N.J.)*

pose that there are *two* independent lamps in the source, say a sodium lamp which emits yellow photons and a mercury lamp which emits blue photons. A photon in a particular single experiment can thus be either yellow or blue, and the color is one of the variables characterizing the photon which we might determine in the experiment. Suppose that we do this for a long sequence of photons. We can then report that the probability is p_1 that the photon in any particular experiment is blue, and that the probability is p_2 that the photon is yellow. We assume that the intensities of the two lamps are kept steady so that these probabilities are reproducible: if we perform several runs of large numbers of repeated experiments we always find the same probabilities p_1 and p_2 in each run.

Are we willing to say, under these circumstances, that the photons are all prepared "in the same way" in the source? Whether this is an appropriate mode of expression or not is not immediately obvious. One might argue that our arrangement with the two lamps introduces an element of chance into the preparation process which could easily be avoided if we would carry out our observations with only *one* lamp operating at a time. Perhaps we should not say that the photons have all been prepared in the same way unless we are assured that the photons are in some sense identical to the highest possible degree?

The difficulty with such a position is that we would then have to decide, for each kind of experiment, whether the particles are "identically prepared to the highest possible degree" or not. This is obviously not a trivial problem. Furthermore, the two-lamp experiment is really just as respectable as the one-lamp experiment in the sense that the probabilities p_1 and p_2, as well as any other probabilities describing the responses of detectors, are *stable and reproducible*. This, of course, is the essential thing for any experiment in which we determine counting rates and probabilities, and unless the source is stable in just this sense the discussion in Sec. 25 would be irrelevant and meaningless.

It is therefore more practical to regard the photons as being all prepared in the same way whenever the source can be kept steady in such a manner that all the relevant probabilities are stable and reproducible. This is the position which we shall adopt in the following.

27 In a certain sense the two-lamp experiment is more realistic than the one-lamp experiment. Ideally we might prefer to do the experiment with only the yellow lamp on, but in the laboratory Nature will see to it that the blue lamp is also on (although its intensity may be very weak). Let us consider two examples to illustrate what we have in mind here.

Fig. 27A shows a semi-realistic electron diffraction experiment in which the objective is to observe the diffraction pattern due to the two slits in the screen S_2. The electrons are emitted by the filament F, and accelerated toward the screen S_1, which is provided with a slit. Let the electrons emerge through this slit with a momentum p. We observe the two-slit diffraction pattern with the help of the counter D, placed at a very large distance from the center of the second screen S_2. This counter can be moved along the circular arc shown in the figure. We assume, for simplicity, that the distance from the counter to the slits is so large that we can regard the rays connecting the entrance slit of the counter with the two slits in S_2 as *parallel*. (This is not what the figure shows, because if we drew the figure correctly it would be hard to see the two slits in the screen S_2. The *essence* of our discussion is, however, not affected by whether the rays are parallel or not.)

Let the separation between the two slits in S_2 be $2a$. The angular distribution $I(\theta, p)$ of the radiation detected by D can then be written, as we found in Sec. 40, Chap. 4

$$I(\theta, p) = 4I_0(\theta) \cos^2 (ap \sin \theta) \qquad (27a)$$

where $I_0(\theta)$ is the angular distribution which we would observe with a *single* slit.[†]

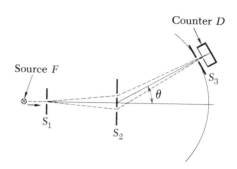

Fig. 27A To illustrate our discussion in Secs. 27–30, of a two-slit electron-diffraction experiment. The counting rate is observed as a function of the angle θ when the counter and the entrance slit S_3 are moved along the circular arc. If the separation between the slits in S_2 is large compared to the wavelength, and if the source produces monoenergetic electrons, the counting rate will be an extremely rapidly varying function of θ. The diffraction pattern will not be observable unless the angular resolution defined by the counter-slit arrangement is very good. If the electrons are *not* monoenergetic, as would be the case if the source is a simple filament, the patterns for the different energies overlap and the diffraction maxima might be smeared out to such an extent that they can no longer be seen.

28 We have written the intensity as $I(\theta, p)$ to emphasize that the angular distribution is a function of p. We shall assume that the widths of the two slits in S_2 are equal, and very small compared to the wavelength of the incident electrons. Within the range of momenta p which come into consideration the intensity $I_0(\theta)$ is therefore *independent* of p. On the other hand, we shall assume that the slit separation $2a$ is very large compared to the wavelength. To be specific, let us assume that for the *mean momentum* p_0 of the beam, we have $ap_0 = \pi \times 10^5$. For this mean momentum we then have the angular distribution

$$
\begin{aligned}
I(\theta, p_0) &= 4I_0(\theta) \cos^2 [(\pi \times 10^5) \sin \theta] \\
&= 2I_0(\theta) \{1 + \cos [(2\pi \times 10^5) \sin \theta]\} \qquad (28a)
\end{aligned}
$$

If we now examine the expression for the intensity we notice that it is a very rapidly varying function of the angle θ. The separation δ between two successive maxima is given approximately by $\delta \cong 10^{-5}/\cos \theta$.

[†] We use units such that $\hbar = c = 1$ in this discussion.

Therefore, if we want to see the diffraction pattern clearly it is essential that the *angular resolution* of our detection equipment be very good. The angle which the entrance slit of the detector D subtends when observed from the center of S_2 must be much less than δ, i.e., much smaller than 10^{-5}. Let us assume that this is the case. If it is *not* the case, i.e., if the angular resolution is much poorer than 10^{-5}, then the second term in the extreme right-hand expression in (28a) will effectively average to zero and we will observe an intensity which is twice the intensity with a single slit.

29   Suppose now that the angular resolution of the detector is very good, so that we can see clearly the two-slit pattern for a beam of electrons all having the momentum p_0. Such a beam is, however, not realistic. The electrons are not all emitted with the *same* energy by the filament F, and they will therefore not emerge with the same momentum from the slit in S_1. The reason for this is the thermal motion of the electrons in the filament. We have said earlier that the random thermal motion is "noise in the symphony of pure quantum mechanics," and we will now see how the noise can prevent us from hearing the music.

In a realistic experiment the momentum p of the emerging electrons will exhibit a finite spread. For simplicity, let us assume that every momentum in the range $(p_0 - q, p_0 + q)$ is equally likely. The quantity q describes the spread in the momentum, and to be specific we shall assume that $q = 10^{-2} p_0$: the momentum is defined to within one per cent.

If we now observe the diffraction pattern with such a beam we will clearly not observe the distribution $I(\theta, p_0)$, but rather an *average* of $I(\theta, p)$ over the range of momenta in the beam. Let us denote this average intensity by $\bar{I}(\theta)$. It is given by

$$\bar{I}(\theta) = \left(\frac{1}{2q}\right) \int_{p_0-q}^{p_0+q} dp\ I(\theta, p) =$$

$$2I_0(\theta) \left(1 + \frac{\cos\ (2ap_0 \sin \theta) \sin\ (2aq \sin \theta)}{2aq \sin \theta}\right) \qquad \text{(29a)}$$

Note that if we let q tend to zero in the expression (29a) we recover the expression (28a).

In accordance with our specific assumptions that $ap_0 = \pi \times 10^5$, and that $q = 10^{-2} p_0$, we can conclude, from (29a), that

$$\left|\bar{I}(\theta) - 2I_0(\theta)\ \right| \leq 2I_0(\theta) \left|\frac{\sin\ [(2\pi \times 10^3) \sin \theta]}{(2\pi \times 10^3) \sin \theta}\right| \qquad \text{(29b)}$$

In the strictly forward direction, i.e., for $\theta = 0$, we see from (29a) that $\bar{I}(\theta) = 4I_0(\theta)$. In this particular direction we always have constructive interference, irrespective of the momentum p. Suppose, however, that we carry out the observations *outside* the strictly forward direction, say for angles θ which satisfy the condition $|\sin \theta| > (2\pi)^{-1} \times 10^{-1} \cong 0.016$. The inequality (29b) then tells us that

$$|\bar{I}(\theta) - 2I_0(\theta)| < 10^{-2} \times 2I_0(\theta) \qquad (29c)$$

For these angles the two-slit pattern is therefore hard to see, because the intensity distribution agrees, within one per cent, with the pattern for a single slit.

30   For a classical billiard ball theory, which we discussed for *photons* in Sec. 41, Chap. 4, we would predict an intensity $I^*(\theta)$, for the two-slit experiment, given by

$$I^*(\theta) = 2I_0(\theta) \qquad (30a)$$

In this model there is no interference, and as we have said this is the wrong prediction; it does not agree with experiments. However, if we compare this prediction with the prediction contained in the inequality (29c) we notice that the prediction (30a) can sometimes *appear* to be right. If the quantum-mechanical interference effects "wash out" for some reason, we end up with the classically predicted observational result.

Our discussion is a very interesting illustration of *one* aspect of the "passage to the classical limit." Suppose that the energy of the electrons in the experiment considered is 10 eV. The slit separation $2a$ is then about 0.04 mm, which we can regard as a *macroscopic* quantity. In spite of this the quantum-mechanical interference effects are certainly there, but in order to see them we must carry out our experiment with an extremely good control over the source of the electrons so that the spread q in the momentum is kept very small. Unless this is the case, the music of quantum mechanics disappears in the noise.

31   As another example of the disappearance of interference effects, consider the observation of interference fringes with a Michelson interferometer, shown schematically in Fig. 31A. Light from a sodium lamp is "split" by a half-silvered mirror, and the essence of the experiment is the observation of interference between the two beams returning from mirrors 1 and 2. We have drawn the two "arms" of the interferometer to be of unequal lengths, L_1 and L_2. The difference in path for the two beams is

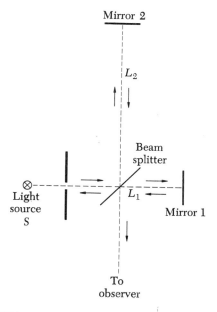

Fig. 31A   Schematic picture of a Michelson interferometer with arms of unequal lengths. (The lengths, i.e., the distances from the mirrors to the beam splitter, are L_1 and L_2, as shown above.) The maximum path difference $2(L_2 - L_1)$ for which interference can be observed depends on the width of the spectral line of the almost monochromatic light source.

thus $d = 2(L_2 - L_1)$. The question arises: can interference fringes be observed for arbitrarily large d?

The answer is: in principle *yes*, but in practice *no*. The precision with which the wavelength of the light is defined sets a limit to the path difference d for which fringes can be seen, and in practice the wavelength is never defined with perfect precision.

Consider a photon of frequency ω emerging from the source. The part of the photon returning from mirror 2 will be retarded in phase relative to the part returning from mirror 1 by an amount $\delta(\omega)$ given by

$$\delta(\omega) = \omega d = 2\pi \left(\frac{d}{\lambda} \right) \tag{31a}$$

where λ is the wavelength. Consider next two different frequencies, ω' and ω''. The difference between the retardations in phase for these frequencies is then given by

$$\delta(\omega') - \delta(\omega'') = (\omega' - \omega'') d \tag{31b}$$

If this difference is numerically very *small*, i.e., $|\delta(\omega') - \delta(\omega'')| \ll \pi$, then the interference fringes will look the same, to a good accuracy, for the two frequencies. On the other hand, if this difference is π, i.e. $|\delta(\omega') - \delta(\omega'')| = \pi$, then *constructive* interference for the frequency ω' corresponds to *destructive* interference for the frequency ω'', and vice versa. The system of fringes for the two frequencies are complementary, and if they are superimposed on each other with equal intensities no fringes will be observed. This leads to a simple criterion for the visibility of the fringes: the frequency spread $\Delta\omega$ in the source must be such that

$$d \, \Delta\omega \lesssim \pi \tag{31c}$$

if the fringes are to be easily observable. For a given source, i.e. for a given $\Delta\omega$, the criterion (31c) gives us the desired upper bound on d.

32 For an approximately monochromatic light source (frequency ω), the quantity $\Delta\omega$ is the linewidth of the emitted light. As we have explained in Chap. 3 several physical effects contribute to the linewidth. One of these is the Doppler effect due to the motion of the atoms in the source. The source is a collection of identical "lamps," but the frequencies emitted by these lamps will not all be the same in the laboratory frame, because the lamps move in a random fashion in the source.

Let us consider the limitation on d arising from the Doppler broadening. The condition for clearly visible fringes is

$$d < \frac{\pi}{\Delta\omega} = \left(\frac{\omega}{\Delta\omega}\right)\left(\frac{\lambda}{2}\right) \tag{32a}$$

In Sec. 44, Chap. 3, we derived an expression for the fractional Doppler broadening, namely

$$\left(\frac{\Delta\omega}{\omega}\right)_D \sim (0.52 \times 10^{-5})\sqrt{\frac{1}{A}\left(\frac{T}{293°\text{K}}\right)} \tag{32b}$$

where T is the effective temperature in the source, and where A is the molecular weight of the emitting atoms, which we assume are present in the form of a gas. Combining (32a) and (32b) we obtain

$$d \lesssim \lambda \sqrt{\frac{A}{(T/293°\text{K})}} \times 10^5 \tag{32c}$$

For $T = 293°\text{K}$ (room temperature), and for $\lambda = 5000$ Å (visible light), and for $A = 100$, we thus obtain $d \lesssim 50$ cm. This estimate is in accordance with observations. The maximum path difference for which interference fringes are seen is of the order of 1 meter for "ordinary" light sources, such as gas discharge tubes (lasers excepted).

33 The two examples which we have considered illustrate how Nature conspires to "keep two lamps on." The background of thermal noise in our surroundings introduces a certain randomness in the preparation of the system before the measurement.

Technical imperfections in our equipment also contribute to the randomness in the preparation process. Suppose, for instance, that we wish to produce a beam of high-energy electrons with a very precisely defined momentum. In order to do this we must be able to control all accelerating potentials very precisely, and the arrangements for the focusing of the beam must be nearly perfect. Furthermore we must be able to maintain a very high vacuum, because the electrons in the beam can lose energy and have their direction of motion changed in collisions with residual gas molecules in the vacuum system. Nothing in the world is perfect, and it is clear that we can never achieve complete control over the preparation stage in practice. It is therefore of interest to see how an "imperfect" preparation process might be described theoretically.

34 Suppose that we have an arrangement for the preparation of a system in a sequence of repeated measurements such that the "system is always prepared in the same manner." As we have agreed before, this means that the *probabilities* and *averages* which we measure in long runs are stable and reproducible. We imagine that we have measured the averages of all possible physical variables. We say that the totality of these averages defines a *statistical ensemble of the system,* and we say that any particular example of the prepared system, as encountered in a single measurement, is an *element of the ensemble.*

A particular method of preparation, whether "perfect" or "imperfect," gives rise to a particular statistical ensemble. From a mathematical point of view an *abstract* statistical ensemble is equivalent to a set of probabilities and averages of physical variables. When we want to consider the concrete physical realization of this abstract concept we can regard the ensemble as a collection of a very large number of prepared systems (elements). We thus describe a beam of light as a statistical ensemble of photons, where the elements of the ensemble are the individual photons.

Another important application of the concept of a statistical ensemble is the description of a quantity of gas in a container as a statistical ensemble of molecules. This description is appropriate if we are studying the average behavior of individual molecules in the gas. Each time we measure, say, the velocity of a molecule we are doing an experiment on an element of the ensemble. The results of a large number of velocity measurements give us the average velocity, which is one of the averages which characterize the ensemble. The conditions under which the gas is kept in the container define, in this case, the preparation procedure. If the temperature and pressure are kept constant the average velocity will also be constant. We can say that the molecules are all prepared in the same way, because they are all subject to macroscopically identical external conditions. This does not, of course, mean that we will find the same velocities in two particular measurements on two individual molecules. The velocity of a molecule (at an instant of time) is a *random variable* from our standpoint: the values which we encounter exhibit a *statistical spread.*

35 Consider a statistical ensemble. To have a concrete example at hand we can think of a beam of electrons emerging from a accelerator operating under conditions which are stable and stationary to the best of our technical abilities. We repeatedly measure a particular physical variable, say the momentum p in the direction of the beam. We denote the average of the values of momentum in a long sequence of measurements by

$$\mathrm{Av}(p;\rho)$$

where the letter ρ symbolizes the particular statistical ensemble, i.e., the particular beam. We call the quantity $\mathrm{Av}(p;\rho)$ the *ensemble*

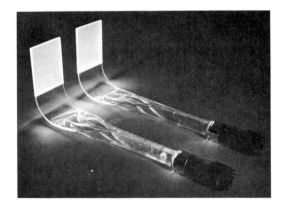

A pair of scintillation counters. When charged particles strike the vertical white panels at left scintillations are produced in the material. The light of these scintillations is "piped" through the lucite guides to the photomultiplier tubes at right. When in use the counter and light pipe are wrapped in aluminum foil and tightly sealed from stray light. *(Courtesy Lawrence Radiation Laboratory, Berkeley.)*

average of p. The average of the *squares* of the values of momentum is denoted $\text{Av}(p^2;\rho)$: this is the ensemble average of the square of momentum. In general $\text{Av}(p^2;\rho)$ will be different from $[\text{Av}(p;\rho)]^2$. Let us study this point. We denote the values of momentum found in the individual measurements by $p_1, p_2, \ldots, p_N$. The two averages are defined by

$$\text{Av}(p;\rho) = \frac{1}{N}\sum_k p_k, \qquad \text{Av}(p^2;\rho) = \frac{1}{N}\sum_k p_k{}^2 \qquad (35a)$$

We can then write the identity

$$\text{Av}(p^2;\rho) - [\text{Av}(p;\rho)]^2 = \frac{1}{N}\sum_k [p_k - \text{Av}(p;\rho)]^2 \qquad (35b)$$

as the reader should immediately convince himself. The right side of (35b) is a sum of non-negative terms, and we can conclude that

$$\text{Av}(p^2;\rho) - [\text{Av}(p;\rho)]^2 \geqq 0 \qquad (35c)$$

where the equality sign applies if and only if all the numbers p_k, $k = 1, 2, \ldots, N$, are equal, in which case their common value equals $\text{Av}(p;\rho)$. In this particular case the particles in the beam all have precisely the same momentum.

The quantity in the left side of (35c) measures the statistical spread in the variable p. In general it will be greater than zero, which we can express by saying that there is an uncertainty in the momentum for the particular ensemble.

36 We can discuss all other physical variables in the same way as we have discussed momentum. For a particular ensemble (beam) we determine their averages, and their *dispersions,* by which we mean their statistical spreads defined in analogy with the expression at left in (35c). The simplest kind of variable is a variable which describes the response of a counter. Let us denote it by D, and let us adopt the convention that D has the value $+1$ in a particular experiment if the counter clicked, and the value 0 if it did not click. $\text{Av}(D;\rho)$ is then simply the probability that the counter will click when we do an experiment with a single element of the statistical ensemble ρ.

At first it might seem that a counter variable D is not of the same kind as a momentum variable p. We might feel that p refers to the *system,* i.e., the particle, whereas D refers to a measuring instrument. We must realize, however, that all our information about the properties of the system is derived by observing the

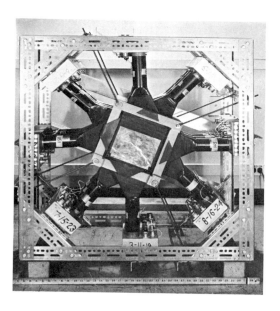

There is a striking contrast between the small and neat "theoretician's counters" which appear in the schematic figures in this chapter, and some of the counters actually used in the laboratory. This photograph shows a stack of 24 scintillation counters, assembled for an experiment in elementary particle physics. The side length of the assembly is about 1 meter. The plastic scintillators are in the middle of the figure, and the photomultiplier tubes are symmetrically placed on the periphery. The direction of the particle beam is normal to the plane of the figure. *(Photograph by courtesy of Lawrence Radiation Laboratory, Berkeley.)*

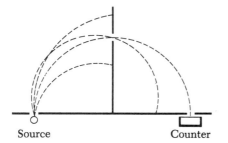

responses of measuring instruments: the *intrinsic* variables of the system are abstractions. If we know the probability that a certain counter, placed in a certain way, will click we do know something about the nature of the statistical ensemble, i.e., something about the particles in the beam. The momentum of particles in a beam is in fact often measured by counters, as illustrated in Fig. 36A.

Fig. 36A To illustrate the principle of a so-called semi-circular beta-spectrograph. The device is used to measure the distribution in momentum (or energy) of the electrons emitted in the decay of beta-active nuclei. The electrons are emitted from the radioactive source at left, and are constrained to move in or close to the plane of the figure. The device is placed in a uniform magnetic field perpendicular to the plane of the figure, and the trajectories are thus circular arcs, the radii of which depend on the momenta of the electrons. The apparatus contains a number of slits so placed that an electron cannot reach the counter at right unless the radius of its trajectory lies within a narrow range. By counting the number of electrons reaching the detector per unit time for different values of the magnetic field we can determine the momentum distribution of the emitted electrons, i.e., the relative number of electrons emitted within different intervals of momentum.

37 Let us discuss further the situation considered in Sec. 26, in which there are two lamps in a source of light: a sodium lamp and a mercury lamp. Let us first consider an experiment in which only the sodium lamp operates, and the beam thus consists of "yellow photons." The source gives rise to a statistical ensemble ρ_1 of photons, and for this ensemble we find the average value d_1 of a certain counter-variable D:

$$\text{Av}(D;\rho_1) = d_1 \qquad (37a)$$

Next we consider an experiment in which only the mercury lamp operates. This arrangement defines a statistical ensemble ρ_2, and the ensemble average of the same counter variable D is found to be

$$\text{Av}(D;\rho_2) = d_2 \qquad (37b)$$

We finally consider the case when *both* lamps operate simultaneously. The two lamps together give rise to the statistical ensemble ρ, and the average of D in this case is

$$\text{Av}(D;\rho) = d \qquad (37c)$$

Suppose now that lamp 1 gives rise to a flux of n_1 photons per unit time in the beam, and that lamp 2 gives rise to a flux of n_2 photons per unit time in the beam. The total flux in the beam is thus $(n_1 + n_2)$ photons per unit time. In any single experiment the photon is either "yellow" or "blue," depending on whether it came from lamp 1 or lamp 2, and we conclude that the *probability* that we find a "yellow" photon in any single experiment is

$$\theta_1 = \frac{n_1}{(n_1 + n_2)} \qquad (37d)$$

whereas the probability that we find a "blue" photon is

$$\theta_2 = \frac{n_2}{(n_1 + n_2)} \qquad (37e)$$

The numbers θ_1 and θ_2 satisfy the conditions

$$1 \geqq \theta_1 \geqq 0, \qquad 1 \geqq \theta_2 \geqq 0, \qquad \theta_1 + \theta_2 = 1 \qquad (37f)$$

as a consequence of the definitions (37d) and (37e). The conditions

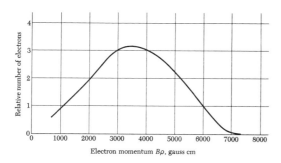

Fig. 36B The beta spectrum of P[32]. The graph shows the relative number of emitted electrons as a function of momentum. The momentum is expressed in terms of the quantity Bρ (in units of gauss centimeters), where ρ is the radius of curvature in the field B. The maximum momentum at 7200 gauss cm corresponds to the maximum kinetic energy 1.7 MeV.

The electrons can emerge with an energy ranging from zero to the upper limit because the total (kinetic) energy released in the decay is shared (in a random fashion) between the electron, the daughter nucleus, and an anti-neutrino.

(37f) are characteristic for the probabilities of two mutually exclusive events, one of which must take place.

38   Consider now a single experiment: i.e., an event involving a single photon. What can we say about the probability $d = \mathrm{Av}(D;\rho)$ that the counter described by the variable D clicks? The photon in question is either yellow or blue. The probability that it is yellow is θ_1: *if* it is yellow then the probability that the counter D clicks is d_1. The probability that the photon is blue is θ_2, and *if* it is blue, then the probability that the counter D clicks is d_2. Since the two cases yellow-blue are mutually exclusive we conclude that the probability d that the counter D clicks must be

$$d = \theta_1 d_1 + \theta_2 d_2 \tag{38a}$$

or

$$\mathrm{Av}(D;\rho) = \theta_1\,\mathrm{Av}(D;\rho_1) + \theta_2\,\mathrm{Av}(D;\rho_2) \tag{38b}$$

The average of D in the ensemble ρ is thus given in terms of the averages of D in the ensembles ρ_1 and ρ_2 and in terms of the probabilities θ_1 and θ_2. These latter probabilities describe how the "combined" ensemble ρ is formed from the ensembles ρ_1 and ρ_2, and they are therefore quantities which characterize the "combined" *source:* the are *independent* of the variable D which describes a particular counter in the region of observation. The formula (38b) therefore holds for *every* counter variable D.

More generally the formula (38b) applies to the averages of an *arbitrary* physical variable. If Q denotes such a variable we must have

$$\mathrm{Av}(Q;\rho) = \theta_1\,\mathrm{Av}(Q;\rho_1) + \theta_2\,\mathrm{Av}(Q;\rho_2) \tag{38c}$$

We say that *the statistical ensemble ρ is an incoherent superposition of the two ensembles ρ_1 and ρ_2 with probabilities θ_1 and θ_2.* We express this statement symbolically in the form

$$\rho = \theta_1 \rho_1 + \theta_2 \rho_2 \tag{38d}$$

The reason for the qualifying attribute "incoherent" is that we must carefully distinguish between this kind of superposition, and the superposition of *waves*, which we discussed in Secs. 36–46, Chap. 5. We shall have more to say about this distinction later.

39   We can clearly generalize the idea of the superposition of two ensembles to include the incoherent superposition of any finite number of ensembles. Let us thus consider the statistical ensembles ρ_k, $k = 1, 2, 3, \ldots, n$. We associate with each one of these ensembles a probability θ_k, such that the numbers θ_k satisfy

$$1 \geqq \theta_k \geqq 0, \qquad \sum_{k=1}^{n} \theta_k = 1 \tag{39a}$$

Let ρ be the incoherent superposition of these ensembles, with probabilities θ_k: we express this symbolically by

$$\rho = \sum_{k=1}^{n} \theta_k \rho_k \tag{39b}$$

This means that the average of any physical variable Q for the ensemble ρ is given by

$$\text{Av}(Q;\rho) = \sum_{k=1}^{n} \theta_k \, \text{Av}(Q;\rho_k) \tag{39c}$$

We shall make the assumption that if $\rho_1, \rho_2, \rho_3, \ldots, \rho_n$ is any set of possible statistical ensembles, then every incoherent superposition of these ensembles is also a possible ensemble. This assumption is mathematical rather than physical, and we make it because we want the set of *all* statistical ensembles to have the property that it is *closed under incoherent superposition*. This means that if the set contains any finite number of ensembles it also contains all possible incoherent superpositions of these ensembles.

40 Note that we have already considered, in our discussion in Secs. 27–29, an incoherent superposition of an infinite number of distinct statistical ensembles. Let $D(\theta)$ be the variable describing the counter D in Fig. 27A for a given angle θ. Let ρ denote the statistical ensemble which a given source, to the left of the screen S_1, gives rise to. We assume that the intensity of the sources which we shall consider is always such that one electron per second enters through the slit in S_1. If the intensity $I(\theta)$, observed with the counter D, is expressed in number of electrons per second, we have

$$\text{Av}[D(\theta);\rho] = I(\theta) \tag{40a}$$

In our discussion in Sec. 27 we first considered the intensity $I(\theta,p)$ for a hypothetical source which produces electrons of an extremely well-defined momentum p. Let us denote the statistical ensemble defined by such a source by $\rho(p)$. We then have

$$\text{Av}[D(\theta);\rho(p)] = I(\theta,p) \tag{40b}$$

We pointed out that if the source is a hot filament with a single

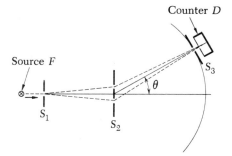

Source F

Counter D

S_1

S_2

S_3

θ

accelerating electrode, then the electrons will *not* emerge through the slit in S_1 with a well-defined momentum. [There is, however, nothing to prevent us from designing a very elaborate source, with "momentum filters," such that the momentum of the emerging electrons is extremely well defined: such a source is described by $\rho(p)$.] Let us denote the statistical ensemble which the simple filament source gives rise to by $\bar{\rho}$. In accordance with our discussion in Sec. 29 we then have

$$\text{Av}[D(\theta);\bar{\rho}] = \bar{I}(\theta) = \left(\frac{1}{2q}\right) \int_{p_0-q}^{p_0+q} dp \, Av[D(\theta);\rho(p)] \quad (40c)$$

This is to be compared with Eq. (39c). Our reasoning in Sec. 29 clearly amounts to regarding the statistical ensemble $\bar{\rho}$, corresponding to the "realistic" hot filament source, as an incoherent superposition of the ideal sources corresponding to the statistical ensembles $\rho(p)$. In other words, in analogy with Eq. (39b), we have

$$\bar{\rho} = \left(\frac{1}{2q}\right) \int_{p_0-q}^{p_0+q} dp \, \rho(p) \quad (40d)$$

Amplitudes and Intensities

41   We can state the difference between coherent and incoherent superposition as follows: in a coherent superposition we add *amplitudes*, but in an incoherent superposition we add *intensities*.

Let us exercise ourselves in the handling of amplitudes and intensities in quantum mechanics. Figure 41A illustrates a semi-realistic double double-slit experiment. Particles with a very well-defined momentum enter through the slit in the screen S, at a rate of, say, one particle per second. We observe the flux of these particles through the other five slits by placing a counter immediately behind the slits, *one slit at a time*. If the observed counting rate is P particles per second for a particular slit we can say that P is the probability that a particle entering through the slit in S will pass through the slit in question.

We assume that the wavelength of the particle is large compared to the widths of the slits, and we furthermore assume that all the slits are of the same width. We can then talk about the (complex) amplitude of the wave *at* a slit.

42   Let A_1 be the amplitude of the wave at the slit *1'* when the amplitude at the slit in S is equal to unity. Similarly, let A_2 be the amplitude of the wave at the slit *2'* when the amplitude at the slit

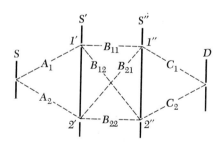

Fig. 41A   To illustrate our discussion in Secs. 41–43 of a somewhat idealized double double-slit experiment. The particles (photons) enter through the slit in S. We are interested in the probabilities that the particles pass through the other slits, and in particular through the slit in D. At each slit we must, of course, add the amplitudes rather than the intensities of the waves which arrive from the preceding slits. The complex numbers A_m, B_{mn}, and C_m are the *transfer amplitudes* between the slits. All the probabilities can be expressed in terms of the transfer amplitudes.

in S is unity. Let B_{11} be the amplitude of the wave at the slit *1″* when the amplitude at the slit *1′* is unity, *but the amplitude at the slit 2′ is zero.* Similarly, let B_{21} be the amplitude of the wave at the slit *1″* when the amplitude at the slit *1′* is zero, but the amplitude at the slit *2′* is unity. C_1 stands for the amplitude at the slit in the screen D when the amplitude at the slit *1″* is unity, but the amplitude at the slit *2″* is zero. The remaining amplitudes are defined in an analogous manner. We can call these amplitudes *transfer amplitudes,* because they describe the propagation of the wave between the slits, from left to right. The dotted lines in Fig. 41A represent this propagation symbolically. A transfer amplitude is associated with each dotted line, as we have explained.

The transfer amplitudes are complex numbers. Their absolute squares define *transfer probabilities* as follows. $P'_1 = |A_1|^2$ equals the probability that a particle which has passed through the slit in S is detected immediately behind the slit *1′*. $P'_2 = |A_2|^2$ equals the probability that a particle which has passed through the slit in S also passes through the slit *2′*. $P_{12} = |B_{12}|^2$ equals the probability that a particle which has passed through the slit *1′* also passes through the slit *2″*. *In this case the slit 2′ must be closed,* to make sure that the particle really passed through the slit *1′*. The absolute squares of the other transfer amplitudes have analogous interpretations. Let us list all the transfer probabilities corresponding to the eight amplitudes:

$$
\begin{aligned}
P'_1 &= |A_1|^2 & P'_2 &= |A_2|^2 \\
P_{11} &= |B_{11}|^2 & P_{12} &= |B_{12}|^2 \\
P_{21} &= |B_{21}|^2 & P_{22} &= |B_{22}|^2 \\
P''_1 &= |C_1|^2 & P''_2 &= |C_2|^2
\end{aligned}
\tag{42a}
$$

The reader should think carefully about how these transfer probabilities can be measured with counters, closing some of the slits when required.

43   Suppose we now ask the question: with all the slits open, what is the probability P that a particle which enters through the slit in S will emerge through the slit in D?

Let us first give a thoughtless answer: since we know all the transfer probabilities between the slits we can find P by compounding these probabilities according to the rules of probability theory. The probability that the particle passes through the slit *1″* should thus be equal to the sum of the probability that it passes through *1″* via the slit *1′* and the probability that it passes through *1″* via the slit *2'*, or, in other words equal to $(P_1'P_{11} + P_2'P_{21})$. This kind of reasoning leads to the final *wrong* result

$$P = (P_1'P_{11} + P_2'P_{21})P_1'' + (P_1'P_{12} + P_2'P_{22})P_2'' \quad (\text{Wrong!}) \quad (43a)$$

What is the correct answer? It is given by

$$P = |(A_1B_{11} + A_2B_{21})C_1 + (A_1B_{12} + A_2B_{22})C_2|^2 \quad (43b)$$

and this is *not* equal to the wrong expression in Eq. (43a). At each slit we must add the *amplitudes* of the waves arriving at the slit, because the waves can interfere. Equation (43b) gives the correct answer according to quantum mechanics, whereas we might regard the expression in Eq. (43a) as the prediction according to a classical billiard-ball theory.

44   How do we find P if we are given only the transfer probabilities, but not the transfer amplitudes? Answer: We cannot find P at all. To find P we must know *both* the phases and the absolute values of the complex transfer amplitudes, but the transfer probabilities only tell us about their absolute values.

Let us discuss further the fallacy in the "compounding-of-probabilities" argument which led us to the wrong prediction (43a). Consider the quantity $P_1'P_{11}$. What does it represent? It represents the probability that a particle entering through the slit in S passes through the slit *1''* *when the slit 2' is closed*. Similarly $P_2'P_{21}$ represents the probability that a particle entering through the slit in S passes through the slit *1''* when the slit *1'* is closed. If *both* slits *1'* and *2'* are open the probability that a particle entering through the slit in S passes through the slit *1''* is *not* given by the sum $(P_1'P_{11} + P_2'P_{21})$. The waves arriving at the slit *1''* from the slits *1'* and *2'* are *coherent* with each other, and we must add their *amplitudes* and not their *intensities*.

45   Consider the slightly modified arrangement shown in Fig. 45A. We insert a phase-retarder R in the path of the wave which passes from the slit in S to the slit *1'*, otherwise the arrangement is exactly as in Fig. 41A. The only effect of the phase-retarder is that the amplitude A_1 becomes replaced by the amplitude $A_1e^{i\theta}$: the phase-retarder retards the phase by an amount θ, but does not affect the amplitude of the wave. If the experiment is done with light we can use glass plates as retarders.

Let $P(\theta)$ be the probability that a particle entering through the slit in S emerges through the slit in D (with all the other slits open). According to Eq. (43b) we have

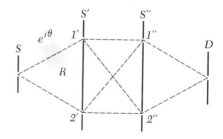

Fig. 45A This figure shows a modification of the double doube-slit experiment illustrated in Fig. 41A. A phase retarder R has been inserted in front of the slit *1'*. It changes the complex amplitude of the wave passing through it by the factor $e^{i\theta}$. The theory of the experiment shown in Fig. 41A applies, provided we replace the transfer amplitude A_1 by the quantity $A_1e^{i\theta}$.

$$P(\theta) = |A_1e^{i\theta}(B_{11}C_1 + B_{12}C_2) + A_2(B_{21}C_1 + B_{22}C_2)|^2$$
$$= |A_1(B_{11}C_1 + B_{12}C_2)|^2 + |A_2(B_{21}C_1 + B_{22}C_2)|^2$$
$$+ U\cos\theta + V\sin\theta \tag{45a}$$

where

$$U = A_1(B_{11}C_1 + B_{12}C_2)A_2^*(B_{21}^*C_1^* + B_{22}^*C_2^*)$$
$$+ A_1^*(B_{11}^*C_1^* + B_{12}^*C_2^*)A_2(B_{21}C_1 + B_{22}C_2) \tag{45b}$$

and

$$V = i\left[A_1(B_{11}C_1 + B_{12}C_2)A_2^*(B_{21}^*C_1^* + B_{22}^*C_2^*)\right.$$
$$\left. - A_1^*(B_{11}^*C_1^* + B_{12}^*C_2^*)A_2(B_{21}C_1 + B_{22}C_2)\right] \tag{45c}$$

as the reader should convince himself.

If we like we can rewrite the expression for $P(\theta)$ in the form

$$P(\theta) = \tfrac{1}{2}[P(0) + P(\pi)] + \tfrac{1}{2}[P(0) - P(\pi)]\cos\theta$$
$$+ \tfrac{1}{2}[2P(\pi/2) - P(0) - P(\pi)]\sin\theta \tag{45d}$$

which means that $P(\theta)$, as a function of θ, is uniquely determined by its values for the three angles $\theta = 0$, $\pi/2$ and π.

46 Next consider the arrangement shown in Fig. 46A. We now have two *separate* sources, 1 and 2, "illuminating" the slits *1'* and *2'*. In all other respects the situation is as in Fig. 41A. Let the two sources be of equal intensity.

What is the probability P_i that a particle which has passed through the screen S passes through the slit in D? It is clearly given by

$$P_i = \tfrac{1}{2}|A_1(B_{11}C_1 + B_{12}C_2)|^2 + \tfrac{1}{2}|A_2(B_{21}C_1 + B_{22}C_2)|^2 \tag{46a}$$

In this case we have to add the *intensities* produced by each source separately at the slit in D to find the intensity with both sources operating together. The expression $|A_1(B_{11}C_1 + B_{12}C_2)|^2$ is the probability that a particle from source *1* passes through the slit in D, and the expresssion $|A_2(B_{21}C_1 + B_{22}C_2)|^2$ is the probability that a particle from source *2* passes through the slit in D. Every particle passing through the slit in D either comes from source *1*, or else from source *2*, with equal probabilities: hence the factors $\tfrac{1}{2}$ in the expression (46a).

47 Let us ask some further questions, concerning Figs. 45A and 46A. Let us regard the screen S', and everything to the left of this screen, as the source. Figures 45A and 46A then show the same experiment performed with two different sources. We can ask: for the situation shown in Fig. 45A, what is the probability $P'(\theta)$

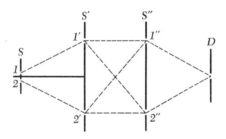

Fig. 46A In this modification of the double double-slit experiment illustrated in Fig. 41A the slits *1'* and *2'* are illuminated by two *independent* sources of equal intensities. The waves from the two sources are incoherent, and the intensity at any particular slit with both sources operating is the sum of the intensities we would observe with only one source operating at a time.

There is an interesting relationship between the experiment illustrated above and the experiment illustrated in Fig. 45A. Any intensity measured in the present experiment is the average over the phase angle θ of the corresponding intensity measured with the arrangement in Fig. 45A. This fact is often expressed by saying that two incoherent sources emit waves of random relative phase.

that a particle coming through the screen S' will also pass through the slit in the screen D? Since every particle coming through the slit in D must have passed through the screen S' it follows that $P'(\theta)$ must be equal to the ratio of the probability $P(\theta)$, given by Eq. (45a), and the probability that a particle coming through the slit in S passes through S'. This latter probability is clearly equal to $[|A_1|^2 + |A_2|^2]$, and we thus obtain

$$P'(\theta) = \big[|A_1(B_{11}C_1 + B_{12}C_2)|^2 + |A_2(B_{21}C_1 + B_{22}C_2)|^2$$
$$+ U \cos\theta + V \sin\theta\,\big] \big[|A_1|^2 + |A_2|^2\big]^{-1} \qquad (47a)$$

which can also be written

$$P'(\theta) = \tfrac{1}{2}[P'(0) + P'(\pi)] + \tfrac{1}{2}[P'(0) - P'(\pi)] \cos\theta$$
$$+ \tfrac{1}{2}[2P'(\pi/2) - P'(0) - P'(\pi)] \sin\theta \qquad (47b)$$

Similarly we can ask: for the situation shown in Fig. 46A, what is the probability P'_i that a particle coming through the screen S' will also pass through the slit in D? We easily see that

$$P'_i = \big[|A_1(B_{11}C_1 + B_{12}C_2)|^2 + |A_2(B_{21}C_1 + B_{22}C_2)|^2\,\big]$$
$$\big[|A_1|^2 + |A_2|^2\big]^{-1} \qquad (47c)$$

Comparing the expression (47c) with the expression (47a) we notice an interesting fact: if we average $P'(\theta)$ over all angles θ between 0 and 2π we obtain P'_i, i.e.,

$$P'_i = \frac{1}{2\pi}\int_0^{2\pi} d\theta\, P'(\theta) \qquad (47d)$$

Actually it is not necessary to average over *all* angles. We also have

$$P'_i = \tfrac{1}{2}[P'(0) + P'(\pi)] \qquad (47e)$$

We can therefore regard the statistical ensemble defined by the source shown in Fig. 46A (the screen S', and everything to the left of S', is the source) as an incoherent superposition of two, or an infinite number of statistical ensembles defined by the sources shown in Fig. 45A, with θ regarded as a variable parameter. (Different values of θ correspond to different sources.)

48   The result (47d) illustrates a general principle concerning an incoherent superposition. If we have two sources which are incoherent we first regard them as coherent and add the amplitudes of the waves from the two sources, but with a variable relative phase factor $e^{i\theta}$. We compute any interesting "intensity" $I(\theta)$, as a function of θ, and we finally average $I(\theta)$ over all angles θ between 0 and 2π. The resulting average, $\bar{I}$, is the appropriate average when the two sources are *incoherent.* Two sources of random relative phase are incoherent.

49 After these exercises on amplitudes, intensities and probabilities, let us continue with our systematic discussion of statistical ensembles.

The set of all statistical ensembles clearly consists of two subsets: those ensembles which can be regarded as incoherent superpositions of two or several other distinct statistical ensembles, and those which cannot be regarded as such superpositions. The statistical ensembles which cannot be regarded as incoherent superpositions of other ensembles are called *pure ensembles,* or *pure states:* the other kind of ensembles are called *mixed ensembles,* or *statistical mixtures.*

Consider now a mixed ensemble. We know that such an ensemble must be an incoherent superposition of other ensembles. Is it also true that the mixed ensemble can be thought of as an incoherent superposition of *pure* ensembles? This question is really a question about the nature of the set of *all* physically realizable statistical ensembles. It might certainly be the case that the set of all physically realizable ensembles does not contain *any* pure states, in which case our question would have to be answered in the negative. On the other hand, we could regard the pure ensembles as limiting cases of mixed ensembles, and we might therefore enlarge our set of statistical ensembles to include not only all physically realizable ensembles but also all limiting cases of these ensembles. If we perform this purely mathematical abstraction, as we shall do, we might expect intuitively that our enlarged set of ensembles will have the property that every statistical ensemble is either a pure ensemble or else an incoherent superposition of pure ensembles.

We shall make this reasonable assumption in the following. As a *physical* assumption it is an idealization: we imagine that it is actually possible to realize all the pure ensembles, and to regard all other ensembles as statistical mixtures of these. In practice we may not be able to realize the ideal of a pure ensemble, but there is no reason why we could not approach this ideal arbitrarily closely.

Can the Outcome of Every Measurement Be Predictable in Principle?

50 It is intuitively clear that we know more about the elements of a pure ensemble than about the elements of a mixed ensemble. Consider, for instance, our example of the light source with two lamps. We obviously know less about the properties of the individual photons emerging from the source when both lamps operate than when only one lamp operates: in particular we know less about the color of the photons.

In order to prepare a pure ensemble we must have perfect control over the preparation stage: we must be able to suppress all sources of statistical fluctuations which can in principle be suppressed.

Now it should be clear that when we perform measurements it is desirable to try to arrange the preparation stage in such a way that the ensemble is as pure as is technically feasible. By so doing we minimize the statistical spread in our data, which means that we increase the accuracy of the results. We can furthermore say that the theoretical interpretation of the experimental results is simpler and more clear-cut for a pure ensemble than for a mixed ensemble. For a pure ensemble we can study the behavior of a system under the best possible conditions, undisturbed by *avoidable* "noise."

51   There now arises a fundamental question. Are the pure states characterized by a complete absence of statistical spread in all physical variables? Differently stated: are the pure states such that the outcome of every measurement can be exactly predicted?

We should understand clearly that this question is a question about the nature of our world which can only be answered on the basis of *experimental* studies. Pure logic tells us nothing about what the answer should be.

The theories of Classical Physics are based on the proposition that the answer to the question is *yes.* Quantum Mechanics is a theory based on the proposition that the answer is *no.* (To prevent misunderstandings we must state here that quantum mechanics is merely a particular one among conceivable theories which answer the question by *no.*) When we accept quantum mechanics as our theory we thus introduce an indeterminacy in our description of nature in the precise sense that no matter how we prepare a pure ensemble there will always be a measurement the outcome of which is not predictable in any particular case. (The measurement which is not predictable depends on the ensemble.) This does not mean that quantum mechanics is "chaotic and vague." The theory is a very definite theory in which we can make precise quantitative statements about *probabilities*, or about *average values* of physical variables.

52   It is in the nature of the question which we have asked that no single set of experiments could possibly decide with finality what the answer should be. Whenever we encounter a phenomenon where we would be tempted to say that the answer is *no*, we might always try to save the situation by the argument that if the measurements are done in a "better way," then the conclusions would be different. In other words: it could always be argued that the indeterminacy arises only from the fact that the experimental arrangements are not the best possible. This kind of argument is hard to refute in an absolute sense. On the other hand, it is fair to ask the person who argues in favor of a deterministic theory of nature, in the classical sense, to show explicitly *how* the measurements are to be performed such that the indeterminacy characteristic of quantum mechanics disappears.

The evidence in favor of the answer *no* is two-fold. First of all, the detailed analysis of a large variety of experiments, in which we take into account observed properties of particles, always seems to lead to the conclusion that the answer should be *no*. Secondly, there is the fact that all the predictions of the theory of quantum mechanics, in which theory the answer *no* is a cornerstone, seem to be in very excellent agreement with all the observed facts: assuming *no* as the answer never seems to lead us to any contradiction with experience.

53 In Chaps. 4 and 5 we have already presented very convincing evidence that the answer must be *no*. The real particles found in nature spread out like waves in space. The waves are divided by half-silvered mirrors and by double slits, and in general diffracted by any obstacle. On the other hand, if we look for the particle with a photo-cell, or some other particle detector, we never find "fractional photons," or "fractional electrons." To describe consistently *all* of these phenomena we are forced to a probability interpretation of the *intensity* of a wave: quantities which are proportional to the absolute square of the wave function must represent *probabilities*. We can only state the probability for a counter to click, but we can never arrange it so that we are certain about the response of every counter in every single experiment.

Consider, for instance, the double-slit experiment. If we want to be certain about the momentum of the incident beam we must arrange it so that the momentum of the particles is extremely well defined. When such a beam is incident on a screen with two slits we will observe the characteristic two-slit diffraction pattern. This pattern can arise only if both slits are open, i.e., only if the particle passes through both slits. However, if we try to catch the particle with a counter placed behind one of the slits we do not detect half of the particle, but the entire particle. In any single experiment the counter may, or may not, click, and we do not know beforehand what will happen with certainty: we can only state the probability that the counter will click. The reader may say: Well, that is only because the ensemble is not pure. But what does the reader suggest that we should do in order to make it purer?

54 The crux is clearly whether it is possible to describe a particle in some other way in more detail than it can be described in terms of the wave theory. If the wave description is correct, and if in addition the particles have the property of indivisibility so that we can never detect "fractional particles," then there seems to be no escape from the probability interpretation of intensities. Let us recall our discussion of the uncertainty relations in the beginning of this chapter. If the *momentum* of the particle is to be accurately known it must be described by a wave which is spread out in space, but then the *position* of the particle cannot be accurately known. A small statistical spread in momentum measurements implies a large statistical spread in position measurements, and as long as we accept the wave description with the probability interpretation of intensities there is just no way of beating the uncertainty relation. On the other hand, there is no experimental material which would suggest that a particle can be described in more detail than the wave description permits: there is absolutely no evidence for any "hidden variables."

These considerations have led to the following basic assumption of Quantum Mechanics: the pure states of a particle are described by waves. *An ensemble of one-particle states is pure if and only if every element of the ensemble is described by the same wave function.* When we can state a wave function which describes all the particles of an ensemble it means that we have *maximum possible* control over the source. Nothing can be purer than a definite wave.

55 It is interesting to compare some aspects of the classical world of fantasy with the real world. The concepts of statistical ensembles, statistical mixtures and pure states are not at all foreign to classical physics. The idea of a statistical ensemble was in fact introduced in classical statistical mechanics long before the discovery of quantum mechanics. A good portion of our discussion of the measurement process remains applicable within the framework of a classical description. A pure state thus arises when we have perfect control over the preparation stage, whereas a statistical mixture arises when our control is less than perfect. The crucial difference between the classical and quantum-mechanical description lies in the nature of the pure state. According to classical ideas a pure state has the property that the outcome of every single measurement is *exactly* predictable. If a given counter clicks in *one* single experiment, it clicks in every subsequent experiment as well. Each time an experiment is repeated the same things happen which have happened before. For a pure state there is no statistical spread in *any* physical variable.

Physicists have recognized for a long time, and long before the development of quantum mechanics, that happenings in the macroscopic world cannot in *practice* be predicted with unlimited precision. Thermal noise and many other kinds of "disturbances" over which we have no control are always present, and in macroscopic situations these causes of uncertainty in the value of a physical variable completely mask the characteristic quantum-mechanical uncertainty. The classical physicist's belief that pure states are characterized by a complete absence of statistical spread in the variables was never really critically tested in macroscopic situations, and this explains why the belief could persist for so long.

56 The recognition of the probabilistic nature of *all* predictions, even in the case of a pure ensemble, was an important step in the development of physical theory. When we consider in retrospect the early history of quantum physics we can appreciate the conceptual difficulties which physicists faced before the emergence of the probabilistic description. That light could exhibit both wave properties and corpuscular properties was a bewildering discovery. This "duality," as it was then called, can now be easily understood along the lines discussed in Chap. 4, but in the early days of quantum physics the situation was different. It had not occurred to anybody to interpret the square of a wave amplitude in terms of probabilities, and without this idea, which is a radical departure from classical ideas, the "duality" of light cannot be understood.

That there is a limit *in principle* to our ability to predict future events has been regarded by many, and in particular by philosophically minded non-physicists, as a most profound and revolutionary idea. Inevitably an enormous amount of nonsense has been written about this issue (and about the uncertainty relation) and the authors have drawn all kinds of far-fetched conclusions about the implications of quantum mechanics for human affairs in general.

The author does not deny that the question of predictability versus non-predictability is an interesting question of principle, and that it is a legitimate subject of philosophical speculation. He wants to mention, however, that professional physicists *today* appear to pay very little attention to this issue. The author cannot recall a single lunch-table discussion in which this issue has been mentioned. (Lunch-table discussions otherwise range over all questions which occupy the thoughts of physicists.) It is, in fact, fair to say that physicists think very little about the theory of measurements in quantum mechanics, *except* when they teach an introductory course on the subject.

Polarized and Unpolarized Light

57 A study of the polarization of light can give us a nice illustration of the distinction between a pure state and a statistical mixture within quantum mechanics. Consider the experimental arrangement shown in Fig. 57A. Almost monochromatic photons of frequency ω pass through a polarization filter F_s, and emerge from the source through the slit in the screen S. The preparation of the statistical ensemble thus takes place to the left of S. The emerging photons are studied by a photo-cell P, fitted with a polarization filter F_p, and the filter together with the photo-cell is to be regarded as a single instrument described by the counter variable D.

It is possible to construct polarization filters which perform to a very high degree of perfection, and which thus have the property that they let through unhindered waves of a definite state of polarization, but completely absorb waves in the opposite state of polarization. We shall assume that the filters F_s and F_p are perfect polarization filters, the properties of which we can select at will.

58 Suppose now that we let the filter F_s be a filter which lets through only left-circularly polarized waves: the emerging photons are elements of a statistical ensemble ρ_L. We first determine the counting rate with the filter F_p removed: this will tell us the number of photons emerging per unit time, and serves to normalize our data. We assume that the counter P has 100 percent efficiency so that it counts every photon which reaches it. Let the counting rate be n photons per unit time.

We consider several different filters F_p, and to every filter-counter combination corresponds a counter variable D. The average of D is defined as the ratio n'/n, where n' is the counting rate in the presence of the filter. If F_p is a filter letting through only left-circularly polarized light, the corresponding counter variable is denoted by D_L; if it lets through only right-circularly polarized light we denote the counter variable by D_R; if it lets through light polarized linearly in the x-direction we denote the variable by D_x; if it lets through light polarized linearly in the y-direction we denote the variable by D_y. Finally, we consider filters which let through

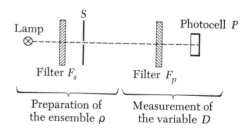

Fig. 57A Schematic figure showing an experiment with polarized light. The filters F_s and F_p are assumed to be ideal polarization filters. (Light passing through an ideal polarization filter emerges in a definite pure state of polarization, and the filter is completely transparent for such light.)

The response of the counter P for an individual photon is not precisely predictable unless the filters F_s and F_p correspond to the *same* (pure) state of polarization.

light linearly polarized along the line which bisects the quadrant bounded by the positive x- and y-axis (counter variable $D_{45°}$) and filters which let through light polarized perpendicular to this bisector (counter variable $D_{135°}$).

For the ensemble ρ_L we find the following averages:

$$\text{Av}(D_L;\rho_L) = 1, \qquad \text{Av}(D_R;\rho_L) = 0 \qquad (58a)$$

$$\text{Av}(D_x;\rho_L) = \text{Av}(D_y;\rho_L) = \text{Av}(D_{45°};\rho_L) =$$
$$\text{Av}(D_{135°};\rho_L) = \tfrac{1}{2} \quad (58b)$$

For this ensemble the two variables D_L and D_R are precisely known, whereas we are in a state of maximum ignorance about the remaining four variables. Is the ensemble ρ_L pure? What this question really means is: can we make it purer? The answer is no. If we require that the variables D_L and D_R be precisely known and have the values given in Eq. (58a) we know that the photons emerging from the source *must* be strictly left-circularly polarized. But every left-circularly polarized wave can be resolved into two linearly polarized waves, of equal amplitude, and polarized perpendicular to each other. If we insert a filter which removes one of the linearly polarized conponents the intensity of the transmitted wave will be $\tfrac{1}{2}$ the intensity of the incident light. The averages of the variables D_x, D_y, $D_{45°}$ and $D_{135°}$ must therefore *necessarily* be as stated in Eq. (58b). If we now combine this *experimental* result about the averages with the *experimental* result that photons cannot be split ("in energy") by a polarization filter, we draw the inevitable conclusion that none of the four variables D_x, D_y, $D_{45°}$ and $D_{135°}$ is precisely predictable in any single experiment. In fact, the uncertainty about these variables is as large as it can be, and this is despite the fact that the ensemble must be regarded as the purest possible ensemble of circularly polarized photons.

59 The reader should note carefully that the conclusion would be entirely different if photons behaved in every respect like classical wave trains. Then the average of the variable D_x would depend on the sensitivity of the detector. If the sensitivity is such that the energy carried by half a wave train is registered, then the counting rate of D_x would be the same as the counting rate of D_L, i.e. $\text{Av}(D_x;\rho_L) = 1$, whereas the average would be zero if the sensitivity is such that the energy carried by half the wave train is not sufficient to trigger the counter. Actual photons *do not* behave like classical wave trains: irrespective of what filter we place in front of the counter we will always find that each photon registered by the counter carries the energy $\hbar\omega$.

The response of the counters D_x, D_y, $D_{45°}$ and $D_{135°}$ is therefore not predictable in any single experiment on the *pure* ensemble ρ_L, and we here have strong evidence in support of the general conclusions stated in Secs. 51–54.

60   What happens if we remove the filter F_s? If we assume that the "lamp" is a spherically symmetric object there will be no preferred directions, and every state of polarization is just as likely as any other state of polarization. We say that the light is *unpolarized*. The corresponding ensemble, ρ_0, is *the most chaotic ensemble* with respect to the polarization-degree-of-freedom, and irrespective of the nature of the ideal polarization filter F_p the counting rate with the filter will be $\frac{1}{2}$ times the counting rate without the filter. We thus observe the averages

$$\mathrm{Av}(D_L;\rho_0) = \mathrm{Av}(D_R;\rho_0) = \tfrac{1}{2} \tag{60a}$$

$$\mathrm{Av}(D_x;\rho_0) = \mathrm{Av}(D_y;\rho_0) = \mathrm{Av}(D_{45^\circ};\rho_0) =$$
$$\mathrm{Av}(D_{135^\circ};\rho_0) = \tfrac{1}{2} \tag{60b}$$

We note that the averages in (60b) agree with the averages in 58b), and the amount of ignorance about the four variables D_x, D_y, D_{45° and D_{135° is therefore the same for the ensembles ρ_L and ρ_0. The ensembles *differ* in the amount of information which we have about the two variables D_L and D_R; for ρ_L we have perfect knowledge about these variables whereas we know as little as is possible about them for the ensemble ρ_0.

We thus expect that the ensemble ρ_0 must be a statistical mixture. To see this explicitly we first consider an experiment when the filter F_s lets through only right-circularly polarized waves. Call the corresponding ensemble ρ_R. The ensemble averages are then given by

$$\mathrm{Av}(D_L;\rho_R) = 0 \qquad \mathrm{Av}(D_R;\rho_R) = 1 \tag{60c}$$

$$\mathrm{Av}(D_x;\rho_R) = \mathrm{Av}(D_y;\rho_R) = \mathrm{Av}(D_{45^\circ};\rho_R) =$$
$$\mathrm{Av}(D_{135^\circ};\rho_R) = \tfrac{1}{2} \tag{60d}$$

As the reader should check in detail, the ensemble averages for ρ_0, ρ_R and ρ_L are such that we can write

$$\rho_0 = \tfrac{1}{2}\rho_L + \tfrac{1}{2}\rho_R \tag{60e}$$

in accordance with our discussion in Sec. 38. We can therefore regard the chaotic ensemble ρ_0 as an *incoherent superposition* of the two *pure* esembles ρ_R and ρ_L.

61   The author wants to mention that he was bothered in his childhood by the difference between unpolarized light and circularly polarized light. The books said that unpolarized light is a mixture of light polarized in two perpendicular directions, and the books also said that circularly polarized light is a superposition of light polarized in two perpendicular directions. The author finally realized that for circularly polarized light one adds the *amplitudes* of the two linearly polarized components, but for unpolarized light one adds the intensities. Circularly polarized light is a *coherent* mixture of light polarized in two perpendicular directions, whereas unpolarized light is an *incoherent* mixture.

References for Further Study

1) It is proper that the reader supplement the theoretical studies in this chapter with some reading on actual counters and related equipment.

  a) Chapter 5, of D. Halliday; *Introductory Nuclear Physics*, (John Wiley and Sons, Inc., 1950), is devoted to a discussion of the detection of charged particles and photons. Various types of counters, and associated electronic equipment, are discussed.

  b) The statistical analysis of counter data is discussed in the above reference. See also, L. J. Rainwater and C. S. Wu: "Applications of Probability Theory to Nuclear Particle Detection," *Nucleonics* vol. 1, no. 2, p. 60, (1947) for a simple and clear discussion.

  c) G. D. Rochester and J. G. Wilson: *Cloud Chamber Photographs of the Cosmic Radiation* (Academic Press, Inc., New York, 1952). It is eminently worthwhile to look at this book and its many interesting pictures.

  d) An elementary discussion of the detection of particles is given in Chap. 3, in D. H. Frisch and A. M. Thorndike: *Elementary Particles* (D. van Nostrand Company, Inc., 1964).

  e) For a collection of stereoscopic bubble chamber pictures, see *Introduction to the Detection of Nuclear Particles in a Bubble Chamber* (Prepared at the Lawrence Radiation Laboratory, The University of California, Berkeley.) (The Ealing Press, 1964.)

2) Note the following articles in the *Scientific American* magazine:

  a) O. M. Bilaniuk, "Semiconductor Particle-Detectors," Oct. 1962, p. 78.

  b) G. B. Collins, "Scintillation Counters," Nov. 1953, p. 36.

  c) G. K. O'Neill, "The Spark Chamber," Aug. 1962, p. 36.

  d) H. Yagoda, "The Tracks of Nuclear Particles," May 1956, p. 40.

  e) D. A. Glaser, "The Bubble Chamber," Feb. 1955, p. 46.

  f) D. E. Yount, "The Streamer Chamber," Oct. 1967, p. 38.

Problems

1 One of the favorite arguments of those who want to refute the uncertainty relation goes as follows. (See adjoining figure.) A monoenergetic beam of electrons, of momentum p, is incident normally on the screen S_1 from the left. This screen has a circular hole, of diameter a. At a distance d

from the screen S_1 we have another screen S_2, which likewise has a circular hole of diameter a. We assume that the two holes are lined up in the direction of the incident beam. Some of the electrons which pass through the first hole might be deflected, but some of them will go on to pass through the second hole. Consider an electron which has passed through the second hole. The uncertainty in its lateral position is of order $\Delta x \cong a$. The *magnitude* of its momentum is p, the same as in the incident beam, because the electrons do not lose or gain energy in this experiment. Since we know that the electron has passed through *both* holes the uncertainty in the direction of the momentum must be less than, or equal to, $\Delta \theta = a/d$. It follows that the uncertainty in the lateral component of the momentum of the electron is of order $\Delta p \cong (a/d)p$. We thus have

$$\Delta x \, \Delta p \cong \left(\frac{a}{d}\right) a \, p$$

for the product of the uncertainties in lateral position and lateral momentum. By making a small and d large we can make this product as small as we please, and hence violate the uncertainty relation, which is one of the cornerstones of quantum mechanics.

Can you demolish this argument? Make sure that you meet all counter-arguments to your arguments.

The above argument is one among many that have been advanced against quantum mechanics via the uncertainty relation. Now it should be clear that there is never any danger that the uncertainty relation will be refuted by this or any similar arguments, *as long as the premises of wave mechanics are accepted*, because with these premises the uncertainty relation can be proved. One might group the "refutations" of wave mechanics into two classes:

(*a*) Arguments in which the ideas of wave mechanics are really denied, although this is not always stated explicitly,

(*b*) Arguments which are "muddled" but which are based on some of the ideas of wave mechanics.

A careful conceptual analysis will clarify the nature of the "refutation." An outright denial of the principles of wave mechanics cannot, of course, be contradicted on logical grounds, but one can always appeal to experimental facts: the "refutation" carried to its logical conclusion might contradict one of these facts. The arguments in the category (*b*) are simply faulty.

2 (*a*) We consider an idealized experiment in which almost perfectly monochromatic light of wavelength 6000 Å passes through an extremely fast shutter. Suppose that the shutter opens and closes periodically in such a way that the shutter is open for a time of 10^{-10} sec, and then closed for 0.01 sec during a period. The light passing through the shutter would then no longer be monochromatic, but would show a certain spread in wavelength.

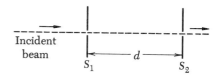

This figure refers to Prob. 1, in which the author fallaciously argues that the uncertainty relation can be violated if we make the slits narrow and the separation d large. It then seems that the product of the uncertainty in lateral momentum with the uncertainty in lateral position at the moment the particle passes through the second slit can be made as small as we like. What is wrong with this idea?

Estimate the magnitude of the uncertainty in the wavelength in Ångström units.

(*b*) We let the light emerging from the shutter pass through a long tube filled with carbon disulfide. This medium is dispersive, and for the wavelength considered the variation of the refractive index n with wavelength is given by

$$\frac{\lambda}{n}\frac{dn}{d\lambda} = -0.075$$

The velocity of the pulse of light let through the shutter could be measured by another shutter placed at a certain distance from the first, and opened at a slightly later time. What is the velocity by which the *pulse* propagates in the carbon disulfide?

3   The author has a new idea for violating the uncertainty principle: this time the time-frequency uncertainty relation. The arrangement is shown very schematically in the adjoining figure. Almost monochromatic light is incident through the slit at left which is fitted with an extremely fast shutter. We shall not be concerned here with purely technical difficulties, and we thus assume that the shutter can be opened for an arbitrarily short interval of time, to admit a sharply defined pulse into the spectrograph indicated symbolically by the prism in the figure. The entering light will, of course, no longer be monochromatic, but will exhibit a spread in frequency as discussed in Prob. 2. We can, however, fit the spectrograph with a suitable narrow exit slit, shown at right in the figure, and thereby select a portion of the incident light whose wavelength falls within an extremely narrow range. The light emerging through the exit slit can therefore be made monochromatic to an arbitrarily high degree: the uncertainty in the frequency can be made as small as we like. On the other hand the duration of the pulse can be made as short as we like with the help of the shutter. The pulse emerging through the exit slit can therefore be of arbitrarily short duration, and be of an arbitrarily precise frequency, contrary to what the uncertainty relation says. Can you find the fallacy in this argument?

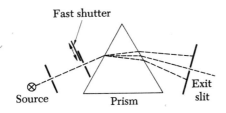

Fast shutter

Source   Prism   Exit slit

This figure refers to Prob. 3. The author again tries to violate the uncertainty relation. The prism symbolizes a spectrograph of very high resolution which is used to select an extremely narrow frequency range of the transmitted light. The incident light is controlled by a fast shutter. The author erroneously maintains that the light pulse emerging from the exit slit can be arbitrarily well defined *both* in frequency and time. What is wrong?

4   With reference to our discussion in Sec. 29, suppose that the temperature of the filament is 1000°C, and suppose that the accelerating potential is 10 volts. Estimate the fractional precision in the momentum of the emerging electrons, i.e., estimate the quantity q/p_0. A crude estimate is enough. Explain your ideas.

5   If we could produce electron beams of very low energy it would be possible to do "macroscopic" electron-diffraction experiments. Suppose that we try to produce a beam of well-defined momentum, with a mean energy of 0.01 eV. Discuss the practical difficulties which we might encounter in this attempt. It is clear that a hot filament with a single accelerating electrode would not do, but there are perhaps other methods which you might think of. If so, state your ideas, and discuss their technical feasibility.

6 Consider the grating shown in the figure illustrating Prob. 5 in Chap. 5. Suppose that the grating is *not* infinitely long, but contains only N openings. In this case the grating is not strictly periodic, and it follows that the diffracted beams exhibit an angular spread. We can express the matter as follows: the characteristic minimum momentum transfer to the grating is no longer precisely $2\pi/a$, but is only defined within an uncertainty of Δq. Try to find a relation between N and Δq. Turn the figure $90°$, and compare with Figs. 5ABCD in the present chapter: this comparison might give you some ideas. Use your result to derive an expression for the uncertainty in the angles at which the various diffracted rays emerge.

7 Consider an almost monochromatic beam of light emerging from a steady source. The problem is to determine the unknown state of polarization of this beam through measurements carried out in the "experimental region."

(*a*) You have available ideal polarization filters and a photomultiplier. What is the smallest number of intensity measurements which you have to carry out in order to determine the polarization state of the beam completely? Explain the basis for your stated conclusion.

(*b*) Suppose you are given a photo-multiplier, two *identical* sheets of polaroid and a quarter-wave plate. How would you determine the state of polarization of the beam, using only the equipment mentioned? In this case you *must not* assume that the polaroid is an ideal polarization filter.

8 The adjoining figure shows a refinement of the double double-slit experiment discussed in Secs. 41–43. Ideal polarization filters are placed (or not placed) in front of the slits and in front of the source and the detector. We assume that the transfer amplitudes discussed in Secs. 41–43 are independent of the state of polarization, and we assume that the light source emits unpolarized light. Derive expressions, analogous to (43b), for the probability that a photon entering through the slit in S will pass through the slit in D, for the various combinations of polarization filters listed below.

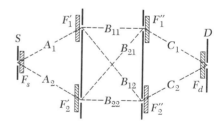

This figure refers to Prob. 8. It illustrates a refinement of the double double-slit experiment shown in Fig. 41A. Ideal polarization filters can cover the various slits. The problem is to determine the probability that a photon entering through the slit in S emerges through the slit in D for various combinations of filters. The numbers A_m, B_{mn}, and C_m are the transfer amplitudes in the absence of the filters. We assume that the transfer amplitudes do not depend on the state of polarization.

F_s	F_1'	F_2'	F_1''	F_2''	F_d
abs	H	V	abs	abs	abs
LC	H	V	abs	abs	abs
LC	H	V	abs	abs	RC
LC	H	V	RC	LC	H
abs	H	abs	abs	H	abs

In this table "abs" means that the filter is not there, H means a horizontal polarizer, V a vertical polarizer, LC a left-circular polarizer and RC a right-circular polarizer.

9 Let us consider the difference between an ideal counter and a practical counter. The practical counter will unfortunately click even if the event under study does not take place, and it will also sometimes fail to register when it ought to. The rate of clicking with the source shut off is known as the *background rate*. One source of background counts is the cosmic radiation which is always present. Furthermore, if two events are separated by too small a time interval the practical counter responds with only a single click. We call the smallest time t_0 for which two events are registered as separate events the *resolving time* of the counter. We can determine the resolving time of a counter as follows. We have two radioactive sources, 1 and 2, which can be placed at certain definite positions near the counter, such that they produce roughly the same counting rate in the counter. Let N_0 be the counting rate with both sources removed. Let N_1 be the counting rate with source 1 present, and let N_2 be the counting rate with source 2 present. Let N_{12} be the counting rate with *both* sources present. We arrange it so that N_{12} is considerably smaller than $1/t_0$, although not completely negligible compared with $1/t_0$. We furthermore assume that N_0 is smaller than N_1, or N_2 or N_{12}. Show that it is possible to determine t_0 from these four rate measurements, and derive an expression for t_0 in terms of N_0, N_1, N_2 and N_{12}.

Note that for an ideal counter, and for the case of no background, we would have $N_{12} = N_1 + N_2$.

Chapter 7

The Wave Mechanics

of Schrödinger

The Wave Mechanics
of Schrödinger

Schrödinger's Non-Relativistic Wave Equation

1   We shall now turn our attention to a phenomenological theory which has played an extremely important role in the development of quantum physics.   This is the theory of the Schrödinger equation, first formulated by Erwin Schrödinger in 1926† shortly after Heisenberg's invention of matrix mechanics.   These two theories were the first quantitative formulation of some of the principles of quantum mechanics.

We discuss the Schrödinger theory in this book because we want to see how a wave theory works out in practice, and see how one can actually compute things within such a theory.   We have selected the non-relativistic Schrödinger theory as our example of a wave theory because it is, in many respects, a particularly simple theory.

2   The theory of the Schrödinger equation (in the most restrictive sense) is based on several drastic approximations, among which we note the following two:

I.   Creation and destruction phenomena of particles are ignored, and in any given physical situation it is thus assumed that the number of each kind of particle remains constant as the process evolves in time.

II.   It is assumed that all relevant velocities are sufficiently small so that a non-relativistic approximation is valid: the discussion is non-relativistic throughout.

We recognize these two assumptions as drastic approximations because we know, empirically, that creation and destruction phenomena *do* occur in nature, and because we also know that any fundamental theory *must* take the facts of special relativity into account.

The two assumptions which we have made are not unrelated to each other.   Consider, for instance, a collision process in which two particles of equal mass collide with each other in such a way that the velocity of each particle in the center of mass system is close to the velocity of light.   Under these circumstances there may be available sufficient kinetic energy for the creation of additional

Erwin Schrödinger.   Born 1887 in Vienna; died 1961. Schrödinger studied physics at the University of Vienna, and obtained his degree in 1910.   After short stays in Stuttgart and Breslau he became professor of physics at Zürich.   In 1927 he was invited to Berlin as Planck's successor.   Schrödinger left Germany in 1933, and eventually accepted a position at the Institute for Advanced Studies in Dublin, as director of its School for Theoretical Physics.   He received the Nobel prize in 1933.

The four papers of Schrödinger mentioned in the beginning of this chapter constitute a monumental contribution to physical theory.   Within a very short time of his discovery of wave mechanics great advances were made in atomic physics.   Schrödinger himself played a very active role in this development. *(Photograph by courtesy of Physics Today.)*

---

† E. Schrödinger, "Quantisierung als Eigenwertproblem," *Annalen der Physik* **79**, 361 (1926); **79**, 489 (1926); **80**, 437 (1926); **81**, 109 (1926).

particles of the same mass, or perhaps of different masses. If, on the other hand, the velocities are small, and the available kinetic energy thus likewise small, then creation phenomena cannot take place: they are forbidden by the law of conservation of energy. There is one notable exception to this statement. Since the rest mass of the photon is zero, photons can always be created or destroyed (i.e., light can be emitted or absorbed) even if all the other particles of non-zero rest masses move at non-relativistic velocities. If we understand the Schrödinger theory in a wider sense, we incorporate the description of light absorption and emission into the theory. We must then amend our assumptions as follows:

I.* Creation and destruction of *material* particles are assumed not to take place, whereas photons may be emitted and absorbed.

II.* All *material* particles are assumed to move with small velocities and they can thus be described non-relativistically. The photon, which can never be described non-relativistically, is accorded a special treatment.

We should mention that there are theories of "relativistic" wave equations in which our second assumption is relaxed. The famous Dirac equation is an example of such an equation. There is also a "relativistic" version of the Schrödinger equation. We shall not discuss these theories here: when we speak of the Schrödinger equation we mean the non-relativistic version, based on our stated assumptions.

3 In the first paragraph of this chapter we referred to the Schrödinger theory as a *phenomenological* theory. It should be so described because it is clearly recognized that the Schrödinger theory cannot claim to be a fundamental theory. We have mentioned some reasons why this is so, and we want the reader to understand this point clearly. The theory of the Schrödinger equation is not the same thing as the theory of quantum mechanics in general.

However, we also want to state explicitly that the Schrödinger theory has proved to be extremely successful when applied to atoms and molecules, and nothing that has been said should be construed as a down-grading of the Schrödinger theory as a useful *approximation*.

4 Before we discuss the Schrödinger equation itself, let us try to understand why the Schrödinger theory, which is based on the two assumptions stated in Sec. 2, should be so successful when applied to atoms and molecules. The basic reason for this is the "small" value of the fine structure constant $\alpha \sim 1/137$. We concluded in Chap. 2 that because α is much smaller than unity, atoms and molecules are loosely bound structures of slowly moving particles. We found, among other things, that to the extent that it makes sense to talk about the velocity of an electron in the hydrogen atom, its velocity will be of the order of $\alpha c \sim c/137$. This velocity is also characteristic of the outermost electrons in other atoms. The nuclei in a molecule will move with a still smaller velocity, and the second assumption on which the Schrödinger theory is based is therefore reasonably well satisfied in the realm of atoms and molecules.

5 Concerning the first assumption we refer to our qualitative discussion in Chap. 2 of the characteristic transition energies in the physics of atoms and molecules. The energies characteristic of molecular binding and of optical transitions are typically of the order of 1–10 eV. The highest energies relevant for atomic structure are the energies of X-rays emitted by the heavy elements, and these energies do not extend beyond 100 keV.

These energies may be contrasted with the rest energy of an electron which is 0.5 MeV. There are no lighter particles than the electron (except the photon, which we agreed should be treated differently), and an electron cannot be created alone in an electromagnetic process but only together with a positron. The creation of a pair requires, however, an energy of 1 MeV, which is much larger than the typical atomic and molecular energies. (The reader may object that neutrinos, which have zero rest mass, are indeed lighter than electrons. The neutrinos interact, however, only *very* weakly with other particles, and in comparison with the electromagnetic interactions the neutrino interactions are completely negligible. In atomic and molecular physics we can forget about the existence of neutrinos.)

6 Quantum electrodynamics, which is a particular example of a so-called *quantum field theory*, has a good claim to being the "correct" theory of atoms and molecules. The Schrödinger theory, applied in this realm, can be regarded as the first approximation to the "correct" theory. If we compare the predictions of quantum electrodynamics with the predictions of the Schrödinger theory we can explicitly study the accuracy of the latter theory. The general result is that the *main* features of atomic and molecular structure are accounted for correctly by the Schrödinger theory. We can express this mathematically as follows. Many theoretical expressions for atomic and molecular quantities, such as energies of

All attempts at a realistic pictorial representation of an atom are doomed to failure. Such a picture suggests to us something which we could actually see with our eyes. The behavior of an atom is so different, however, from the behavior of any familiar macroscopic object that a *direct* visualization is out of the question. This does not prevent us from representing *some* aspects of an atom by a picture. A schematic picture of this kind is somewhat analogous to a cartoonist's representation of complex human activities. If the conventions upon which the picture is based are generally understood, the picture does convey a message.

The above representation of the helium atom is *not* intended to be frivolous. The reader is here reminded that the electrons move slowly in (light) atoms, which is why the non-relativistic Schrödinger theory is applicable. There is another purpose behind the figure. Whenever the reader sees a picture which is supposed to represent an atom, nucleus or molecule, he should remember the Wichmann model and the above remarks about pictorial representations.

stationary states, wavelengths of emitted lines, lifetimes of excited states, geometric parameters of molecules, etc., can be expanded in powers of the fine structure constant α. In such expansions the Schrödinger theory gives the leading term correctly. The terms of higher order can be thought of as "relativistic corrections." These corrections are generally small because of the smallness of α.

7   Let us now try to formulate the Schrödinger theory for a very simple physical situation, namely the motion of a particle, say an electron, in an external field of force. The Schrödinger theory is to be sure much more general than that, and it may be employed to describe the motion of any number of particles interacting mutually with each other. To understand the general features of this theory we should, however, begin with the simplest physical situation.

Let us first consider an even simpler case, namely the case of a single particle moving in the absence of any external forces: in this case we speak of a *free particle*. The Schrödinger theory is concerned with a wave equation, known as the Schrödinger equation, which describes the de Broglie waves associated with the particle. In Sec. 37, Chapter 5, we have already derived such a wave equation, namely the Klein-Gordon equation. This equation is relativistically invariant: it holds irrespective of whether the particle moves slowly or rapidly, and it has the same form in every inertial frame of reference. We now wish to modify this wave equation in accordance with the principles on which the Schrödinger theory is built, which means that we wish to carry out a non-relativistic approximation. Furthermore we will give a definite physical interpretation to the wave function $\psi(\mathbf{x},t)$ which describes the de Broglie wave.

8   In Chap. 5 we have already given a *rough* interpretation of the wave function: the "particle is most likely to be found in those regions of space in which the amplitude $\psi(\mathbf{x},t)$ is large." We shall now make a specific assumption through which this idea becomes precise.

The Schrödinger wave function $\psi(\mathbf{x},t)$, i.e., the amplitude of the de Broglie wave in the Schrödinger theory, describes the probability distribution of the particle in space and time, as follows. If we attempt to locate the particle through a measurement of its position at a given instant of time t the *probability* that we find the particle in a small region of volume $d^3(\mathbf{x})$ containing the point $\mathbf{x}$ is proportional to $|\psi(\mathbf{x},t)|^2 d^3(\mathbf{x})$. The *probability density* is thus proportional to the absolute square of the wave function.

This assumption is characteristic of, and fundamental to, the Schrödinger theory. If we wish to be able to make precise computations we naturally have to give *some* interpretation to the wave function, and the probability interpretation which we have formulated above is both convenient and physically transparent. This profound and important idea was first stated by Max Born[†]

9 The Schrödinger wave function is a complex-valued function of position and time which satisfies the (linear) Schrödinger equation which we shall presently write down. Every definite wave function corresponds to a definite state of motion of the particle. Now we should note that if $\psi(\mathbf{x},t)$ is a possible wave function, so is the function $e^{i\theta}\psi(\mathbf{x},t) = \psi_1(\mathbf{x},t)$ where θ is any real constant. Furthermore, and what is most important, the probability distributions defined by ψ and by ψ_1 are *identical*. This means that the two wave functions $\psi(\mathbf{x},t)$ and $\psi_1(\mathbf{x},t)$ describe the *same* state of motion of the particle. We may express this as follows: to every wave function corresponds a *unique* state of motion of the particle. The converse is not true: a given state of motion of the particle defines a Schrödinger wave function only up to a *constant* complex factor of modulus unity, i.e., up to a complex factor of absolute value 1. Two wave functions which differ only in such a factor correspond to the *same* physical state.

10 Let the mass of the particle be m. We consider a plane wave of momentum $\mathbf{p}$. The energy of the particle is then given by[‡]

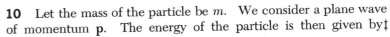

$$E = \sqrt{m^2c^4 + c^2p^2} \qquad (10a)$$

Let us now go to the non-relativistic approximation in which we assume that the velocity of the particle is much smaller than the velocity of light. This means that in Eq. (10a) the term $(cp)^2$ is much smaller than the term $(mc^2)^2$, and we shall therefore expand the square root in (10a) keeping only the first two terms:

$$E \cong mc^2 + \frac{p^2}{2m} \qquad (10b)$$

The first term in (10b) is the rest energy of the particle, and the second term is the non-relativistic expression for the kinetic energy of the particle.

The corresponding de Broglie wave function, which we denote

Max Born. Born 1882 in Breslau, Germany (now in Poland). Born first studied mathematics at Breslau, Heidelberg, Zürich and Göttingen, but later changed to physics. In 1921 he was appointed professor of theoretical physics at the University of Göttingen. Born left Germany in 1933, and after three years at Cambridge he was appointed professor of natural philosophy at the University of Edinburg. After his retirement in 1953 he returned to Germany. He was awarded the Nobel prize in 1954.

Born made many significant contributions to the development of both matrix mechanics and wave mechanics, as well as in other domains of physics. His statistical interpretation of quantum mechanics is particularly noteworthy. It was an essential step that had to be taken before the theory could really be physically interpreted in a consistent manner. *(Photograph by courtesy of Physics Today.)*

[†] M. Born, "Quantenmechanik der Stossvorgänge," *Zeitschrift für Physik* **38**, 803 (1926).

[‡] We use MKS or cgs units in this chapter.

by $\psi_B(\mathbf{x},t)$, is then given approximately by

$$\psi_B(\mathbf{x},t) = \exp\left(\frac{i\mathbf{x}\cdot\mathbf{p}}{\hbar} - \frac{itp^2}{2m\hbar}\right)\exp\left(-\frac{itmc^2}{\hbar}\right) \qquad (10c)$$

We have written the wave function as a product of two factors. The first of these we shall denote by $\psi_S(\mathbf{x},t)$:

$$\psi_S(\mathbf{x},t) = \exp\left(\frac{i\mathbf{x}\cdot\mathbf{p}}{\hbar} - \frac{itp^2}{2m\hbar}\right) \qquad (10d)$$

and we then have

$$\psi_B(\mathbf{x},t) = \psi_S(\mathbf{x},t)\exp\left(-\frac{itmc^2}{\hbar}\right) \qquad (10e)$$

and hence

$$|\psi_B(\mathbf{x},t)|^2 = |\psi_S(\mathbf{x},t)|^2 \qquad (10f)$$

As we see from equation (10f) the two wave functions ψ_S and ψ_B differ only in a complex factor of modulus unity and this factor is *independent* of the state of motion of the particle, i.e., it is independent of $\mathbf{p}$. The absolute squares of the two wave functions are identical everywhere and at all times. To describe the probability distribution of the particle we can employ the wave function ψ_S just as well as the "correct" de Broglie wave function ψ_B. This is precisely what is done in the Schrödinger theory, and ψ_S, as given by (10d), is thus taken to be the Schrödinger wave function describing a free particle moving with the small momentum $\mathbf{p}$. This convention is purely a matter of convenience: why should we carry along the factor $\exp\left(-itmc^2/\hbar\right)$ in our calculations when it ultimately has "no physical effect?"

11   An arbitrary Schrödinger wave can be obtained as a superposition of plane Schrödinger waves of the form (10d). To find the wave equation satisfied by every Schrödinger wave we proceed as in Sec. 37, Chap. 5. We find, in other words, the simplest *linear* wave equation satisfied by every *plane*-wave function. The derivation is entirely analogous to our discussion in Chap. 5 and we obtain

$$i\hbar\frac{\partial}{\partial t}\psi(\mathbf{x},t) = -\frac{\hbar^2}{2m}\boldsymbol{\nabla}^2\psi(\mathbf{x},t) \qquad (11a)$$

where we have dropped the subscript S from the wave function: henceforth we will only deal with the Schrödinger wave function $\psi_S(\mathbf{x},t) = \psi(\mathbf{x},t)$ and the subscript is therefore superfluous.

The equation (11a) is the Schrödinger wave equation for a free particle. It describes the motion of such a particle in the *non-relativistic* approximation. Comparing (11a) with the relativistic equation (37e) in Chap. 5 we note that (11a) involves only the *first* derivative with respect to time. Also, the velocity of light does not appear in Eq. (11a) which is in accordance with the non-relativistic nature of the Schrödinger theory.

12 Consider the plane wave solution (10d) of the Schrödinger equation (11a). The *phase velocity* v_f' of this wave is

$$v_f' = \frac{\omega}{k} = \frac{p}{2m}, \quad \text{where } \omega = \frac{p^2}{2m\hbar}, \quad k = \frac{p}{\hbar} \quad (12a)$$

The phase velocity v_f of the de Broglie wave given (in the non-relativistic approximation) in (10c) is, on the other hand,

$$v_f \cong \frac{mc^2}{p} + \frac{p}{2m} \quad (12b)$$

The reader may be bothered by the fact that the two phase velocities v_f' and v_f are not equal, although the two kinds of waves ψ_B and ψ_S are supposed to describe exactly the same physical situation. However, there is no cause for alarm: the phase velocity is not the same thing as the velocity of the particle, and it does not correspond to anything directly observable. The *group* velocity v, on the other hand, is given by

$$\frac{1}{v} = \frac{dk}{d\omega} = \frac{m}{p} \quad (12c)$$

for the Schrödinger wave, and this velocity is indeed equal to the velocity of the particle as it should be. We have already shown, in Chap. 5, that the group velocity of a de Broglie wave is also equal to the velocity of the particle, and the two kinds of waves therefore do propagate with the same group velocities.

13 Let us now try to go one step further and consider the motion of the particle in an external field of force derivable from a potential. We shall denote the potential energy of the particle by $V(\mathbf{x})$: the potential is a function of position but not of time.

The reader may have some doubts about the idea of introducing a potential in quantum mechanics to describe the forces acting on a particle. The forces on a particle are, of course, due to the presence of other particles, and consistency requires that these other particles should also be described quantum mechanically. *All* the particles in a given physical situation should be described as waves, and a *fundamental* theory of particle interactions must therefore be a theory describing the interactions between the de Broglie waves of the particles. *Quantum field theory* is a theory in which such a fundamental description is attempted. According to this theory the de Broglie wave describing an electron interacts with the quantized electromagnetic field, and this field may in turn interact with the de Broglie wave describing a proton. The electromagnetic interaction between an electron and a proton thus arises indirectly; it is mediated by the quantized electromagnetic field. We express this by saying that the interaction arises through an *exchange of photons*. (This is a nice figure of speech.)

In this chapter we are, however, working within the framework of the approximations characteristic of the Schrödinger theory: we are not working with a fundamental but with a phenomenological theory. We are only interested in the motion of a *single* particle, and it is then reasonable to try to represent the effect of all the other particles by an *effective potential* $V(\mathbf{x})$, and it is furthermore reasonable to be guided by the classical analog in the selection of this potential.

The justification for the introduction of a potential function is particularly transparent if we consider the motion of a charged particle in a *macroscopic* electric field, defined by a number of conductors connected to batteries. In this case we know that we can describe the motion of an electron to a high accuracy within classical theory, and the nature of the trajectories is determined by the electrostatic potential defined by the system of conductors. In the language of quantum field theory the electron exchanges photons with all the charged particles in the conductors. It should be intuitively obvious, however, that the net effect of all these "photon exchanges" can be described in terms of an electrostatic potential which the electron "sees" in space.

14　The idea of introducing an effective potential function in the Schrödinger theory is in many respects quite analogous to the introduction of a refractive index in classical optics. We know very well that on the microscopic scale glass is not a homogeneous substance, but is made up of atoms. If we wish to describe the propagation of a light wave (a photon) through the glass in a *fundamental* way we would have to consider the interaction of the light wave with all the individual atoms in the chunk of glass. If, on the other hand, we are contented with a *phenomenological* description of how the light propagates through the chunk of glass (which may be a component of an optical system) then we may describe the effect of all the elementary interactions by an effective refractive index. As we have said there is a certain analogy between this refractive index and the potential in the Schrödinger

Fig. 15A We "derive" the Schrödinger equation by first finding the equations which a wave might reasonably obey in the regions I, II and III within which the potential is constant. We easily see that the equations (16c), (16e) and (16f) must hold, and then, hocus-pocus, we combine these into the single equation (17a) which is the Schrödinger equation.

In the figure the potential is represented by the solid line. The energy E is assumed larger than the potential in the three regions. The magnitude of the energy is represented by the heavy dashed line, which thus lies above the potential energy curve.

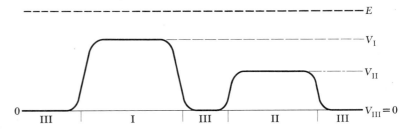

theory, and it will aid our understanding of the Schrödinger theory if we keep this analogy in mind. We should also remember that the description of the electromagnetic properties of a solid through a refractive index has its limitations. Similarly there are physical situations in which the interactions between elementary particles cannot be described at all by a potential function: the potential function makes sense only in those physical situations in which the two basic premises of the Schrödinger theory hold true.

15 Consider now a situation in which there is a bounded region in space, region I, in which the potential energy of the particle is V_{I}. Let there be another bounded region, region II, in which the potential energy is V_{II}. Let us furthermore assume that at the boundaries of these regions the potential falls rapidly to zero. We designate the outside as the region III, and we thus assume that $V_{\text{III}} = 0$. The situation is shown schematically in Fig. 15A, where the heavy solid line represents the potential as a function of position.

Suppose now that a particle of *total non-relativistic* energy E moves around in this potential field of force. Since our discussion is non-relativistic E stands for the sum of the kinetic and potential energies of the particle; the rest energy mc^2 is *not* now included. According to classical mechanics the *kinetic* energy of the particle is then E in region III, $(E - V_{\text{I}})$ in region I and $(E - V_{\text{II}})$ in region II. The kinetic energy, E_{kin}, is related to the momentum p of the particle through

$$E_{\text{kin}} = \frac{p^2}{2m} \tag{15a}$$

The total energy is indicated by the dashed line in Fig. 15A. We assume for the moment that the total energy is everywhere larger than the potential energy.

16 Let us now consider the behavior of a Schrödinger wave associated with the particle. The frequency ω of the wave is related to the energy E through $E = \hbar\omega$, and the wave function therefore depends on the time t through the factor $\exp(-itE/\hbar)$ only. It follows that a Schrödinger wave associated with a particle moving with a definite energy E satisfies the equation

$$i\hbar \frac{\partial}{\partial t}\, \psi(\mathbf{x},t) = E\psi(\mathbf{x},t) \tag{16a}$$

The *spatial* dependence of the wave is determined by the momentum of the particle: the momentum p and the wavelength λ being related through the de Broglie equation $\lambda = h/p$. Consider a wave of energy E in the region III. We imagine that we resolve this wave into a superposition of plane waves. The spatial dependence of these plane waves will be given by the exponential factor $\exp(i\mathbf{x}\cdot\mathbf{p}/\hbar)$ where the magnitude of $\mathbf{p}$ is given by

$$E = \frac{p^2}{2m} \tag{16b}$$

It follows that each one of these plane waves satisfies the differential equation

$$-\frac{\hbar^2}{2m}\, \boldsymbol{\nabla}^2\, \psi(\mathbf{x},t) = E\psi(\mathbf{x},t) \tag{16c}$$

Therefore, the Schrödinger wave corresponding to a particle of energy E must satisfy the differential equation (16c) throughout the region III.

Consider now the wave in the region I. If we resolve the wave in *this* region into plane waves of the form $\exp(i\mathbf{x}\cdot\mathbf{p}/\hbar)$ the magnitude of the momentum $\mathbf{p}$ will now be determined, in accordance with (15a), by

$$\frac{p^2}{2m} = E_{\text{kin}} = E - V_{\text{I}} \tag{16d}$$

and we conclude that the Schrödinger wave in the region I must satisfy the equation

$$-\frac{\hbar^2}{2m}\, \boldsymbol{\nabla}^2\, \psi(\mathbf{x},t) = (E - V_{\text{I}})\, \psi(\mathbf{x},t) \tag{16e}$$

Similarly we conclude that the Schrödinger wave function in the region II must satisfy the differential equation

$$-\frac{\hbar^2}{2m}\, \boldsymbol{\nabla}^2\, \psi(\mathbf{x},t) = (E - V_{\text{II}})\, \psi(\mathbf{x},t) \tag{16f}$$

17 Our reasoning, leading to the three equations (16c), (16e) and (16f), satisfied by the wave function in the regions I, II and III, is certainly plausible, and it is very tempting to summarize these three equations into the single equation

$$-\frac{\hbar^2}{2m}\nabla^2 \psi(\mathbf{x},t) = [E - V(\mathbf{x})] \psi(\mathbf{x},t) \qquad (17a)$$

where $V(\mathbf{x})$ is the potential function which assumes the values V_I, V_{II} and $V_{III} = 0$ in the three regions. It should be noted, however, that we presented no arguments for what the "correct" differential equation should be in the boundary-regions in which the potential changes rapidly, and it is therefore not self-evident that the equation (17a) must hold everywhere. In fact the author now wants to confess that he arranged the arguments leading to our equations, and drew the picture in Fig. 15A in a manner deliberately designed to lead the reader to believe that the equation (16e), for instance, *must* be true. There is actually a flaw in our argument. As long as the region II is very *large* compared to the wavelength of the de Broglie wave in this region we can safely accept our conclusion (16e) as extremely plausible. The *local* behavior of the wave in the region should not depend on the potential elsewhere, and the relation between wavelength and kinetic energy must then be in accordance with our assumption. The situation is different, however, if the region II is *small* compared to the wavelength, i.e., if the potential $V(\mathbf{x})$ varies considerably over a wavelength. In this case it is not so clear what the spatial dependence of the wave function should be, because the "wavelength" at the point $\mathbf{x}$, as defined through the de Broglie relation in terms of the kinetic energy $[E - V(x)]$, is a function of position.

It is therefore not self-evident that the equation (17a) is the correct equation everywhere in space, and for every potential function $V(\mathbf{x})$. Nevertheless we shall *assume*, following Schrödinger, that the equation (17a) *is* correct. As an equation describing the behavior of the Schrödinger waves it is at least a reasonable equation, and we should give it a fair trial. We wanted to make it clear, however, that our discussion is not a *proof* of the correctness of equation (17a) but rather a plausibility argument in its favor. Actually one can do a bit better. One possible approach is to start from quantum electrodynamics, in which case one can show that the equation (17a), as applied to non-relativistic problems involving atoms and molecules, arises as an approximation to the field theory formulation. Another approach is to study systematically what the possible wave equations are which allow a sensible physical interpretation, including the probability interpretation discussed in Sec. 8. We wish to preserve this interpretation of the wave function in the case when the particle is subject to forces. One may then show that the equation (17a) is, in a certain sense, the simplest wave equation for the quantum-mechanical problem which "corresponds" to the classical problem of a particle moving in the potential field of force $V(\mathbf{x})$. It would take us too far to examine these arguments in detail, and we must therefore accept the equation (17a) as a working hypothesis on the basis of the arguments which we have presented.

18   The equation (17a) refers to a wave of definite energy E. For such a wave the relation (16a) holds, and we may therefore rewrite (17a) in the form

$$-\frac{\hbar^2}{2m}\,\nabla^2\,\psi(\mathbf{x},t)\,+\,V(\mathbf{x})\,\psi(\mathbf{x},t)\,=\,i\hbar\,\frac{\partial}{\partial t}\,\psi(\mathbf{x},t) \qquad (18a)$$

In this equation E no longer occurs, and (18a) therefore holds for *every* E, and hence for *every* Schrödinger wave.

The equations (17a) and (18a) are the celebrated Schrödinger equations. The equation (18a) is known as the *time-dependent Schrödinger equation*, whereas (17a) is known as the *time-independent Schrödinger equation*. We should keep in mind that (18a) is the equation holding for *all* Schrödinger waves, whereas (17a) (for a given value of E) holds only for Schrödinger waves describing a particle of total energy E.

The best possible justification for the equations (17a) and (18a) comes, of course, from a comparison of predictions based on these equations with experimental facts. Within a short time after Schrödinger's great discovery the equation was applied to many problems within atomic and molecular physics with spectacular success, and these branches of physics took a great leap forward. In this development Schrödinger himself played an active role, and in the next chapter we shall see how he was able to account for the quasi-stable states of atoms. We have every reason to admire Schrödinger's insight in writing down just the equation (18a) which has turned out to be the correct one for the situations for which it was intended.

It is not our intention to discuss the general theory of the solution of Eq. (18a) in this course: that must be reserved for more advanced courses. We merely wish to consider a few very simple applications of the Schrödinger theory to see how it works out.

Some Simple "Barrier Problems"

19   We have assumed that the Schrödinger equations (17a) and (18a) are valid for an arbitrary potential function $V(\mathbf{x})$. In our "derivation" of Eq. (17a) we considered, however, only the case when the potential $V(x)$ is everywhere smaller than the total energy E. Let us now see what happens in the case when there are regions in space in which the potential is *larger* than E. According to classical mechanics these regions are not accessible to the particle, but we will see that the situation is different in quantum mechanics.

For simplicity we shall restrict our discussion to a one-dimensional world: the particle can move along a line, and its position is given by the coordinate x. The one-dimensional model has the great advantage that the time-independent Schrödinger equation is an ordinary differential equation instead of a partial differential equation, and the mathematical discussion is therefore an order of magnitude simpler. The essential features are, however, already present in this simple model.

20 We consider the Schrödinger equation for the case when the energy of the particle is $E > 0$, and the one-dimensional analog of Eq. (17a) then reads

$$-\frac{\hbar^2}{2m}\frac{\partial^2}{\partial x^2}\,\psi(x,t) = [E - V(x)]\,\psi(x,t) \tag{20a}$$

The time-dependence of the wave function $\psi(x,t)$ is given by the factor $\exp(-itE/\hbar)$, and if we like we may write

$$\psi(x,t) = \varphi(x)\exp\left(-\frac{itE}{\hbar}\right) \tag{20b}$$

in which case the time-independent factor $\varphi(x)$ satisfies the same equation (20a), namely

$$-\frac{\hbar^2}{2m}\frac{d^2}{dx^2}\varphi(x) = [E - V(x)]\,\varphi(x) \tag{20c}$$

which is an ordinary differential equation. If we can solve this equation for $\varphi(x)$ we obtain the Schrödinger wave function $\psi(x,t)$ from (20b).

21 Consider now the situation shown in Fig. 21A. The dashed line indicates the total energy E, whereas the solid line represents the potential function $V(x)$. We assume that as we go to the left in the figure the potential tends to the constant value zero, whereas as we go to the right it tends to the constant value $V_0 > E$. The point x_0, where the kinetic energy has the value zero, is known as the *turning point*. According to classical mechanics a particle incident from the left would stop and turn around at this point. The region to the right of x_0 is inaccessible to the classical particle.

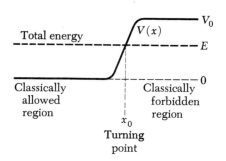

Fig. 21A To illustrate discussion in Sec. 21. The solid line represents the potential, and the heavy dashed line indicates the magnitude of the total energy E. The point x_0 at which the potential equals E is the classical turning point. According to quantum mechanics there is a finite probability that the particle will be found in the classically forbidden region.

We should now solve the equation (20c) for the potential shown in Fig. 21A. The solution $\varphi(x)$ is some function of x which is continuous and has a continuous first derivative. Without really solving the equation explicitly we may guess that the wave function $\varphi(x)$ will not vanish to the right of x_0, which, according to our probability interpretation of the wave function, means that there is a certain non-vanishing probability that we will find the particle to the right of x_0. Quantum mechanics therefore predicts that a particle can penetrate into a region forbidden to it according to classical mechanics.

22 Let us try to study this phenomenon more explicitly. For this purpose we simplify our problem further and replace the smoothly rising potential in Fig. 21A by the step potential in Fig. 22A. For convenience we also select the turning point x_0 as the origin, $x_0 = 0$, on the x-axis. We thus have

$$V(x) = 0 \quad \text{for } x < 0 \qquad V(x) = V_0 > E \quad \text{for } x > 0 \quad (22a)$$

The potential shown in Fig. 22A can be regarded as a limiting case of potentials of the kind shown in Fig. 21A. The potential rises more and more steeply until we reach the idealized situation shown in Fig. 22A. As long as the potential is a continuous function the wave function will be continuous and have a continuous first derivative, and this property is also preserved in the limiting case of a step potential. In this latter case, however, the *second* derivative of the wave function will, in general, exhibit a "jump." These statements, it should be noted, are *mathematical* statements about the differential equations which arise in the Schrödinger theory. As physicists we should always regard the step potential as an idealization of the actual potential, and then there will never be any doubt that the physical wave function must satisfy the continuity properties mentioned.

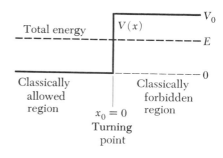

Fig. 22A To simplify the mathematical discussion the continuously varying potential in Fig. 21A is replaced by a step-potential, as shown above.

23 Let us consider the wave equation in the region $x > 0$. In this region it assumes the form

$$-\frac{\hbar^2}{2m}\frac{d^2}{dx^2}\varphi(x) = (E - V_0)\varphi(x) \qquad (23a)$$

and we can immediately find two linearly independent solutions, namely

$$\exp(-xq), \quad \exp(+xq), \quad \text{where } q = \sqrt{\frac{2m(V_0 - E)}{\hbar^2}} \qquad (23b)$$

The solution $\exp(+xq)$ increases exponentially as x increases, and so does its absolute square. According to our probability interpretation of the wave function this means that the probability density for finding the particle increases without limit as x increases. A solution of this nature is not physically acceptable. We here encounter another example of the boundary conditions which physically meaningful solutions of a wave equation must satisfy: a solution which increases indefinitely as we go to infinity must be ruled out on physical grounds. This leaves us with the solution $\exp(-xq)$ as the only possibility, and if we denote the wave function in the region $x > 0$ by $\varphi_R(x)$ we have

$$\varphi_R(x) = \exp(-xq) \tag{23c}$$

24 We next consider the region $x < 0$. In this region the Schrödinger equation assumes the form

$$-\frac{\hbar^2}{2m}\frac{d^2}{dx^2}\varphi(x) = E\varphi(x) \tag{24a}$$

and two linearly independent solutions of this equation are

$$\exp(ixk), \qquad \exp(-ixk), \qquad \text{where } k = \sqrt{\frac{2mE}{\hbar^2}} \tag{24b}$$

These solutions are oscillatory solutions: they do not increase indefinitely as x tends to $-\infty$. Both solutions are physically acceptable,† and if we denote the wave function in the region $x < 0$ by $\varphi_L(x)$ we conclude that this wave function must be of the form

$$\varphi_L(x) = A \exp(ixk) + B \exp(-ixk) \tag{24c}$$

where A and B are constants.

How do we determine the constants A and B? We have said that *the wave function must be continuous and have a continuous first derivative.* This means that the functions $\varphi_R(x)$ and $\varphi_L(x)$ must match at the origin in such a way that

$$\varphi_R(0) = \varphi_L(0), \qquad \varphi'_R(0) = \varphi'_L(0) \tag{24d}$$

since they both represent the *same* wave function but in two different regions which meet at the turning point $x_0 = 0$. The two conditions (24d) give us two equations, namely

$$A + B = 1, \qquad ik(A - B) = -q \tag{24e}$$

---

† If the reader is bothered by this statement, see Sec. 51 of this chapter.

and these two equations determine the constants A and B. The solution is simply

$$A = \frac{(1 + iq/k)}{2}, \qquad B = \frac{(1 - iq/k)}{2} \tag{24f}$$

25 To interpret our solution it is convenient to multiply our wave function (everywhere) by the constant $1/A$: this we can do since the Schrödinger equation is a linear equation. We may then write our solution explicitly in the form:

$$\varphi(x) = e^{ixk} + \left[\frac{1 - i\sqrt{V_0/E - 1}}{1 + i\sqrt{V_0/E - 1}} \right] e^{-ixk}, \qquad \text{for } x < 0 \tag{25a}$$

and

$$\varphi(x) = \frac{2e^{-xq}}{1 + i\sqrt{V_0/E - 1}}, \qquad \text{for } x > 0 \tag{25b}$$

where

$$k = \sqrt{\frac{2mE}{\hbar^2}}, \qquad q = \sqrt{\frac{2m(V_0 - E)}{\hbar^2}} \tag{25c}$$

Consider now the wave function in the region $x < 0$, as given by (25a). It is a superposition of two waves. The first term, $\exp(ixk)$, represents a wave traveling to the *right*. The second term, proportional to $\exp(-ixk)$, represents a wave traveling to the *left*. The coefficient standing in front of $\exp(-ixk)$ in the second term is of modulus unity:

$$\left| \frac{1 - i\sqrt{V_0/E - 1}}{1 + i\sqrt{V_0/E - 1}} \right| = 1 \tag{25d}$$

and the two waves therefore have amplitudes of the same magnitude. The absolute square of the amplitude of a wave must somehow be proportional to the "flux" of the particle, and we can conclude that the wave function in (25a) describes the situation in which a particle incident from the left is reflected back to the left by the potential "hill." This interpretation is in accordance with our classical picture of what goes on.

The wave function in the region $x > 0$, as given by (25b), describes the penetration of the Schrödinger wave into the region forbidden to the classical particle. The amplitude of the penetrating wave decreases exponentially as we go further into the forbidden region, and at large distances from the barrier the amplitude is for

Fig. 25A The upper part of the figure shows the potential $V(x)$. The total energy E is indicated by the heavy dashed line. The lower part of the figure shows the absolute square of the wave function $\varphi(x)$. As we see the wave penetrates into the classically forbidden region. To the left of the barrier we have a standing wave pattern, which arises because the reflected wave interferes with the incident wave. Note that the wave function and its derivative are continuous at the turning point.

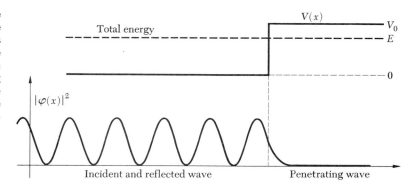

Total energy

$V(x)$

V_0

E

$|\varphi(x)|^2$

0

Incident and reflected wave   Penetrating wave

all practical purposes zero, in accordance with the classical picture. These features are illustrated in Fig. 25A.

26 It is interesting to consider the limiting case when the height of the potential barrier tends to infinity, i.e., when $V_0 \rightarrow +\infty$. (The energy E is kept constant.) By inspection of (25c) we see that as V_0 tends to infinity, q will also tend to infinity, which means that the rate of decrease of the wave function with the distance (from the classical turning point) tends to infinity. The wave function penetrates less and less into the forbidden region. By inspection of (25b) we see that the amplitude of the penetrating wave tends to zero as V_0 tends to infinity. In the limiting case of an infinitely high potential "hill" we thus obtain:

$$\varphi(x) = e^{ixk} - e^{-ixk} \qquad \text{for } x < 0 \qquad (26a)$$

$$\varphi(x) = 0 \qquad \text{for } x > 0 \qquad (26b)$$

Our conclusion is that if the potential barrier is infinitely high, then the wave function must vanish *at* the barrier, i.e., for $x = 0$, and to the right of the barrier, i.e., for $x > 0$.

Fig. 26A shows the behavior of the absolute square of the wave function, i.e., the probability density for the particle. Note that the probability density shows an oscillatory behavior to the left of the barrier. This is a quantum-mechanical interference effect which has no counterpart in classical mechanics. The same feature can, of course, also be seen in Fig. 25A.

27 We have considered the case of the step potential in such detail in order to give the reader confidence that the Schrödinger equation can be solved, and that the solutions can be interpreted physically. Given any reasonably continuous, or stepwise continuous, potential, we can be confident that the solution exists. To

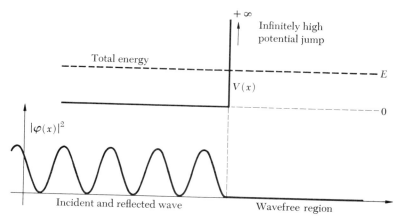

Fig. 26A   This figure illustrates the limiting case of an infinitely high potential step. (Compare with Fig. 25A.) The upper part of the figure shows the potential. The heavy dashed line indicates the total energy E. The lower part shows the absolute square of the wave function $\varphi(x)$. The wave function, but not its derivative, vanishes at the turning point. The derivative of the *square* of the wave function will, of course, vanish at this point.

find it *explicitly* is, however, often not an easy matter, but the complications are only of a technical-mathematical nature. Even without knowing the precise explicit solution we can often say quite a lot about the *nature* of the solution, and hence make general statements about the behavior of the physical system.   On the basis of our studies so far we can conclude that it is a general feature of quantum mechanics that the Schrödinger wave can penetrate into regions forbidden to the particle within classical mechanics.

28   To increase the reader's understanding of the Schrödinger equation, let us consider the following.   Figure 28A shows a "potential step."   We want to study the motion of a particle of energy $E > V_0$ in this potential.   (The *detailed* study is left to the reader as an exercise: Prob. 1 at the end of this chapter.)

The reader will note that we can find *two* physically acceptable solutions of the wave equation (20c) in the region to the left of the step, and we can also find *two* physically acceptable solutions in the region to the right of the step.   How do we know which ones to pick?   That depends on the physical situation which we wish to study.   Suppose that we want to consider the case when a particle is incident on the step from the left.   The wave will perhaps be partially reflected at the step, but part of the wave will continue to travel to the right past the step.   This means that the correct wave function for this problem must be such that it represents a particle traveling to the right in the region to the right of the step: it must be of the form $\exp(ixk')$ for $x > 0$.   In the region to the left of the step the wave function can be of the form $[A \exp(ixk) + B \exp(-ixk)]$, where the first term describes a wave traveling to the right, and the second term describes a wave traveling to the left.   This second term describes the *reflected* wave, whereas the

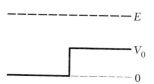

Fig. 28A   To illustrate discussion in Sec. 28.   The energy E of the particle is here larger than the height of the potential barrier.   According to classical theory a particle would not be reflected by this barrier, but according to quantum mechanics the incident wave will be partly transmitted and partly reflected.

first term describes the *incident* wave. How do we find *A* and *B*? *A* and *B* are determined by the two conditions that the wave function and its first derivative must be continuous everywhere, and in particular at the step. This gives us two equations for the two unknowns *A* and *B*, and after we have found these amplitudes we can also find the intensities of the incident, reflected, and transmitted waves, and hence the reflection coefficient of this kind of "barrier."

Suppose that we want to consider instead what happens when the particle is incident from the right. In this case we know that the wave function to the left of the barrier must be of the form exp $(-ixk)$, because to the left of the barrier we only have a wave traveling to the left. To the right of the barrier the wave function is of the form $[A' \exp (ixk') + B' \exp (-ixk')]$. We again find A' and B' by imposing the conditions that the wave function and its first derivative must be continuous at the step. The choice of wave function thus depends on the particular physical problem which we want to consider.

The lesson to be learned from a consideration of the behavior of a particle moving in a potential of the kind shown in Fig. 28A is that in general the particle will be partly reflected by any discontinuity in the potential, and it will be partly able to penetrate past the region of the discontinuity.

29   Let us next consider the situation shown in Fig. 29A in which the potential is discontinuous at the *two* points $x = 0$, and $x = a$. In view of what we learned in the preceding section, a wave will be partly reflected and partly transmitted at both discontinuities.

Suppose we want to consider the case when the particle is incident on this barrier from the left. The reader might feel that this is a difficult problem which should be solved as follows. We consider a wave incident from the left, and we find the part of this wave which is reflected, and the part of this wave which is transmitted, at the first discontinuity at $x = 0$. The transmitted wave hits the second discontinuity, at $x = a$, and is partly reflected and partly transmitted. The reflected part returns to the discontinuity at $x = 0$, and is again partly reflected and partly transmitted. To find the wave which is emerging toward the right from the barrier we thus have to consider an infinite number of reflections back and forth between the two discontinuities, and then add the amplitudes of all the partial waves which are transmitted to the right of the point $x = a$. Can we really solve this problem? The answer is that the problem can indeed be solved in this manner, but there is a much easier way of finding the answer. All we have to do is to

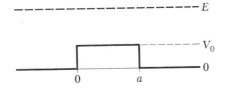

Fig. 29A   To illustrate discussion in Sec. 29. This problem could be solved by taking into account all the repeated partial reflections and transmissions at the discontinuities at $x = 0$ and $x = a$. It is, however, much easier to find the *global* solution of the Schrödinger equation directly: all the multiple reflections are then taken into account in one sweep.

find that solution of the Schrödinger equation (20c) which is continuous everywhere, and has a continuous derivative everywhere, and which is of the form exp (ixk) for $x > a$. The last condition means that the part of the incident wave which is able to penetrate the barrier *must* travel to the right in the region $x > a$: this corresponds to the physical situation which we wanted to consider.

For $x > a$ the wave function is thus of the form exp (ixk). For $a > x > 0$ the wave function is of the form [A exp (ixk') + B exp $(-ixk')$], and to find A and B we impose the conditions that the wave function and its first derivative must be continuous at $x = a$. In the region $0 > x$ the wave function is of the form [A' exp (ixk) + B' exp $(-ixk)$], and we can then determine A' and B' by imposing the conditions that the wave function and its first derivatives must be continuous at $x = 0$. In this manner we find the *global solution* of the Schrödinger equation (20c) which corresponds to the physical problem which we wish to study, and this solution is *unique* (except for an overall constant factor). With a finite effort we can clearly solve this problem.

30   The important thing to understand is that to solve a barrier problem of this kind all we have to do is to find the solution of the Schrödinger equation (20c) which is valid *everywhere*, and which is subject to *boundary conditions* determined by the physical problem under study, i.e., conditions such as the condition that the wave must be of the form exp (ixk) to the right of the barrier. This procedure automatically takes into account all the "multiple reflections" which we are led to think about on the basis of our physical intuition. It is not wrong to try to solve the problem by considering the multiple reflections, but it is *much* easier to find the global solution of the Schrödinger equation directly.

Consider the potential barrier shown in Fig. 30A. *Where* does the reflection of the particle take place? The answer is that it "takes place" throughout the *entire* region in which the potential is changing. If we like we can approximate the continuously varying potential function $V(x)$ by a function which increases by a large number of very small steps, as shown in Fig. 30B. At each step a wave is partly transmitted and partly reflected, and we can again regard our problem as a "multiple reflection problem." The Schrödinger equation (20c) describes concisely all these multiple reflections, and if we like we can interpret the equation in this manner. If we find the global solution of Eq. (20c) we have in effect taken into account all these infinite *local* reflections and transmissions in one sweep.

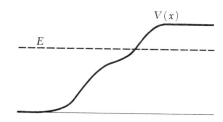

Fig. 30A   The particle (wave) is reflected by this barrier because the energy E is less than the limiting value of the potential at right. (The total energy is indicated by the dashed line, and the potential by the solid line.) *Where* does the reflection take place? The answer is that it takes place throughout the region in which the potential is changing.

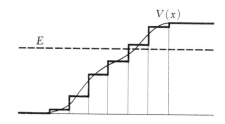

Fig. 30B   The potential shown in Fig. 30A is approximated by a potential which changes stepwise. At each discontinuity a wave is partly reflected and partly transmitted. The solution of the Schrödinger equation takes into account all the "multiple reflections."

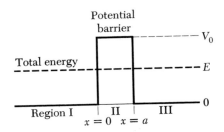

Potential barrier — — — — V_0

Total energy — — — — — — — — E

Region I | II | III   — 0

$x = 0$ $x = a$

Fig. 31A The solid line represents the potential, and the heavy dashed line indicates the total energy. According to classical theory a particle incident from the left cannot pass through this barrier. According to quantum mechanics there is a finite probability that the particle "leaks through." This phenomenon is known as the *tunnel effect*.

31 Let us now consider another problem which readily suggests itself. What happens if the potential is of the form shown in Fig. 31A, when the height V_0 of the barrier is greater than E?

It is easy to guess the answer: a wave incident from the left will be partially reflected by the barrier, and will be partially able to penetrate the barrier into the region III. Classically a particle originally in region I would be reflected at the point $x = 0$, and it could not penetrate into the regions II and III. That a particle can "leak through" a potential barrier which is absolutely opaque classically is one of the striking features of quantum mechanics. The phenomenon is known as the *tunnel effect*.

To solve the Schrödinger equation for the situation shown in Fig. 31A we could proceed just as we explained in Secs. 28–30. We find the general solution in each one of the three regions I, II and III, and we then impose the condition that the wave function, as well as its first derivative, shall be continuous everywhere, and in particular at the two turning points $x = 0$ and $x = a$. The barrier problem of Fig. 31A is therefore not difficult in principle, but to find the detailed solution is somewhat laborious. Fortunately we can understand the essential features of this problem without actually solving the Schrödinger equation completely, and we can therefore leave the detailed solution for a later course. (Or as a homework problem: See Prob. 2.)

32 Let us consider the solution in the following particular case: a particle is incident from the left. It is partially reflected by the barrier, and it is partially able to penetrate. This means that we are interested in a solution of the Schrödinger equation for which the wave function is of the form exp (ixk) in the region III, which represents a particle propagating to the right in this region. In the region I we will necessarily have two waves: one propagating to the left and one to the right. The first of these represents the reflected wave and the second the incident wave. In the region I the wave function is therefore of the form

$$\varphi(x) = e^{ixk} + Ae^{-ixk}, \qquad \text{where } k = \sqrt{\frac{2mE}{\hbar^2}} \qquad (32a)$$

and where A is a constant which describes the amplitude of the reflected wave. Its absolute value will be less than 1 since part of the incident wave penetrates the barrier.

Inside the barrier the wave function will be essentially an exponential function of the form

$$\varphi(x) \cong B \exp{(-xq)}, \qquad q = \sqrt{\frac{2m(V_0 - E)}{\hbar^2}} \qquad (32b)$$

where B is a constant. The above wave function is only approximate, but the approximation is good if the potential barrier is not too low.

Let us suppose that aq is large compared to unity. In that case the ratio $\varphi(a)/\varphi(0) \cong \exp(-aq)$ for the wave function given by (32b) is a small number. If we recall how we matched the two solutions at the turning point in the discussion in Sec. 24, we may conclude that the absolute value of the ratio of the amplitude of the wave in the region III to the amplitude of the wave traveling to the right in region I must be *roughly* given by the ratio $\varphi(a)/\varphi(0) \cong \exp(-aq)$. It is certainly true that the ratio in question is not simply the exponential factor, but, what is most important, this factor completely dominates the situation when aq is large compared to unity, i.e., when the barrier is high and thick.

33 We have assumed that the incident wave is of unit amplitude. The amplitude of the wave penetrating into region III is smaller. Its magnitude, or more correctly its *order* of magnitude, is approximately equal to $\exp(-aq)$. The (absolute) square T of this amplitude has a simple physical interpretation. It is equal to the *probability* that a particle hitting the barrier passes through the barrier. This probability is thus given by

$$T = |\varphi(a)|^2 \sim \exp(-2aq) \tag{33a}$$

or, in view of the second expression (32b),

$$T \sim \exp\left\{-2a\sqrt{\frac{2m(V_0 - E)}{\hbar^2}}\right\} \tag{33b}$$

The quantity T is known as the *transmission coefficient* of the barrier. Our approximate derivation of the expression (33b) for this quantity is, as we see, based on the very simple fact that the amplitude of the wave decreases roughly exponentially as we go to the right *inside* the barrier. We are primarily interested in the case when aq is large, which means that T is very small. We could of course, derive an exact expression for T, in which case an additional factor would appear in the expression (33b). The exponential factor given above is, however, the crucial factor, and the approximate expression (33b) is completely adequate for our purposes.

Figure 33A shows the barrier effect *schematically*. The upper portion of the figure represents the potential, whereas the lower portion shows the absolute square of the wave function. The transmitted wave is a single complex wave traveling to the right. It is therefore of constant modulus as shown in the figure.

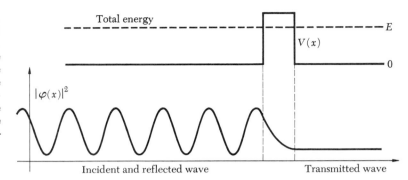

Fig. 33A Schematic illustration of the tunnel effect. The upper portion of the figure shows the potential (with the total energy indicated by the dashed line). The lower portion shows the absolute square of the wave function. Note the transmitted wave and note the exponential decrease of the wave function inside the barrier. To the left of the barrier we have an imperfect standing wave pattern. The amplitude of the reflected wave is smaller than the amplitude of the incident wave, and the combined amplitude is therefore nowhere zero.

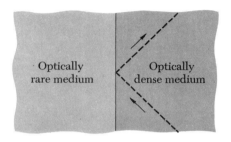

Fig. 34A Total reflection of a plane electromagnetic wave at the plane interface of two media of different refractive indices. The dashed line shows the reflection of a ray.

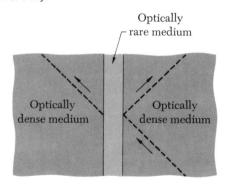

Fig. 34B Frustrated transmission. Classical electromagnetic theory predicts that a wave incident on a thin slab at an angle larger than the critical angle of total reflection will be partly transmitted and partly reflected. This phenomenon is analogous to the quantum-mechanical tunnel effect. A transmitted and a reflected ray are indicated by the dashed lines.

34 Before we consider a physical application of our theory of the quantum-mechanical tunnel effect we wish to point out an analogy to this effect in classical electromagnetic theory. This has to do with the reflection of a plane electromagnetic wave at the plane interface of two regions of different refractive indices.

Suppose a plane wave traveling in some medium (represented by the shaded regions in Fig. 34A) is incident on a plane boundary between an optically rare medium and an optically dense medium. (The refractive index is larger in the dense medium than in the rare medium.) Suppose further that the angle of incidence is *larger* than the angle of total reflection, and that the rare medium extends to infinity to the left of the boundary. The wave will then be totally reflected. This is indicated schematically in Fig. 34A, in which the dotted line represents a "ray," i.e., a normal to the local wave front. Whereas it is true that the wave cannot propagate in the rare medium, the electric field in the neighborhood of the interface is not zero: the field penetrates into the rare medium. As we go further to the left from the interface the electric field amplitude decreases exponentially. This situation is entirely analogous to the quantum-mechanical problem considered in Secs. 22–25.

Consider now the situation shown in Fig. 34B, in which the rare medium is merely a thin slab. In this case the wave incident on the boundary from the right is *partially* reflected. Part of the wave will however be able to penetrate through the "forbidden region" and this part will then propagate in the dense medium to the left. This situation is analogous to the quantum-mechanical barrier penetration. Note that we have not drawn the "ray" in the forbidden region, and the reason for this is that "ray-optics" is not applicable in this region: the wave vector is a complex vector.

The phenomena just described can be completely accounted for by classical electromagnetic theory. The transmission coefficient for the situation shown in Fig. 34B is very small whenever the thickness of the slab of the optically rare material is large compared to the wavelength of the incident radiation. As the thickness decreases the transmission coefficient increases, and it reaches the value one when the thickness is zero.

35 Let us now generalize our discussion of the quantum-mechanical tunnel effect. Instead of the rectangular potential barrier shown in Fig. 31A we consider a barrier of arbitrary shape, as shown in Fig. 35A. Suppose a wave of energy E is incident from the left. This wave will be partially reflected and partially transmitted. We are primarily interested in the overall transmission coefficient T for the barrier, and to find this coefficient precisely we must solve the Schrödinger equation for the potential $V(x)$. We can, however, obtain an approximate expression for T by a different method, based on our discussion in Secs. 32–33. This approximation is better the smaller the wavelength is compared to the width of the barrier.

To derive an approximate expression for the transmission coefficient T we imagine the region of the potential barrier divided into several subregions, as indicated in Fig. 35B. In each one of the subregions we replace the actual potential by a constant potential, as shown in the figure. We have already found the transmission coefficient for a rectangular barrier. Let the transmission coefficients for the 5 rectangular barriers shown in Fig. 35B be $T_1, \ldots, T_5$. The overall transmission coefficient T must then be approximately equal to the product of the transmission coefficients of the subregions, and we have

$$T \cong T_1 T_2 T_3 T_4 T_5 \qquad (35a)$$

or

$$\ln T \cong \ln T_1 + \ln T_2 + \ln T_3 + \ln T_4 + \ln T_5 \qquad (35b)$$

36 Consider now Eq. (33b). If dx_n denotes the thickness of one of the rectangular barriers, and if $V(x_n)$ is the height of the barrier, then the transmission coefficient T_n of this barrier is given by

$$\ln T_n \cong -2 \sqrt{\frac{2m[V(x_n) - E]}{\hbar^2}} \, dx_n \qquad (36a)$$

The logarithm of the overall transmission coefficient is obtained, according to (35b), by summing over all the subregions, and if we now go to the limit of an infinitely fine subdivision we can replace the sum by an integral, and we finally obtain

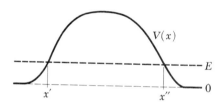

Fig. 35A The solid line represents the potential, and the heavy dashed line indicates the total energy E. How can we derive an expression for the transmission coefficient of this barrier?

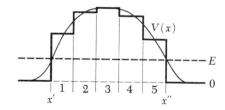

Fig. 35B To illustrate our derivation of an approximate expression for the transmission coefficient for the barrier shown in Fig. 35A. We imagine the continuously varying potential approximated by a set of rectangular barriers. The overall transmission coefficient is the product of the transmission coefficients for all the rectangular barriers. Note that this method is only approximately correct: multiple reflections have not been taken into account.

$$\ln T \cong -2 \int_{x'}^{x''} dx \sqrt{\frac{2m[V(x) - E]}{\hbar^2}} \qquad (36b)$$

We remind the reader that this formula is an *approximate* expression for the transmission coefficient. It is, however, a highly useful formula as it gives us a good qualitative picture of the barrier penetration phenomenon. Note that the integral is evaluated between the two classical turning points x' and x''.

The dependence of the transmission coefficient on the parameters occurring in the expression (36b) should be carefully noted. Other parameters remaining fixed, the transmission coefficient decreases as the mass of the particle increases. Similarly the transmission coefficient increases as the total energy E increases, and for two reasons. The integrand, which is always positive, becomes smaller, and furthermore the range of integration becomes smaller as the turning points move toward each other. The transmission coefficient naturally increases as the width of the barrier decreases.

Theory of Alpha-Radioactivity

37   Let us now try to apply our theory of barrier penetration to an actual physical situation.

In Prob. 3 at the end of Chap. 2 we noted that the half-life of the alpha-particle-emitting radium nucleus $_{88}Ra^{226}$ must be regarded as being "unnaturally long." The half-life is 1622 years, and this seems very long indeed on any sensible *nuclear* time scale. As a characteristic time for nuclear processes we might select the time it takes for light to pass through a nucleus, and this time is of the order of 10^{-23} seconds. The half-life of radium, is, however, 5×10^{10} seconds, or something like 10^{33} "natural nuclear time units." We are thus faced with the problem of explaining the enormous number 10^{33}. Admittedly the "natural nuclear time unit" is a somewhat loose concept, but our problem would not be any simpler even if we made the nuclear time unit 1000 times larger.

There is a further experimental fact which we must note: some alpha-radioactive nuclei have very much shorter lifetimes. For instance the alpha-active polonium isotope $_{84}Po^{212}$ has a half-life of only 3×10^{-7} sec. At the other extreme we note the uranium isotope $_{92}U^{238}$ which is also an alpha-emitter: its half-life is 4.5×10^9 years. The real problem is therefore the enormous range encountered in the lifetimes of alpha-emitters.

The energies of the emitted alpha-particles typically lie in the range 4–10 MeV. Each alpha-active isotope is in general characterized by a definite energy of the emitted alpha-particle, although there are cases in which a nucleus can emit alpha-particles of several different discrete energies. Let us forget about the latter complication, which we discussed briefly in Sec. 40, Chap. 3. Empirically it has been found that there is a strong correlation between the lifetime of the nucleus and the energy of the emitted alpha-particle: the larger the energy the shorter the lifetime.

38 Let us now see whether we can explain the observed facts.† As long as the alpha-particle is inside the nucleus it is acted upon by the strong nuclear forces. These forces, as we have explained, have a short range, and we may imagine that they are inoperative outside the nuclear surface, of radius R. Outside the nuclear surface the dominant force is the electrostatic repulsion between the alpha-particle, which carries the charge $+2e$, and the *daughter nucleus* remaining after the decay. It carries the charge $+Z'e$ if Z' is the atomic number of the daughter. The original nucleus, the parent, carries the charge $+Ze$, where $Z = (Z' + 2)$ is its atomic number. We have shown the situation schematically in Fig. 38A. The distance from the center of the nucleus increases toward the right. The solid curve represents the potential energy of the alpha-particle in the presence of the daughter nucleus. Outside the nuclear surface, i.e., for $r > R$, this potential is simply the Coulomb potential

$$V(r) = \frac{2e^2Z'}{r}, \qquad \text{for } r > R \tag{38a}$$

As we reach the nuclear surface the strongly attractive nuclear force becomes operative, which means that the potential must dip sharply. In Fig. 38A we have idealized the situation by assuming that we actually have a step potential. We have not drawn the potential curve inside the nucleus because it is not well known: it is in fact not well defined since the alpha-particle may lose its individuality as a particle in the strong nuclear force field.

The dashed line represents the total energy of the alpha-particle. This energy, E, is also the energy with which the alpha-particle finally emerges at large distances from the nucleus, where the electrostatic potential energy is effectively zero.

Fig. 37A Early cloud chamber photograph showing tracks of alpha-particles emitted by a radioactive substance. The photograph is taken from L. Meitner: "Über den Aufbau des Atominnern," *Die Naturwissenschaften* **15:1**, 369 (1927).

An alpha-particle of a given energy has a quite well-defined range in matter in bulk. The particle loses energy by ionizing the atoms in the material. The track ends when it has lost all of the initial kinetic energy. The range R in cm in air at standard pressure and temperature is very roughly given by $R = 0.32 \times E^{3/2}$, where E is the energy in MeV.

The radioactive source, located at the bottom of the picture, emitted alpha-particles of two different energies. We can clearly see the well-defined range of the more energetic group. The slower particles reach only about half the distance of the faster particles. *(Courtesy of Springer Verlag.)*

---

† The justification for trying to find the explanation on the basis of the Schrödinger theory is that the velocity of the alpha-particle *outside* the nuclear surface is "non-relativistic," as the reader can estimate for himself. Remember that the energy of the alpha-particle does not exceed 10 MeV.

Fig. 38A Schematic representation (by the solid line) of the potential which an alpha-particle sees in the vicinity of a nucleus. Outside the nucleus, i.e., beyond the distance R, the potential is the Coulomb potential. Inside the nucleus the forces are strongly attractive. The precise form of the potential is not known, but the attractive force is represented by the sudden drop in the potential at R. The dashed line represents the total energy of the alpha-particle. According to quantum mechanics the alpha-particle can penetrate the potential barrier. This takes place in the alpha-decay of heavy nuclei.

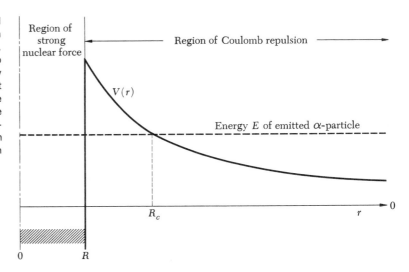

39 We have drawn Fig. 38A in a manner which suggests that the alpha-particle has to penetrate a potential barrier, in the region from R to R_c, before it can be emitted. Let us check immediately whether this is the correct picture. *If the picture is correct, then the classical turning point R_c given by*

$$R_c = \frac{2e^2 Z'}{E} \tag{39a}$$

must satisfy the condition $R_c > R$.

Inserting numerical values relevant for $_{88}Ra^{226}$, for which $Z = 88$, $Z' = 86$ (the atomic number for the noble gas radon), $E = 4.78$ MeV, we obtain $R_c \cong 50 \times 10^{-13}$ cm $= 50$ fermi. (To simplify the numerical work we may write $R_c = (e^2/m_e c^2) \times (2Z') \times (m_e c^2/E) \cong (2.8 \times 10^{-13}$ cm$) \times (172) \times (0.5$ MeV$/4.78$ MeV$) \cong 50$ fermi, where m_e is the electron mass.)

In Sec. 36, Chap. 2, we stated that the radius R of a nucleus of mass number A is given by

$$R \cong r_0 A^{1/3} \qquad r_0 = 1.2 \times 10^{-13} \text{ cm} \tag{39b}$$

and for $_{88}Ra^{226}$, for which $A = 226$, we thus obtain $R \cong 7.3$ fermi.

The picture is therefore qualitatively correct: the alpha-particle indeed has to penetrate a potential barrier. Quantitatively the picture is wrong: the barrier should have been drawn *much* thicker. Our execution of the figure has been motivated by aesthetic considerations: the important features of the situation are nevertheless reproduced qualitatively.

The inequality $R_c > R$ is true *generally* for the alpha-active nuclei. These radioactive elements are all heavy elements with a large atomic number Z. The isotope $_{88}Ra^{226}$ can be regarded as a typical alpha-emitter. The potential barrier which the alpha-particle has to penetrate is therefore an essential feature of the alpha-decay process, and we may hope to understand the extraordinary variation of the lifetime as a function of the energy E in terms of our simple theory of the tunnel effect.

40 Let us therefore compute the transmission coefficient T for the potential barrier shown in Fig. 38A. According to our results (36b) T is given by

$$\ln T \cong -2 \int_R^{R_c} dr \sqrt{\frac{2m_\alpha(2e^2Z'/r - E)}{\hbar^2}} \tag{40a}$$

where we note that in view of (39a) the integrand vanishes at the upper limit R_c. To discuss this integral we introduce a new variable of integration defined by $x = r/R_c$. As r varies from R to R_c the new variable x varies from $x_c = R/R_c$ to $+1$. Taking the relation (39a) into account we may then write the integral (40a) in the form

$$\ln T \cong -\frac{4e^2Z'}{\hbar} \sqrt{\frac{2m_\alpha}{E}} \int_{x_c}^1 dx \sqrt{\frac{1}{x} - 1} \tag{40b}$$

The integral occurring in (40b) can be evaluated in closed form easily enough. Since, however, the quantity $x_c = R/R_c$ is in general a fairly "small" quantity, it will suffice for our purposes to carry out an approximate evaluation in which we retain only the first two terms in an expansion in x_c. We proceed as follows:

$$\int_{x_c}^1 dx\sqrt{\frac{1}{x} - 1} = \int_0^1 dx \sqrt{\frac{1}{x} - 1} - \int_0^{x_c} dx \sqrt{\frac{1}{x} - 1}$$
$$\cong \int_0^1 dx \sqrt{\frac{1}{x} - 1} - \int_0^{x_c} dx \sqrt{\frac{1}{x}} = \int_0^1 dx \sqrt{\frac{1}{x} - 1} - 2 \sqrt{x_c} \tag{40c}$$

The first term in the extreme right-hand side of (40c) can be evaluated trivially if we make the substitution $x = \sin^2 \theta$, and we get

$$\int_0^1 dx \sqrt{\frac{1}{x} - 1} = 2 \int_0^{\pi/2} d\theta \cos^2 \theta = \frac{\pi}{2} \tag{40d}$$

The integral in (40b) is therefore given approximately by

$$\int_{x_c}^{1} dx \sqrt{\frac{1}{x} - 1} \cong \frac{\pi}{2} - 2\sqrt{\frac{R}{R_c}} \tag{40e}$$

and if we substitute this expression into (40b) taking into account (39a) we obtain

$$ln\ T \cong -\frac{2\pi e^2 Z'}{\hbar}\sqrt{\frac{2m_\alpha}{E}} + \left(\frac{8}{\hbar}\right)\sqrt{e^2 Z' R m_\alpha} \tag{40f}$$

41 To obtain a useful and transparent formula we shall make some further approximations. We shall set $Z' = 86$, and $R = 7.3$ fermi, which are the values of these parameters for the case when the radium isotope $_{88}\text{Ra}^{226}$ is the parent. We thus regard these values of Z' and R as "typical" for *all* alpha-active nuclei. The alpha-emitters are all heavy nuclei, and the variation in Z' actually found for this class of nuclei is not very large. The important parameter in (40f) is the energy E, which, as we said varies from 4–10 MeV. Our approximation is therefore reasonably justified. especially in view of the other approximations which we have made.

Inserting now the appropriate numerical values for the physical constants in (40f) and setting $Z' = 86$, $R = 7.3$ fermi, we finally obtain

$$\text{Log}\ T \cong -\frac{148}{\sqrt{E/\text{MeV}}} + 32.5 \tag{41a}$$

Notice that (41a) gives the common logarithm (for which 10 is the base) of T. To go from natural logarithms to common logarithms we make use of the relation $\text{Log}\ x = (\text{Log}\ e)\ (\ln x) \cong 0.434 \ln x$.

We have now derived the general expression (41a) for the transmission coefficient T, as a function of the energy E, for the potential barrier which the alpha-particle has to penetrate in alpha-emission. Let us see how we can use this result to find the lifetime of the alpha-emitter.

42 For this purpose we consider a very naive model of the process. We assume that before the emission the alpha-particle bounces back and forth inside the nucleus along a diameter. Let the time between two successive collisions with the "walls" be τ_0. In each collision there is a certain chance that the particle will leak through the potential barrier, and in fact the probability for emission in any single collision is just equal to the transmission coefficient T. It follows that the alpha-particle has to make of the order of $1/T$ collisions before it gets out, and we may therefore write the lifetime τ as

$$\tau = \frac{\tau_0}{T} \tag{42a}$$

or

$$\text{Log } \tau = \text{Log } \tau_0 + \frac{148}{\sqrt{E/\text{MeV}}} - 32.5 \tag{42b}$$

To estimate τ_0 we may assume, on the basis of our native model, that the alpha-particle inside the nucleus moves with the same velocity v as it will have after emission. We then have

$$\tau_0 = \frac{2R}{v}, \qquad v = \sqrt{\frac{2E}{m_\alpha}} \tag{42c}$$

If we apply this to the case of the radium isotope $_{88}\text{Ra}^{226}$, which we have used as a "standard" alpha-emitter, we obtain $\tau_0 \cong 10^{-21}$ sec.

As we see from (42c) the time τ_0 does depend on the energy E as well as on the nuclear radius R. The quantity τ_0 occurs, however, as the argument of a logarithm in (42b) and the variation of the first term with E is completely insignificant compared with the variation of the second term with E. To see this explicitly we consider what happens when E changes from 9 to 4 MeV. The increment of the first term in (42b) is then equal to Log $(\frac{3}{2}) \cong 0.18$. The increment of the second term in (42b) is *much* larger, namely $148 \times (\frac{1}{2} - \frac{1}{3}) \cong 25$. We may therefore well assume that the value $\tau_0 = 10^{-21}$ sec is approximately valid for *all* alpha-emitters, and we shall do so. We may state the matter as follows: The *dominant* factor in alpha-emission is the barrier penetration phenomenon. What goes on inside the nucleus before the emission we do not know too well, but we may say that the inside process defines a time τ_0, which can be interpreted as the time between successive attempts by the particle to penetrate the barrier. This time certainly depends on the parent nucleus in question, but it can reasonably be assumed to be of roughly the same order of magnitude for all alpha-emitters. Anyway the variation in the first term in (42b) can for any reasonable model be expected to be small compared to the variation in the second term, and for this reason our naive model, which should at least give the right order of magnitude for τ_0, is not as bad as it may seem at first, or, more correctly stated: it may be bad but it does not matter very much if it is bad.

We have thus reached our final goal, which is a general relation between the lifetime τ and the energy E of an alpha-emitter:

$$\text{Log } (\tau/\text{sec}) \cong \frac{148}{\sqrt{E/\text{MeV}}} - 53.5 \tag{42d}$$

Fig. 43A Half-life of alpha-emitters versus energy. The small circles in this graph show a selection of alpha-radioactive nuclei. The ordinate is the logarithm of the half-life, and the abscissa is the quantity $-1/\sqrt{E}$ where E is the kinetic energy of the emitted alpha-particle. Our simple theory predicts that the points should lie on a straight line, shown dashed in the graph. As we can see, the agreement is far from perfect in the details, but the general trend in the dependence of half-life on energy is correctly reproduced. Considered as a whole this graph is a most impressive confirmation of the ideas of quantum mechanics.

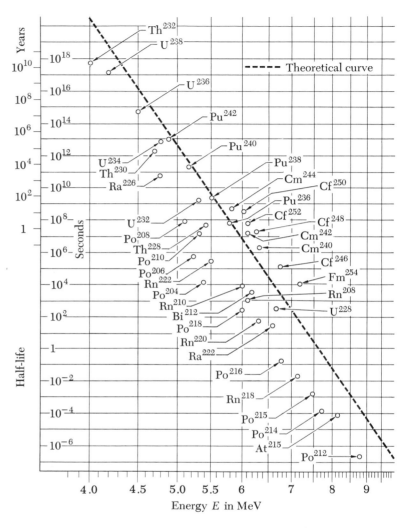

43 In Fig. 43A we have plotted the half-life of an alpha-emitter versus the energy E: the dashed line represents Eq. (42d). In this graph the ordinate is $\text{Log}\,(\tau/\text{sec})$ and the abscissa is $-1/\sqrt{E/\text{MeV}}$, and the relation (42d) is thus represented by a straight line. In the same graph we have plotted a large number of known alpha-emitters in order to compare our theory with the observed facts. We note that the experimental points by no means all fall on the theoretical curve, but it is also clear that the general trend of the observational data *is* correctly reproduced by our curve. We can regard it as a spectacular triumph of quantum mechanics that our simple and naive theory should take us to this level of understanding of alpha-radioactivity, which at first sight seems to be such a hopelessly complicated phenomenon.

The theory of quantum-mechanical barrier penetration was first given by Gamow, and independently by Condon and Gurney, in 1928.† Since then many refinements have naturally been added to the theory of alpha-disintegration, and with these it is possible to account for the observational results in much more detail.

44 The "lifetime" given in Fig. 43A is the *half-life* of the radioactive nucleus. As the reader undoubtedly knows, radioactive decay is governed by an exponential law such that if there are initially present, at time $t = 0$, N_0 nuclei of the species in question, then the average number of nuclei present at a later time t is given by

$$N(t) = N_0 \exp(-\lambda t) \tag{44a}$$

The constant λ is known as the *decay constant*, or *decay rate*. Its inverse, $1/\lambda$, is known as the *mean life* of the nucleus. The half-life is defined as the time at which $N(t) = N_0/2$: at this time half the nuclei originally present have decayed, on the average. Denoting the mean life by τ_m, and the half-life by $\tau_{1/2}$, we have

$$\tau_{1/2} = \frac{1}{\lambda} \ln 2 = \tau_m \ln 2 \tag{44b}$$

We may wonder whether our formula (42d) gives us the mean life, half-life, or maybe some other "lifetime." Our reasoning actually gives us the mean life, but within the accuracy of our discussion it makes absolutely no difference whether we speak of the mean life or the half-life. As we can see in Fig. 43A our result is correct only within a factor of 100 or 1000.

45 Consider again Fig. 38A. This figure is also relevant to the "inverse" process in which a charged particle of energy E, *less* than the height of the barrier, collides with a nucleus. The particle in question may be an alpha-particle or a proton, or maybe a deuteron. If the particle can get inside the barrier, i.e., into the region in which the strong nuclear forces are active, a nuclear reaction will

---

† G. Gamow, "Zur Quantentheorie des Atomkernes," *Zeitschrift für Physik* **51**, 204 (1928). See also G. Gamow, "Quantum theory of nuclear disintegration," *Nature* **122**, 805 (1928); R. W. Gurney and E. U. Condon, "Wave mechanics and radioactive disintegration," *Nature* **122**, 439 (1928).

in general take place.   According to classical mechanics the particle cannot penetrate the barrier, but we now know that the situation is different in quantum mechanics.   If the energy E is very small the transmission coefficient T is also small, and it is unlikely that a nuclear reaction will take place in any particular collision.   As the particle energy is increased the transparency of the barrier increases, and the chance for a nuclear reaction likewise increases. This increase is furthermore roughly represented by an exponential function of the energy.   The phenomenon of barrier penetration is thus an important feature of many nuclear reactions involving *charged* incident particles of not too high energy.   The situation is entirely different when the incident particle is a *neutron*.   The Coulomb barrier is then absent, and the neutron may freely enter the nucleus no matter how small its energy is.   In fact many nuclear reactions take place with a large yield for *thermal neutrons,* by which we understand neutrons of an energy corresponding to room temperature, i.e., about $1/40$ eV.

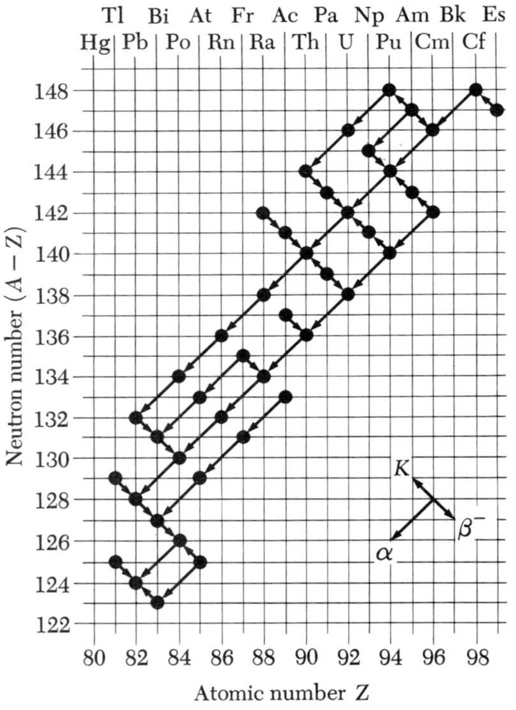

46   The heavy radioactive nuclei can be arranged into four groups corresponding to four different radioactive series, or decay-chains. In alpha-emission the mass number A of the nucleus changes by -4 units, and the charge number Z by -2 units.   In beta-decay, in which an electron (positron) and an anti-neutrino (neutrino) are emitted, the mass number does not change, but the charge number changes by $+1$ (-1).   Some heavy nuclei decay through alpha-emission, and some decay through beta-emission.   There is a further possibility: a nucleus can capture one electron from the cloud of electrons surrounding it, and at the same time emit a neutrino. This process is known as *K-capture*.   It is closely related to beta-decay.   The basic interaction responsible for K-capture and beta-decay is the universal *weak interaction,* which we have mentioned before.   The electron, positron, and neutrino do not participate in the strong interactions of which the "nuclear force" is an example, whereas the alpha-particle does.   The reason for the long lifetimes encountered in beta-decay or K-capture is not a barrier penetration effect, but simply the *intrinsic* weakness of the weak interactions.

In alpha-decay, beta-decay or K-capture, the mass number A either changes by four units, or not at all.   The radioactive nuclei can thus be grouped into four sets, and the mass number within a set is of the form $A = (4n + r)$, where n is variable, but r is fixed. The four sets correspond to the four different values $r = 0, 1, 2$ or 3. One such radioactive decay series, for $r = 2$, is shown in Figs. 46A–B.

A naturally occurring radioactive element either has a very long

Fig. 46A   Heavy radioactive nuclei, whose mass numbers are of the form $A = (4n + 2)$.   The arrows indicate radioactive decays, the type of decay being shown by the directions of the arrows as in the small insert in the lower right of the figure.   The symbol α indicates α-decay, the symbol β^- indicates beta-decay (through the emission of an electron and an anti-neutrino), and the symbol K indicates K-capture.

Note that some nuclei decay in two different ways. Note also that the end product of all the above decay chains is the stable lead isotope Pb^{206}.

lifetime, or else it is a member of a decay chain originating from a long-lived element. Among heavy nuclei with long lifetimes we note U^{238} with half-life 4.5×10^9 years; Th^{232} with half-life 1.4×10^{10} years, and U^{235} with half-life 7.13×10^8 years. The most long-lived member of the $(4n+1)$-family is a neptunium isotope, Np^{237} with half-life 2.2×10^6 years. This is a short time measured on a geological time scale, and the $(4n + 1)$-family therefore does not occur naturally.

A few of the naturally occurring light nuclei are also radioactive. Examples are the beta-active nuclei K^{40} with half-life 1.3×10^9 years and Rb^{87} with half-life 4.7×10^{10} years.

47 The phenomenon of natural radioactivity makes it possible to determine the ages of rocks, i.e., the time which has passed since the rocks were last chemically transformed. The principle is simple. We determine the relative amounts of a long-lived radio-active isotope and the stable end product in the disintegration chain which are present in a sample. Consider, for instance, the uranium-radium decay chain which originates with U^{238} and ends with the stable lead isotope Pb^{206}. Suppose that we find, in a given sample, an amount of Pb^{206} corresponding to N_{Pb} atoms, and an amount of U^{238} corresponding to N_U atoms. If we assume that *all* of the Pb^{206} atoms come from the decay of uranium, we can write

$$N_U = N_0 \, e^{-\lambda T}, \qquad N_{Pb} = N_0 \left(1 - e^{-\lambda T}\right) \qquad (47a)$$

where N_0 is the number of U^{238} atoms present originally, λ is the decay rate of uranium, and T is the age of the sample. Since $N_0 = N_U + N_{Pb}$ we have

$$e^{\lambda T} = \frac{(N_{Pb} + N_U)}{N_U} \qquad (47b)$$

and since λ is known we can find T. This method actually gives only an *upper* limit on T, because some of the Pb^{206} atoms present today may have been present when the mineral was formed. A more sophisticated approach is therefore necessary, and we have to compare the isotopic composition of lead in minerals which contain no uranium with the composition in minerals which contain uranium. Our example was therefore oversimplified, but it does illustrate the principle involved.

Another method depends on comparing the helium content of a rock with its uranium content. In each alpha-disintegration in the decay chain a helium nucleus is produced, and if we can be sure that the helium did not escape from the interior of the rock

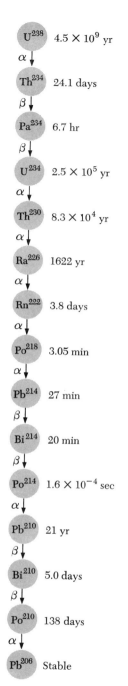

Fig. 46B The uranium-radium-lead radioactive chain. Half-lives are shown at right and decay modes at left. These isotopes occur naturally (in uranium minerals) because they originate from the long-lived uranium isotope 238. No transuranic elements belonging to this series (mass numbers of the form $4n + 2$) have appreciable half-lives on a geological time scale.

Fig. 48A Graph showing the estimated relative abundances of the elements in the solar system. The data are taken from a table (on pages 192–193) in L. H. Aller: *The Abundance of the Elements* (Interscience Publishers, Inc., New York, 1961), and the graph is inspired by a similar graph appearing on page 191 in the same book.

The ordinate is the logarithm to base ten of the relative abundance, i.e., *relative number of atoms*. The points representing neighboring elements have been joined for ease of reading. A graph such as this one is based on a variety of very different kinds of measurements, as well as on specific theoretical ideas. The data for the lighter elements derive mostly from spectroscopic studies of the sun, whereas the estimates for the heavier elements take into account studies of the composition of meteorites. The graph is a reasonable summary of present knowledge, but it must be understood that some of the data shown are quite uncertain and tentative.

It is believed that the abundances of the elements in the entire (visible) universe is *roughly* similar to the abundances in the solar system. The abundances found in our immediate surroundings differ markedly from the "cosmic" abundances (see Table 48B).

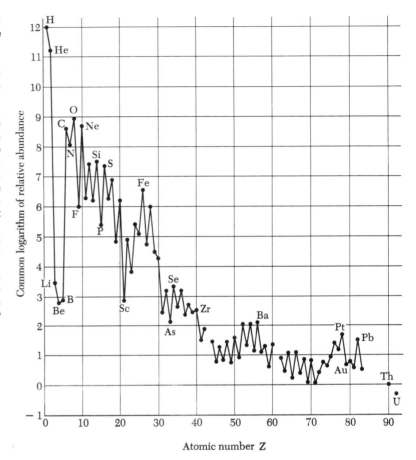

Atomic number **Z**

we can find out how many uranium atoms have disintegrated since the rock was formed.†

Through methods such as these it has been found that the oldest rocks in the crust of the earth are about 3×10^9 years old. This is definitely a *lower* limit on the age of the earth, because the crust has undergone many chemical transformations in the past. Meteorites have also been studied, and they have been found to be about 4.6×10^9 years old. How the meteorites were formed is not known with any certainty, but there is good evidence that they

† The first estimate of the age of the earth on the basis of radioactivity was made by Rutherford. See E. Rutherford, "The Mass and Velocity of the α particles expelled from Radium and Actinium," *Philosophical Magazine* **12**, 348 (1906). See pp. 368–369, where Rutherford arrives at the estimate of 400 million years for the minerals he studied.

were formed (crystallized) at about the same time as the other solid bodies in the solar system. The age of the earth as a body would thus be about 4.6×10^9 years. It is furthermore possible to estimate, using radioactive "clocks," the time that elapsed between the latest formation of the chemical elements in the meteorite and its crystallization. According to one such estimate,[†] it appears that this time was about 0.35×10^9 years. This implies that the final building of the chemical elements contained in planets and meteorites took place about 5 billion years ago. This is thus the estimated age of our solar system.

48   It is natural to speculate further. How old is the universe? How were the chemical elements formed? We cannot discuss here the ideas which lead to estimates of the age of the universe. It is believed that the age of the universe might be something like 10 billion years, roughly of the same order of magnitude as the age of the solar system.

It is believed that the chemical elements were formed from hydrogen in nuclear reactions in the stars. Fig. 48A shows the estimated abundances of the chemical elements in the solar system. The dots which represent individual chemical elements are not experimental points in the sense that they would all derive from measurements on a single "standard sample." The dots represent estimates based on a large number of different kinds of measurements, such as spectroscopic determinations of relative abundances in the solar atmosphere, relative abundance measurements in meteorites and estimates of the chemical composition of the earth's crust. Notice that hydrogen is by far the most abundant element. Notice also the peaks in the abundance curve corresponding to particularly stable elements. There is a clearly visible systematic trend in that elements of even atomic number are more abundant than neighboring elements of odd atomic number, which reflects the fact that nuclei which have an even number of protons and an even number of neutrons tend to be more stable than other nuclei.

To explain this curve in all details and thereby trace out the early history of the solar system is a fascinating problem. The main features of the abundance curve are believed to be fairly well understood at this time.

The author has absolutely nothing to say about the question of where the *hydrogen* came from originally.

TABLE 48B   *The Eight Most Common Elements in the Earth's Crust*

Element	Number of Atoms percent
Oxygen	62.6
Silicon	21.2
Aluminum	6.5
Sodium	2.64
Calcium	1.94
Iron	1.92
Magnesium	1.84
Potassium	1.42

This table shows the estimated composition of the ten outermost miles of the earth's crust together with the oceans and the atmosphere. These eight elements make up nearly 99 percent of the *mass* of this domain. The weak gravitational field of the earth cannot retain the light elements hydrogen and helium, which explains their low abundances compared to the "cosmic" data. The abundances for the heavier elements in the earth can be expected to be similar to the cosmic abundances, but geological processes on the earth have led to a chemical segregation of the elements, and the data for the crust are not representative for the earth as a whole.

[†] J. H. Reynolds, "Determination of the age of the elements," *Physical Review Letters* **4**, 8 (1960). See also C. M. Hohenberg, F. A. Podesek, and J. H. Reynolds, "Xenon-Iodine Dating: Sharp Isochronism in Chondrites," *Science* **156**, 202 (1967), for some results which indicate that the time might be very considerably shorter.

Advanced Topic:
Normalization of the Wave Function†

49 Let us consider the Schrödinger wave function, and for simplicity, let us restrict ourselves to the one-dimensional case, in which case the wave function $\psi(x,t)$ is a function of x and t. We have said that the absolute square of the wave function is proportional to a probability density. This means that the probability of finding the particle at the time t in the interval $x_2 > x > x_1$ is

$$P(x_1,x_2) = N \int_{x_1}^{x_2} dx \, |\psi(x,t)|^2 \tag{49a}$$

where N is some constant which is *independent of* x. How do we determine the constant N? By the simple requirement that the probability of finding the particle *somewhere* must be unity, i.e., we must have

$$1 = N \int_{-\infty}^{+\infty} dx \, |\psi(x,t)|^2 \tag{49b}$$

Now it could conceivably happen that the integral in (49b) does not converge. If so the constant N must be zero, and it follows from Eq. (49a) that the probability of finding the particle in any finite interval must also be zero. This cannot correspond to anything physically meaningful, and we draw the important conclusion that *the Schrödinger wave function $\psi(x,t)$ must, for all values of t, be a square-integrable function of x.* By the term "square-integrable" we understand the condition that the integral in (49b) converges.

Suppose therefore that the wave function $\psi(x,t)$ is indeed square-integrable. We can then define a new wave function $\psi_n(x,t)$ by

$$\psi_n(x,t) = \sqrt{N} \, \psi(x,t) \tag{49c}$$

where N is given by Eq. (49b). This wave function has the nice property that

$$\int_{-\infty}^{+\infty} dx \, |\psi_n(x,t)|^2 = 1, \qquad P(x_1,x_2) = \int_{x_1}^{x_2} dx \, |\psi_n(x,t)|^2 \tag{49d}$$

and the probability density is thus *equal* to the absolute square of the wave function.

A wave function which satisfies the first condition in (49d) is called a *normalized wave function,* or is said to be *normalized to unity.* It is clearly convenient to work with such a wave function since its absolute square is equal to the probability density.

---

† Can be omitted in a first reading.

50   We must now settle an important question: does the constant
N, as defined by (49b), depend on the time t?   We have assumed
that $\psi(x,t)$ is an actual solution of the Schrödinger equation

$$-\frac{\hbar^2}{2m}\frac{\partial^2}{\partial x^2}\,\psi(x,t)\,+\,V(x)\,\psi(x,t)\,=\,i\hbar\,\frac{\partial}{\partial t}\,\psi(x,t) \qquad (50a)$$

and the new wave function $\psi_n(x,t)$ would then also be a solution
of this equation *provided* the constant N is independent of the time.

We shall prove the following theorem: if $\psi(x,t)$ satisfies the Eq.
(50a) and if $\psi(x,t)$ tends to zero "sufficiently rapidly" as x tends to
$+\infty$ or $-\infty$, then

$$\frac{d}{dt}\int_{-\infty}^{+\infty} dx\,|\psi(x,t)|^2 = 0 \qquad (50b)$$

Here "sufficiently rapidly" means, among other things, that
$\psi(x,t)$ is square-integrable.

To prove this theorem we differentiate inside the integral sign:

$$\frac{\partial}{\partial t}\,|\psi(x,t)|^2 = \frac{\partial}{\partial t}\,\psi^*(x,t)\,\psi(x,t)$$

$$= \psi^*(x,t)\,\frac{\partial\psi(x,t)}{\partial t} + \frac{\partial\psi^*(x,t)}{\partial t}\,\psi(x,t) \qquad (50c)$$

Eq. (50a) gives us an expression for the time derivative of $\psi(x,t)$.
To obtain a similar expression for the complex conjugate $\psi^*(x,t)$ of
the wave function we simply form the complex conjugate of (50a)
which gives us the equation

$$i\hbar\,\frac{\partial}{\partial t}\,\psi^*(x,t) = \frac{\hbar^2}{2m}\frac{\partial^2}{\partial x^2}\,\psi^*(x,t)\,-\,V(x)\,\psi^*(x,t) \qquad (50d)$$

We have here assumed that $V(x)$ is a *real* function, which is justified in the Schrödinger theory since this potential is to correspond
to the potential in the "corresponding" classical problem.   That the
potential is real is essential for our argument, and this assumption is
always made in the Schrödinger theory.

Eliminating now the time derivatives from (50c) using (50a) and
(50d) we obtain

$$\frac{\partial}{\partial t}\,|\psi(x,t)|^2 = \frac{i\hbar}{2m}\left(\psi^*\,\frac{\partial^2\psi}{\partial x^2} - \psi\,\frac{\partial^2\psi^*}{\partial x^2}\right)$$

$$= \frac{i\hbar}{2m}\frac{\partial}{\partial x}\left(\psi^*\,\frac{\partial\psi}{\partial x} - \psi\,\frac{\partial\psi^*}{\partial x}\right) \qquad (50e)$$

We therefore obtain

$$\frac{d}{dt}\int_{-\infty}^{+\infty} dx \, |\psi(x,t)|^2 = \int_{-\infty}^{+\infty} dx \frac{\partial}{\partial t}\, |\psi(x,t)|^2 =$$

$$\frac{i\hbar}{2m}\left[\psi^*\frac{\partial\psi}{\partial x} - \psi\frac{\partial\psi^*}{\partial x}\right]_{-\infty}^{+\infty} \quad (50f)$$

If, however, the derivative of the wave function (with respect to *x*) stays bounded, then the expression at right in (50f) will vanish since we assumed that the wave function will vanish at infinity. The relation (50b) therefore holds, and it follows immediately, from (49b), that *N* is indeed a constant, independent of *t*. The function $\psi_n(x,t)$ is therefore also a bona-fide wave function, i.e., a solution to the Schrödinger equation (50a). *We may always form a normalized wave function from a given physical wave function, and if we like we could work entirely with wave functions normalized to unity.*

These important conclusions also hold in the three-dimensional case. We shall not show this here, but merely state that the proof is quite analogous to the proof in the one-dimensional case.

51 At this point the reader may be quite alarmed since our *firm* conclusion that every physically meaningful wave function must be square-integrable seems to cast doubts on our discussion of plane monochromatic waves earlier in this chapter. It is clear that a wave function of the form exp $(ixp/\hbar - itp^2/2m\hbar)$ is *not* square-integrable and therefore cannot be normalized to unity. We are forced to the conclusion that a wave of *precisely* defined momentum *p*, which depends on the coordinate *x* only through the factor *exp* $(ixp/\hbar)$, does not correspond to a physically realizable state of motion of the particle. On the other hand we are not forbidden to consider a wave which over a very *large* interval on the *x*-axis depends on *x* through the factor exp $(ixp/\hbar)$, provided that this wave function does tend to zero as *x* tends to $+\infty$ or $-\infty$. We can therefore resolve our difficulty by agreeing that when we discuss "waves of precisely defined momentum" we do not *really* mean that the wave is *everywhere* of the form exp $(ixp/\hbar)$. We realize that the wave function *must* tend to zero at infinity, but we assume that the wave function is of this form in a very large interval of the *x*-axis which includes the region in which we are primarily interested. Our "monochromatic waves" are thus to be understood as "almost monochromatic waves." With this understanding we can safely continue to talk about waves which depend on the co-ordinates through the factors exp $(ixp/\hbar)$, or exp $(ix \cdot \mathbf{p}/\hbar)$, as is done in almost all texts on quantum mechanics. We can regard the non-normalizable waves as limiting cases of normalizable waves, and if we like we may call the former kind of wave functions *improper* *wave functions.* This term will also serve to placate the mathematicians whose sensibilities are rightly offended by the frequently careless manner in which physicists talk about "plane waves" as if these would be bona-fide Schrödinger wave functions.

References for Further Study

1) Schrödinger's original papers on quantum mechanics have been published in translation. E. Schrödinger: *Collected Papers on Wave Mechanics* (Blackie and Son, Ltd., Glasgow, 1928).

For the history of the subject we also refer to the books mentioned at the end of Chap. 1. (Items 3 and 5.)

2) Some readers may feel a strong desire to learn more about the Schrödinger equation immediately. We therefore mention the following books:

a) R. M. Eisberg: *Fundamentals of Modern Physics* (John Wiley and Sons, New York, 1961).

b) E. Merzbacher: *Quantum Mechanics* (John Wiley and Sons, New York, 1961).

c) L. I. Schiff: *Quantum Mechanics*, 3rd edition (McGraw-Hill Book Company, New York, 1968).

The third of these books is the most advanced, and they are all more advanced than our book. The author mentions them only because the reader may be interested in seeing a complete discussion of some particular topic. Simple barrier problems are discussed in detail in the first two references.

3) Radioactivity and nuclear reactions are, of course, discussed in all books on nuclear physics. Among the many books we mention the following:

a) D. Halliday: *Introductory Nuclear Physics* (John Wiley and Sons, New York, 1955).

b) E. Segrè: *Nuclei and Particles* (W. A. Benjamin, New York, 1964).

4) For the questions of the formation of the chemical elements, and the ages of the solar system and universe, we refer to:

a) E. M. Burbidge, G. R. Burbidge, W. A. Fowler, and F. Hoyle: "Synthesis of the Elements in Stars," *Reviews of Modern Physics* **29**, 547 (1957).

b) W. A. Fowler and F. Hoyle: "Nuclear Cosmochronology," *Annals of Physics* **10**, 280 (1960).

c) J. H. Reynolds: "The Age of the Elements in the Solar System," *Scientific American*, Nov. 1960, page 171.

Problems

1 Consider the barrier shown in Fig. 28A for the case when $E > V_0$.

(*a*) Consider first the case of a particle incident from the left. It, i.e., the wave packet, will be partly reflected and partly transmitted by the step. To discuss this case we desire a solution which in the right-hand region describes a wave traveling to the right. Find this solution everywhere, and derive an expression for the reflection coefficient R, i.e., the probability R that the particle is reflected. The transmission coefficient T, i.e., the probability that the particle is transmitted, is then equal to $(1 - R)$.

(b) Consider next the case when the particle is incident from the right. In this case we desire a solution of the Schrödinger equation which in the left-hand region represents a wave traveling to the left. Find this solution everywhere, and derive an expression for the reflection coefficient R' and the transmission coefficient $T' = (1 - R')$. Note that a *classical* particle would not be reflected at all by the step.

2 Derive an *exact* expression for the transmission coefficient T of the potential barrier shown in Fig. 31A and compare your exact result with our approximate formula (33b). This comparison is best carried out by forming the logarithm of both expressions for T. Our approximate result arises in the limiting case of a "high and thick" barrier.

3 It is of interest to consider a specific example of the optical barrier penetration effect illustrated in Fig. 34B. At the wavelength 6000 Å (in air) the refractive index of flint glass is 1.75. Suppose the optically denser medium in Fig. 34B is flint glass, and suppose the rarer medium is air. Let the angle of incidence be 45°, and let the separation between the plates be 0.01 mm. Estimate the fraction of light which is able to pass through the barrier. (It is not necessary to carry out a precise computation: An estimate in the spirit of our discussion of barrier penetration will suffice.)

Note that the intensity of the transmitted light decreases exponentially with the thickness of the air space between the two glass prisms. The important quantity is the ratio of the thickness to the wavelength. Note that the component of the wave vector *parallel* to the interfaces is the same in the air space as in the glass. (Why?)

4 Let us fuss about a small detail: Is the Fig. 34B appropriately drawn? Consider the relationship between the transmitted ray and the incident ray. Perhaps the transmitted ray should have been drawn as a continuation of the incident ray, and not as in the figure? To find out how this picture should really be drawn we could perhaps perform some experiments. Suppose that the thickness of the optically rare medium is of the order of a wavelength of the light used. By an arrangement with slits we select an *extremely* narrow pencil of light as the incident beam, and this pencil is represented by the dotted line in the lower right part of the figure. We can then study the transmitted pencil of light, and find out whether it actually follows the dotted line in the upper left part of the figure. You do not really have to do this experiment in the laboratory: you can do it as a mental experiment instead, because there is nothing in this experiment which cannot be accurately predicted within electromagnetic theory.

After a consideration of this experiment, state your opinions as to whether Fig. 34B is correctly drawn.

5 Consider the motion of a particle in an "arbitrary" potential, such as the one shown in the figure on this page. As x tends to $+\infty$, or $-\infty$, the potential function $V(x)$ tends to zero.

Suppose that a particle of energy E is incident from the left. The wave function $\varphi(x)$ must then be of the form: $\varphi(x) = [e^{ixk} + Ae^{-ixk}]$ for very large negative x, and of the form: $\varphi(x) = B\,e^{ixk}$ for very large positive x. To actually find the two constants A and B we would have to solve the Schrödinger equation for the potential $V(x)$.

We have been led to interpret $|A|^2$ as the reflection coefficient of the barrier, and $|B|^2$ as the transmission coefficient. If this interpretation is to make sense we must clearly have:

$$|A|^2 + |B|^2 = 1 \qquad (a)$$

This raises an interesting question of principle. Is it really true that the above relation holds for *all* potential functions $V(x)$?

Prove this relation in general. *Hint:* Consider the function

$$F(x) = \varphi^*(x)\,\frac{d\varphi(x)}{dx} - \varphi(x)\,\frac{d\varphi^*(x)}{dx}$$

and show that $dF(x)/dx = 0$ if $\varphi(x)$ satisfies the Schrödinger equation.

This problem illustrates the fact that one can sometimes prove general statements about the *nature* of the solutions, without actually finding the solutions explicitly. In this particular case it is clear that we have discovered an important general property of the Schrödinger equation and its solutions. If the theory is to make sense the equation (a) *must* hold, and it is comforting that we can prove it.

6 There are further interesting questions which we can ask about the situation illustrated in the figure associated with the preceding problem. For instance: Is the transparency of the barrier the same in both directions?

Theorem: The transmission coefficient when the particle is incident from the left is the same as when the particle is incident from the right, provided that the energies are the same in both cases.

Prove this theorem. *Hint:* Note that if $\varphi(x)$, as discussed in the preceding problem, is a solution of the Schrödinger equation, so is $\varphi^*(x)$, and so is every linear combination of $\varphi(x)$ and $\varphi^*(x)$. Consider a *suitable* linear combination of $\varphi(x)$ and $\varphi^*(x)$.

7 Many unstable nuclei decay through the emission of a positron and a neutrino. The positrons emerge with energies ranging from 10 keV to a few MeV. This kind of decay is a consequence of the weak interactions, as we have stated before. We have also said that the reason for the long lifetimes sometimes found for beta-active nuclei is the intrinsic weakness of this interaction. This does not exclude the possibility that the barrier penetration effect might also play an important role. Investigate this point with the help

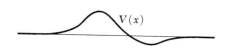

This figure refers to Prob. 5. Can you prove, for an *arbitrary* barrier of this kind, that the reflection and transmission coefficients, defined in terms of the amplitudes of the waves traveling to the right and to the left, actually add up to unity?

of some explicit numerical examples, i.e., estimate the transmission coefficients for "typical" potential barriers which the positron has to pass through, and convince yourself that the barrier penetration phenomenon is not here the dominant factor determining the lifetime.

8 L. Meitner and W. Orthmann [*Zeitschrift für Physik* **60**, 143 (1930)], once carried out a calorimetric measurement of the energy released in the beta-decay of RaE. (This is the old name for the nucleus Bi^{210}.) In this experiment they used a sample of RaE, placed in a suitable calorimeter, and they measured the rate at which heat evolved in the calorimeter. From a knowledge of the half-life of RaE (5.0 days), and the size of the sample, they were able to find the number of disintegrations per second, and hence the amount of heat energy evolved per disintegration. This amount of heat turned out to be (0.337 ± 0.020) MeV/disintegration.

On the other hand it was known that the maximum kinetic energy with which the electron can be emitted is 1.170 MeV. There thus appears to exist a disturbing discrepancy between the known maximum energy and the calorimetrically determined energy, and it is not surprising that physicists were quite worried at that time. Since we believe that the decay takes place between two definite energy levels we must assume that the energy 1.170 MeV is the kinetic energy released per decay, and the question then arises why this energy partly "disappears" in the calorimeter. Physicists were in fact so disturbed that some people, including Bohr, contemplated the possibility that the principle of conservation of energy might not hold in microphysics.

In the light of your knowledge of beta-decay, explain in detail the above mentioned circumstances (including the worried feelings of the physicists at that time).

9 In naturally occurring uranium the abundance of the isotope 235 is 0.71 percent and the abundance of the isotope 238 is 99.28 percent. The half-life of U^{235} is 7.1×10^8 years and the half-life of U^{238} is 4.50×10^9 years.

(*a*) The abundances quoted above hold for all terrestial samples, and also for uranium-containing meteorites. What conclusion can you draw from this circumstance?

(*b*) If you make the simplifying assumption that the original amounts of the two isotopes of uranium in the solar system were equal, what estimate of the age of the solar system do you then arrive at?

10 (*a*) Compute the amount of radium which you expect to find in an amount of uranium ore which contains 1 ton of uranium. Does it make any difference whether the mineral is 1 million years old, or 500 million years old?

(*b*) How much lead do you expect to find if the mineral is 500 million years old?

Chapter 8

Theory of Stationary States

Chapter 8   Theory of Stationary States

Quantization as an Eigenvalue Problem

1 Our subheading above is the English translation of the common title of Schrödinger's four famous papers† on wave mechanics, in which he showed how the existence of discrete energy levels in atoms can be understood in terms of the wave picture, and specifically in terms of the Schrödinger equation.

Preceding Schrödinger's theory Bohr had formulated, in 1913, a semi-classical theory of atoms. We say semi-classical because he assumed a planetary model, to be described by the laws of classical mechanics, but with the added assumption that not every orbit permitted classically is actually realized. The actual orbits are restricted by a number of *quantum conditions* of a definitely non-classical nature. As an example we mention the rule that the total angular momentum due to the orbital motion of the particles in the atom must be an integral multiple of $\hbar$. The values of total energy associated with the orbits permitted by the quantum conditions form in many cases (but not always) a discrete set. In this way Bohr created a theory of the discrete energy levels of an atom. His procedure can be called the *quantization* of the motion in the atom. This is the historical origin of the term "quantization."

2 Bohr's quantum conditions were of an ad hoc nature, and they could hardly be regarded as satisfactory. By the time of Schrödinger's papers it had become evident that whereas Bohr's theory did explain some of the observed facts, the theory also had definite shortcomings and outright failures. The time was therefore ripe for new ideas.

Schrödinger's great contribution was to show that if the wave picture of matter is taken seriously, then there is a systematic and natural way to "quantize." He noted that his wave equation would have, under suitable conditions, solutions describing *standing waves*, and he associated these solutions with the stationary states of atoms. The standing wave solutions are all characterized by a time dependence of the form exp $(-i\omega t)$, where the possible frequencies form a discrete set, say ω_1, ω_2, ω_3, . . . , and the energy of the nth stationary state is then given by $E_n = \hbar\omega_n$. In this chapter we shall follow in Schrödinger's footsteps and explore this idea.

---

† E. Schrödinger, "Quantisierung als Eigenwertproblem," *Annalen der Physik* **79**, 361 (1926); **79**, 489 (1926); **80**, 437 (1926); **81**, 109 (1926).

3 In Chap. 7 we arrived, via a number of plausibility arguments, at the Schrödinger equation†

$$-\frac{\hbar^2}{2m}\boldsymbol{\nabla}^2\,\psi(\mathbf{x},t) + V(\mathbf{x})\psi(\mathbf{x},t) = i\hbar\frac{\partial}{\partial t}\psi(\mathbf{x},t) \qquad (3a)$$

which describes a particle of mass m, moving in a field of force derivable from the potential function $V(\mathbf{x})$. In our "derivation" we recognized that this equation is clearly an approximation: the motion of the particle is treated non-relativistically, and all creation and destruction phenomena are ignored. We gave reasons why this equation should be very useful in atomic and molecular physics, and in some situations in nuclear physics. In the latter realm we scored a great success in being able to explain the dependence of the lifetime of an alpha-emitter on the energy of the emitted particle in terms of the quantum-mechanical tunnel effect.

Just as in our discussion in Chap. 7 it will be instructive here to consider the simplified version of the Schrödinger theory which describes one-dimensional problems. The Schrödinger equation for such problems is of the form

$$-\frac{\hbar^2}{2m}\frac{\partial^2}{\partial x^2}\,\psi(x,t) + V(x)\psi(x,t) = i\hbar\frac{\partial}{\partial t}\psi(x,t) \qquad (3b)$$

The equation (3b) is much simpler to discuss mathematically than the three-dimensional equation (3a). Since the essential features of the phenomena in which we are now interested show up in pretty much the same way in the two equations, we can gain a real understanding of how Schrödinger's theory works out through a study of the simpler equation (3b). Furthermore, it should be stated that this equation is not as unphysical as one may think at first: many problems involving motion in three dimensions can be reduced to equivalent one-dimensional problems.

4 Let us begin with a very simple problem. A particle is confined inside a "box" of length a, with infinitely high walls. The solid curve in Fig. 4A represents the potential $V(x)$ for this problem. The potential $V(x)$ is zero for x in the interval $(0,a)$, and $+\infty$ outside this interval.

In Sec. 26, Chap. 7, we considered the case when there is only *one* infinitely high potential wall. In that case we found monochromatic standing wave solutions describing the reflection of a particle of any positive energy E by the wall. The new element in the present situation is that the particle is confined between *two* infinitely high potential walls.

---

† In this chapter we employ cgs or MKS units.

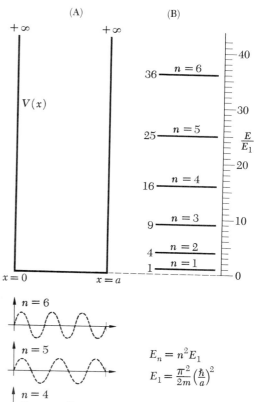

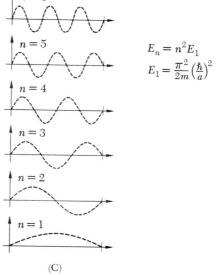

$$E_n = n^2 E_1$$

$$E_1 = \frac{\pi^2}{2m}\left(\frac{\hbar}{a}\right)^2$$

Figs. 4ABC The somewhat unphysical situation of a particle confined to a one-dimensional box provides us with a very simple illustration of the essence of Schrödinger's theory of stationary states. (A) shows the potential, which becomes infinite at the points $x = 0$ and $x = a$. The wave function corresponding to a stationary state must vanish at these points. This is possible only if the (total) energy has one of the values shown in (B) the term scheme (only the first six levels are shown). (C) shows the corresponding wave functions (eigenfunctions) for the first six stationary states.

Let us now try to solve the Schrödinger equation (3b) assuming that the wave function $\psi(x,t)$ depends on the time through a factor which is a simple exponential function of t, i.e.,

$$\psi(x,t) = \varphi(x) \exp\left(-\frac{itE}{\hbar}\right) \tag{4a}$$

Inserting a wave function of this form into (3b) we obtain the time-independent Schrödinger equation

$$-\frac{\hbar^2}{2m}\frac{d^2}{dx^2}\varphi(x) = [E - V(x)]\,\varphi(x) \tag{4b}$$

In our discussion in Sec. 26, Chap. 7, we concluded that the wave function must vanish in a region in which the potential is infinite, as well as *at the boundary* of such a region. In the present problem the wave function must thus vanish at the points $x = 0$ and $x = a$, as well as outside the interval $(0,a)$.

Inside the box the general solution of the equation (4b) is of the form

$$\varphi(x) = A \exp(ixk) + B \exp(-ixk) \tag{4c}$$

where

$$k = \sqrt{\frac{2mE}{\hbar^2}} \tag{4d}$$

and where A and B are constants. If we now first impose the condition that the wave function must vanish at $x = 0$, we find that the physically acceptable solution must be of the form

$$\varphi(x) = C \sin(xk) \tag{4e}$$

where C is a non-zero constant. The wave function must also vanish at $x = a$, and we obtain the further condition

$$C \sin(ak) = 0 \quad \text{or} \quad ak = n\pi \tag{4f}$$

This is a condition on k, and thereby a condition on the energy E. Taking into account the relation between E and k, as expressed by (4*d*), we have

$$E = \frac{n^2\pi^2(\hbar/a)^2}{2m} \tag{4g}$$

where n is a *positive* integer: our problem has *no* physically acceptable solution unless E is of this form. The case $n = 0$ is ruled out because it corresponds to an identically vanishing wave function, which is not acceptable physically. Since we assumed that k is non-negative it follows that n is positive.

5 We have thus found that for our particle in a box the Schrödinger equation (3b) has stationary solutions with a simple exponential time-dependence, i.e., solutions of the form $\psi(x,t) = \varphi(x) \exp{(-itE/\hbar)}$, *only* if the energy E assumes one of a discrete set of values $E_1, E_2, E_3, \ldots, E_n, \ldots$, given by

$$E_n = \frac{n^2\pi^2(\hbar/a)}{2m} \tag{5a}$$

where n is any positive integer. The *normalized*† wave function $\psi_n(x,t)$ corresponding to the nth possible energy E_n is then of the form

$$\psi_n(x,t) = \sqrt{\frac{2}{a}} \sin{\left(\frac{n\pi x}{a}\right)} \exp{\left(-\frac{itE_n}{\hbar}\right)} \tag{5b}$$

inside the interval $(0,a)$, and zero outside. (That this wave function is correctly normalized to unity we see trivially by integrating $|\psi_n(x,t)|^2 = (2/a)\sin^2{(n\pi x/a)}$ from 0 to a: the result is 1.)

We have represented the energies E_n in the form of a term scheme for our system in Fig. 4B, which shows the first six energy levels. In Fig. 4C we have drawn the corresponding wave functions $\varphi_n(x)$. These functions are, of course, equal to the functions $\psi_n(x,t)$ at the particular time $t = 0$.

See also Fig. 5A for a composite picture.

6 Let us now study the difference between stationary and non-stationary solutions of the Schrödinger equation (3b).

Consider first the nth stationary solution, given by (5b). Since the solution is normalized to unity, the absolute square of the wave function gives us the probability density $P_n(x)$ of finding the particle anywhere on the x-axis. We find

$$P_n(x) = |\psi_n(x,t)|^2 = \left(\frac{2}{a}\right)\sin^2{\left(\frac{n\pi x}{a}\right)} \tag{6a}$$

inside the interval $(0,a)$, and $P_n(x) = 0$ *outside* this interval. As we see, the probability density *does not depend on the time* for a stationary solution.

Let us next consider a non-stationary solution. Since the Schrödinger equation (3b) is a linear differential equation the linear combination of any two solutions gives a new solution. This new solution will satisfy the same boundary conditions $\psi(0,t) = \psi(a,t) = 0$, provided that the two original solutions satisfy these

---

† For a discussion of the normalization of the Schrödinger wave function, see Sec. 49, Chap. 7.

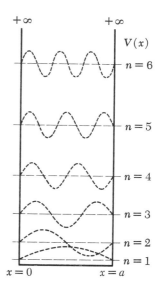

Fig. 5A It is common practice in texts on quantum mechanics to show figures such as the one above. The three figures 4ABC have been collapsed into a single figure. This is perhaps a deplorable practice, but since the author has never been misled by such pictures, he assumes that the reader will not be misled either.

The energy levels are indicated by the thin broken lines. Each one of these lines also serves as the x-axis in a superimposed figure showing the corresponding wave function.

conditions. We can conclude that any linear combinations of the stationary solutions (5b) give us new physically acceptable solutions, in accordance with the Principle of Superposition.

To see what happens in such a superposition of two solutions, let us consider the particular linear combination

$$\psi(x,t) = \sqrt{\frac{1}{2}} \, [\psi_{n'}(x,t) + \psi_{n''}(x,t)] \tag{6b}$$

where we assume that $n' \neq n''$. We claim that this new solution of the Schrödinger equation is normalized to unity (for all times t). The probability density $P(x,t)$ corresponding to the solution (6b) is given by

$$P(x,t) = |\psi(x,t)|^2 = \left(\frac{1}{a}\right) \left\{ \sin^2\left(\frac{n'\pi x}{a}\right) + \sin^2\left(\frac{n''\pi x}{a}\right) \right.$$

$$+ \left. 2 \sin\left(\frac{n'\pi x}{a}\right) \sin\left(\frac{n''\pi x}{a}\right) \cos\left[\frac{t(E_{n'} - E_{n''})}{\hbar}\right] \right\} \tag{6c}$$

The reader should immediately verify, by integrating this expression for $P(x,t)$ between 0 and a, that the wave function given in (6b) is indeed normalized to unity.

As we see, the probability density $P(x,t)$ is *not* time-independent: the last term in the expression (6c) shows an oscillatory behavior, and the frequency of this oscillation is given by

$$\omega_{n'n''} = \frac{(E_{n'} - E_{n''})}{\hbar} \tag{6d}$$

7 A moment's reflection tells us that this behavior must be exhibited by *all* superpositions of the stationary solutions (5b) as long as at least two different stationary solutions occur in the superposition. (The superposition may contain any number of stationary solutions: even an infinite number could occur.) We furthermore easily see that if the stationary solutions $\psi_{n'}$ and $\psi_{n''}$ occur in the superposition, then there must be a term in the probability density which oscillates with the frequency $\omega_{n'n''}$, given by equation (6d). This term comes from the "cross-terms" $\psi_{n'}^*\psi_{n''}$ and $\psi_{n''}^*\psi_{n'}$ occurring in the expansion of the absolute square of the wave function

$$\psi(x,t) = \sum_n c_n \psi_n(x,t) \tag{7a}$$

where the c_n are constants.

Now it is actually possible to prove a *theorem*: every physically acceptable solution of the Schrödinger equation for the particle-in-the-box problem can be written in a unique way in the form of an expansion, as given in (7a), in terms of the *stationary* solutions (5b) of the problem. We shall not prove this theorem here, but we accept it as very plausible: mathematically the theorem is a theorem about Fourier series. Accepting this theorem we can conclude that the only solutions of the Schrödinger equation which correspond to a time-independent probability density are the stationary solutions.

8 We have now learned the essence of the Schrödinger theory of the stationary states and energy levels of a quantum-mechanical system. The stationary states correspond to stationary solutions of the Schrödinger equation, and for these the probability density is independent of the time. For the non-stationary states the probability density shows an oscillatory dependence on the time, and the frequencies of oscillation which can occur are given by equation (6d) in terms of the energy *differences* of the various stationary levels. These frequencies are clearly the characteristic frequencies of the system at which we might expect radiation to be emitted and absorbed: these are the frequencies at which the system will resonate. The transition frequencies $\omega_{n'n''}$ in turn determine the location of the energy levels, except for a common additive constant which we can fix by assigning some suitable energy to the ground state. (In our example we selected the "bottom of the well" as the zero-point.)

We might now propose an ambitious program: to solve the Schrödinger equations (properly generalized to apply to many-particle systems) for all cases of physical interest in which the Schrödinger theory can be expected to be a good approximation. In particular we should look for the stationary solutions which can be normalized to unity: they give us the stationary states and the corresponding energy levels. Needless to say this grandiose program is very far indeed from its completion: our mathematical abilities are totally inadequate for solving the Schrödinger equation exactly for a complicated system, although we can handle simple systems quite well.

9 Considering the above-mentioned program we may ask whether it is really what we want. As we carefully discussed in Chap. 3 the "stationary" states are, strictly speaking, not stationary at all. Our theory of the particle in the box, on the other hand, did give us strictly stationary states. The program which we have outlined would also give us strictly stationary states, in contradiction to known observational facts. We here encounter a clear shortcoming of the Schrödinger equation: it does not describe radiative transitions. The Schrödinger equation is therefore not the whole story: something has been left out. In this respect the Schrödinger theory is analogous to a classical theory in which all the electrostatic interactions between the electrons and the nucleus are taken into account, but in which the radiation of electromagnetic waves from the moving particles is ignored. We can nevertheless hope that the Schrödinger theory is a good approximation in atomic and molecular physics. We can thus expect that a stationary state predicted by the Schrödinger equation would correspond to an *almost* stationary state in the "true" theory, and that the "mean energy" of the latter state would be very close to the precise energy predicted by the Schrödinger equation.

10 Let us explain some commonly employed terms before we go on. The time-independent Schrödinger equation (4b) is typical of the equations we have to consider when we wish to find the energy levels of a system. Let us write this equation in the symbolic form

$$H\varphi(x) = E\varphi(x) \tag{10a}$$

where H stands for the *differential operator*

$$H \equiv -\frac{\hbar^2}{2m}\frac{d^2}{dx^2} + V(x) \tag{10b}$$

We wish to find the solutions $\varphi(x)$ to the differential equation (10a). This equation will always have solutions, for any E, but not all of these will be physically acceptable. We should therefore regard the conditions for physical acceptability, namely that the wave function be square-integrable,† as an *essential* part of the problem. If we do this we find that E cannot be arbitrary. Those values of E for which the equation (10a) has a physically acceptable solution are called the *eigenvalues of the differential operator H.* The corresponding wave functions are called the *eigenfunctions* of the operator.‡

We can now understand why Schrödinger called his papers "Quantisierung als Eigenwertproblem."

11 The problem of a particle confined to a potential well with infinitely high walls is a bit unrealistic. Let us now consider the one-dimensional eigenvalue problem more generally. Let us assume that the potential $V(x)$ is nowhere infinite, but rather of the form shown in Fig. 11A. We assume that the potential function

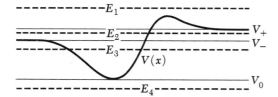

Fig. 11A A particular kind of potential function, which tends to the constant values V_+ and V_- as x tends to $+\infty$ or $-\infty$. We shall study the nature of the solutions of the Schrödinger equation for various values of the total energy E. The dashed horizontal lines indicate four energies representative of the cases which can occur.

---

† In the case of a "potential well" with infinitely high walls this condition leads to the condition that the wave function vanishes outside the well, as well as on the boundary, as we found in Sec. 26, Chap. 7.

‡ The words "eigenvalue" and "eigenfunction" are German-English hybrids which have become firmly established in physics. The German terms are "Eigenwert" and "Eigenfunktion."

tends to the constant value V_+ as $x \to +\infty$, and to the constant value V_- as $x \to -\infty$. We denote the minimum of the potential by V_0. This particular potential is, of course, a special case, but it is very instructive to consider this special case. We assume that $V_+ \geqq V_-$.

We shall examine the nature of the solutions of the time-independent Schrödinger equation (4b) for the potential $V(x)$. Let us write this equation in the form

$$\frac{d^2}{dx^2}\varphi(x) = -\left(\frac{2m}{\hbar^2}\right)[E - V(x)]\varphi(x) \tag{11a}$$

We consider this differential equation for various values of the energy-parameter E: namely for $E \leqq V_0$; for $V_- \geqq E > V_0$; for $V_+ \geqq E > V_-$; and for $E > V_+$. It should be clearly understood that the differential equation (11a) will have solutions for *all* values of E, but these solutions will in general not all be physically acceptable.

The graphical representation of the (in general) complex wave function presents some problems. One possibility is to plot the absolute value of the wave function. Another possibility is to consider real solutions of equation (11a). We note that if $\varphi(x)$ is any (complex) solution of (11a), so is $\varphi^*(x)$, since both E and $V(x)$ are real. The real part $[\varphi(x) + \varphi^*(x)]/2$, and the imaginary part $[\varphi(x) - \varphi^*(x)]/2i$, of a solution $\varphi(x)$ are also solutions, and we may imagine that we plot these *real* functions.

12 Let us first consider the *local* behavior of the real solutions in a region throughout which $[E - V(x)] < 0$. Inspection of the Schrödinger equation (11a) shows that in such a region the second derivative of the wave function has the *same* sign as the wave function. It follows that if the wave function does not vanish in an interval, then it "bulges toward the x-axis," as illustrated by the two segments shown in Fig. 12A. If the wave function passes through the axis it will "grow away" from the axis on both sides of the zero, as shown in Fig. 12B. The wave function might also approach the axis asymptotically either on the left side, or else on the right side, as illustrated by the two segments shown in Fig. 12C.

We conclude that if $V(x) > E$ for *all* values of x, then the solutions of (11a) cannot be physically acceptable, because the absolute value of the wave function grows without bound either on the left side or on the right side, or possibly on both sides. With reference to Fig. 11A the physical system cannot have an energy E less than V_0.

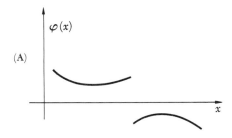

(A)

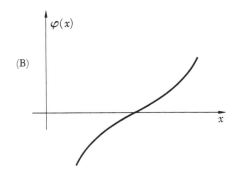

(B)

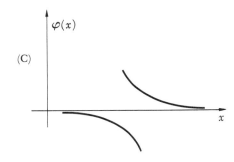

(C)

Figs. 12ABC The segments shown in the figures above illustrate the local behavior of the (real) wave function in a region throughout which $E < V(x)$. In such a region the second derivative has the *same* sign as the wave function.

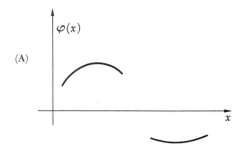

(A)

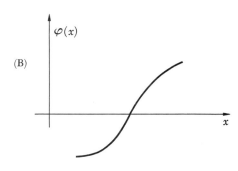

(B)

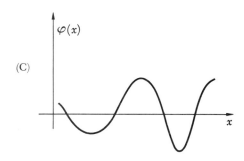

(C)

Figs. 13ABC The segments shown in the figures above illustrate the local behavior of the (real) wave function in a region throughout which $E > V(x)$. In such a region the second derivative and the wave function are of *opposite* sign. The reader should carefully compare the above figures with the figures 12ABC.

13 We next consider the behavior of the wave function in a region throughout which $[E - V(x)] > 0$. In this case the second derivative of the wave function and the wave function are of *opposite* sign. It follows that in a region in which the wave function does not vanish it must "bulge away from the x-axis." This is illustrated by the two segments shown in Fig. 13A. If the wave function crosses the axis it will turn toward the axis on both sides of the point of intersection, away from the tangent to the curve at the zero point. This behavior is illustrated in Fig. 13B, which should be compared to Fig. 12B.

A longer segment of the curve representing the wave function might cross the axis several times, and we then have the "oscillatory" behavior illustrated in Fig. 13C.

14 Finally we consider the case when $[E - V(x)] = 0$ *throughout* a region. (This very special situation can only arise if the potential function $V(x)$ is constant throughout the region.) The second derivative of the wave function must be zero, and it follows that the first derivative must be a constant. The curve representing the wave function is a straight line, as illustrated by the segments shown in Figs. 14AB.

Let us here note that for a potential of the kind shown in Fig. 11A a physically meaningful wave function and its first derivative cannot *both* vanish at the same point, because if this happens then the wave function must vanish everywhere. This statement is a theorem in the theory of ordinary differential equations. For this reason the curve segments shown in Figs. 12ABC, 13ABC and 14AB never *touch* the x-axis, although they may cross the axis, or approach the axis asymptotically.

15 Let us use our knowledge of the *local* behavior of the wave function to discuss its *global* behavior, for *all* values of x, when the potential function is as shown in Fig. 11A. We must now impose on the solutions of the differential equation (11a) the conditions which must be satisfied by physically meaningful wave functions.

With reference to Fig. 11A we first consider the case when the energy E satisfies the condition $E > V_+$. Such an energy E_1 is represented by the dashed line so labeled in the figure. This case is actually somewhat special because we have $[E - V(x)] > 0$ for all x. The solutions are oscillatory everywhere, and in particular at $+\infty$ and $-\infty$. The solutions are also oscillatory at $+\infty$ and $-\infty$ in the case when the energy E is below the maximum of the potential $V(x)$, provided $E > V_+$: in this case we have a barrier penetration problem. We can therefore find two linearly independent solutions, oscillatory at infinity, for *every* $E > V_+$, and these solutions describe traveling particles (or waves). We have already discussed such solutions, and their physical interpretation, in Chap. 7. The solution for a fixed E is not normalized to unity, but we can construct normalizable solutions in the form of a (continuous) superposition of the traveling-wave solutions. We agreed, in Sec. 51, Chap. 7, to call the solutions corresponding to a definite E *improper* wave functions, and for any $E > V_+$ we thus have *two* linearly independent improper wave functions. These wave functions, or rather the normalized wave packets which can be formed from them, may describe, for instance, a particle incident from the left of the "barrier." This particle is partly reflected back to the left, and partly transmitted past the barrier to the right. Similarly the particle may be incident from the right.

16 Next suppose that $V_+ > E > V_-$. In this case we have a region to the right in which $[E - V(x)] < 0$, and a region to the left in which $[E - V(x)] > 0$. This is the same kind of problem as we considered in Sections 21–25, Chap. 7. In this case only *one* of the two linearly independent solutions in the right region is physically acceptable, namely the solution which tends to zero as x tends to $+\infty$ (corresponding to the segment at right in Fig. 12C). This solution, when continued to the left, has an oscillatory behavior in the region in which $[E - V(x)] > 0$. (The wave function, and its first derivative, are, of course, everywhere continuous, otherwise the wave function does not correspond to a *global* solution of the Schrödinger equation.) For every E such that $V_+ > E > V_-$ we thus have *one* (improper) wave function, and this wave function describes the reflection of a particle incident from the left by the potential "hump," just as in the problem considered in Chap. 7.

17 Let us next consider the case when $V_- > E > V_0$. An example of this is the energy E_3, indicated by the dashed line so labeled in Fig. 11A. In this case we have a region to the left and a region to the right in which $[E - V(x)] < 0$, and a region in the middle in which $[E - V(x)] > 0$. The two boundary points separating these regions from each other are the *classical turning points:* we shall denote them by x_1 and x_2.

To the left of x_1 the wave function must tend asymptotically to the x-axis, and the behavior must be as indicated by the left segment in Fig. 12C (except for the sign of the wave function, which is immaterial). Unless the wave function behaves in this way it would grow as x tends to $-\infty$, and a steadily growing wave func-

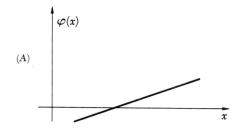

(A)

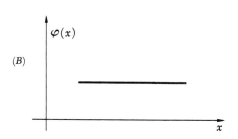

(B)

Figs. 14AB The segments shown in the figures above illustrate the local behavior of the (real) wave function in a region throughout which $E = V(x)$. This is a very special case which can occur only if the potential $V(x)$ is constant in the region. The second derivative of the wave function vanishes, and the wave function is represented by a straight line.

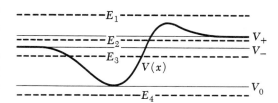

For the convenience of the reader we show Fig. 11A once more. The potential function tends to the constant values V_+ and V_- as x tends to $+\infty$ or $-\infty$. The dashed horizontal lines indicate four energies representative of the cases which can occur.

tion cannot be physically acceptable. To the right of x_2 the wave function must behave like the right segment in Fig. 12C. In the middle region between x_1 and x_2 the wave function shows an oscillatory behavior, and in this region we have two linearly independent physically acceptable solutions. The problem is now to "match" these different types of solutions so that we obtain a physically acceptable wave function which is everywhere continuous and has a continuous first derivative everywhere.† *For an arbitrary E this cannot be done: a physically acceptable solution (which is square integrable) can only be found for certain discrete values of the energy E. To each one of these energies corresponds a bound stationary state of the system.*

18   This phenomenon can be readily understood with the help of Figs. 18ABC. Suppose we pick an arbitrary energy E such that $V_- > E > V_0$. We satisfy the physical conditions "to the left" by picking a solution which tends asymptotically toward the x-axis as x tends to $-\infty$. At the turning point x_1 this solution must be "matched" to the oscillatory solution between x_1 and x_2. Since both the wave function and its first derivative must be continuous we obtain a *unique* solution for this region. This solution must be "matched" to the solution to the right of x_2, and we again obtain a *unique* solution to the right of x_2. This solution will not exhibit the behavior of the right segment in Fig. 12C unless the energy E is just right, but will instead grow away from the axis in which case the overall solution is not physically acceptable. The conditions that the wave function shall decrease to the left, as well as to the right, are in general not compatible, *except* if E has one of a discrete set of values. These values must be greater than V_0. We have already concluded that there can be *no* physically acceptable solution for $E < V_0$.

For the potential problem shown in Fig. 11A the term scheme thus consists of a (possibly empty) set of discrete levels between V_0 and V_-, and a continuum above the energy V_-.

19   A one-dimensional problem of the kind just discussed, and which is comparatively simple to handle analytically, is shown in Fig. 19A. In this case we have $V_+ = V_-$, and the potential function $V(x)$ is piecewise constant. The term scheme is shown to the right, and we see that there are four bound states below the continuum. The wave functions corresponding to these bound states

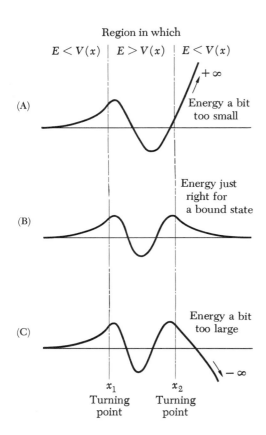

Region in which

$E < V(x)$ | $E > V(x)$ | $E < V(x)$

$+\infty$

(A)   Energy a bit too small

(B)   Energy just right for a bound state

(C)   Energy a bit too large

$-\infty$

x_1 x_2
Turning   Turning
point   point

Figs. 18ABC Schematic representation of the behavior of the solutions of the Schrödinger equation which tend to zero asymptotically as x tends to $-\infty$. The three curves represent solutions for three different energies. Unless the value of the energy parameter is "just right," the solution will diverge to $+\infty$ or $-\infty$ as x tends to $+\infty$. The unbounded solutions of the differential equation are not physically acceptable: they are not solutions of the Schrödinger *problem*. The energy is "just right" for curve *B*: the wave function tends to zero asymptotically as x tends to $+\infty$. This curve represents the wave function for a bound state.

---

† The "matching" is, of course, done automatically if we find a *global* solution of the wave equation.

are shown in the left part of the figure. Note that the first wave function has one extremum (and no node), the second wave function has two extrema (and one node), and the fourth wave function, corresponding to the highest *discrete* energy level, has four extrema (and three nodes). For a deeper potential well we would have more bound states, and in the extreme case of an infinitely deep well, which is the problem discussed in Sec. 4, we have an infinite number of bound states. The term schemes in Figs. 4B and 19A should be compared: the locations of the first four bound state levels are similar, although not identical, in the two cases.

The reader may wish to try to solve the problem of finding the bound states for the situation shown in Fig. 19A: it is not particularly difficult.

We have now learned that we can understand, on the basis of the Schrödinger theory, why a quantum-mechanical system will have bound states, and why there will in general be a continuum of possible energies above a certain limit. The beginning of the continuum is simply the energy above which the system can dissociate, which in our simple examples means that the particle can behave like a propagating wave packet far away from the "central region."

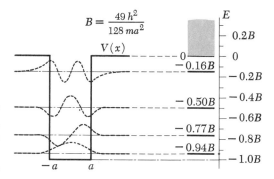

Fig. 19A The case of a particle in a potential well of depth B. This figure is based on an example given in R. B. Leighton, *Principles of Modern Physics*, p. 154 (McGraw-Hill Book Co., New York, 1959). The potential well is shown at left, and the term scheme at right. There are four bound states (four discrete energy levels). The corresponding eigenfunctions are shown at left, superimposed on the graph of the potential function. The continuum begins at the top of the well, as indicated in gray on the term scheme.

20 Let us next indicate how one can understand the phenomenon which we encountered in Sec. 38, Chap. 3, of energy levels *above* the beginning of the continuum. (See the term scheme in Fig. 38A, Chap. 3.)

We consider the one-dimensional potential problem shown in Fig. 20A. It differs from the problem in Fig. 19A in that the potential does not stay constant outside the well, but decreases to the value $-B_\infty$, in a stepwise fashion, at some distance from the well. We assume that outside these steps the potential remains constant at the value $-B_\infty$.

In accordance with our theory the continuum will now begin at the energy $-B_\infty$, as indicated in the term scheme to the right in Fig. 20A. For a not too small b there will be three bound states. These energy levels, E_1, E_2, E_3, are very close to the three first energy levels in the term scheme in Fig. 19A whenever the constant b is *large*, i.e., when the two potential barriers shown in Fig. 20A are very thick. Let us restrict our considerations to the case of a very large b. If b were infinite, then the problem in Fig. 20A would become identical with the problem in Fig. 19A. The continuum would begin at the energy 0, and the fourth bound state at the energy E_4 would be present. For any finite b, no matter how large, we have only three strictly stationary states, and the

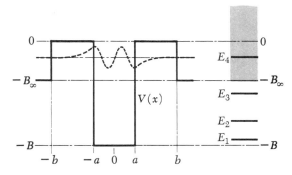

Fig. 20A This figure illustrates a modification of the situation shown in Fig. 19A. The potential functions are identical within the interval $(-b, +b)$, but outside this interval the potential shown above has the constant value $-B_\infty < 0$. The continuum therefore begins at $-B_\infty$ in the present case, and there are only three strictly stationary states. However, if b is very large, corresponding to a very thick "barrier," then there will be a fourth *almost* stationary state. This *virtual level* is designated by E_4 above. It corresponds to the fourth stationary level in Fig. 19A.

continuum begins at $-B_\infty$. However, suppose that the width of the well is of typical atomic size, that its depth is of the order of 10 eV, that the particle is an electron, and that b is larger than one kilometer. Under these conditions it is difficult to see how the situation in Fig. 20A can really differ from the situation in Fig. 19A. Common sense tells us that the behavior of the particle *in the neighborhood of the well* must be very similar in the two cases, and we therefore expect that the fourth bound state present in the term scheme in Fig. 19A must somehow manifest itself in the problem in Fig. 20A also. A careful mathematical examination of the situation, which we cannot undertake here, bears this out. Let us, however, try to indicate a possible line of analysis.

21 We compare the behavior of a particular Schrödinger wave function $\psi(x,t)$ with time in the two cases. Suppose that at time $t = 0$ the wave function is identical with the fourth eigenfunction shown in Fig. 19A, corresponding to the fourth energy level $E_4 \cong -0.16B$. We have, in other words,

$$\psi(x,0) = \varphi_4(x) \tag{21a}$$

where the wave function $\varphi_4(x)$ is the wave function represented by the dotted line about the level of E_4 in Fig. 19A. This same wave function is also shown by a dotted line in Fig. 20A. Note that this wave function tends to zero fairly rapidly outside the well.

For the problem in Fig. 19A it is easy to solve the time-dependent Schrödinger equation (3b), subject to the initial condition given by (21a). Since $\varphi_4(x)$ is an eigenfunction of the Schrödinger differential operator we simply have

$$\psi(x,t) = \varphi_4(x) \exp\left(-\frac{itE_4}{\hbar}\right) \tag{21b}$$

which expresses the stationary nature of $\psi(x,t)$. We may now find the probability, $P(t)$, that the particle is found *within* the well:

$$P(t) = \int_{-a}^{a} dx \, |\psi(x,t)|^2 = P(0) \tag{21c}$$

and, as we see, this probability is independent of the time t, which again reflects the stationary nature of the wave function $\psi(x,t)$. Note that the integral in (21c) extends over the well only, from $-a$ to a.

22 If we now try to solve the same problem for the situation shown in Fig. 20A, i.e., with the same initial condition (21a), then the solution is *not* of the form (21b), although one may say that it is of approximately this form. If we actually find the time-dependent wave function $\psi(x,t)$ for the problem in Fig. 20A, and then compute the probability $P(t)$ for finding the particle *in* the well, it is possible to show that we have, instead of (21c), an approximate relation of the form

$$P(t) = \int_{-a}^{a} dx \, |\psi(x,t)|^2 \cong P(0) \exp\left(-\frac{t}{T}\right) \qquad (22a)$$

where T is a positive constant. We emphasize that the relation (22a) is an *approximate* relation: it is valid only for times t which are not "too large." The detailed proof of this result would take us too far, but we shall try to make it plausible.

The interpretation of the result (22a) is that if the particle is placed "in the well" at time $t = 0$, with an energy roughly equal to E_4, then the particle will eventually leak out of the well. If T is large, which is the case when b is large, it takes a long time for the particle to leak out, and we have an *approximately stationary state*. The time T is the mean-life of the state. If we let b tend to infinity, T will tend to infinity, and we obtain a *strictly* stationary state, as in the problem of Fig. 19A. If we let b tend to a, T becomes smaller, and in the limit $b = a$ the "state" of energy E_4 loses its meaning as a quasi-stationary state.

In view of this result we are justified in drawing the fourth energy level E_4 in the term scheme in Fig. 20A *within* the continuum: it corresponds to an approximately stationary state. Such levels are often called *virtual energy levels*.

The result (22a) can be understood in a qualitative fashion as a consequence of a barrier penetration phenomenon, like the ones we discussed in Chap. 7. A particle confined, at an energy E_4, inside the well would stay within the well forever if classical mechanics applied. Within the framework of quantum mechanics this is not so: the particle can leak through the barriers on both sides of the well. The wider the barriers the longer it will take, and the larger will be the constant T. For a very large T the particle has to bounce back and forth very many times inside the well, and the particle behaves approximately as if it were in a stationary state.

23 In each case in our discussion so far the problem of finding the stationary states can be regarded as the problem of fitting an oscillatory wave function between two classical turning points. For the ground state the wave function has one maximum, and no node. For the next state the wave function has two extrema, and one node. Generally the wave function corresponding to the mth state will have m extrema, and $(m - 1)$ nodes. Let us employ the quantum number n to label the stationary states, where n is the number of nodes (zeros) of the wave function. The ground state is thus assigned the quantum number $n = 0$, and the nth *excited* state is assigned the quantum number n. The wave function corresponding to the quantum number n has $(n + 1)$ extrema.

Let us now try to find a method to determine the *approximate* energy levels of a particle in a potential valley. For this purpose we consider Fig. 23A, which shows the potential for a typical problem of this kind.

The heavy solid line represents the potential. The heavy dashed line represents the energy, E_6, of the sixth excited state, and the oscillatory dotted line represents the corresponding wave function. The wave function is drawn only between the turning points x_1 and x_2 [which are defined by $V(x_1) = V(x_2) = E_6$]. Outside this interval the wave function tends asymptotically toward the axis.

24 Suppose we were to try to represent a wave function of the kind shown in Fig. 23A by an expression of the form

$$\varphi(x) = A(x) \sin [f(x)] \tag{24a}$$

where $A(x)$ is a positive amplitude, and where $f(x)$ is a phase function which increases monotonically with x. Every time the phase function $f(x)$ assumes a value $k\pi$, where k is an integer, the wave function has a node. Let us consider the change Δf in the phase function between the turning points,

$$\Delta f = f(x_2) - f(x_1) \tag{24b}$$

Looking at Fig. 23A we see that for the wave function shown in the figure the change in the phase is about $(6 + \frac{1}{2})\pi$. We shall be guided by this suggestive picture and *assume* that the wave function for the nth excited state is such that the phase function $f(x)$ changes by the amount

$$\Delta f_n \cong (n + \tfrac{1}{2})\pi \tag{24c}$$

between the turning points.

We make the assumption (24c) for convenience, in order to have a definite formula. If we wish to be more correct we might at most state an *inequality*, namely

$$(n + 1)\pi \geqq \Delta f_n > n\pi \tag{24d}$$

as the reader can easily convince himself. If we look at Fig. 4C we see that in that case the upper limit in (24d) is assumed, whereas

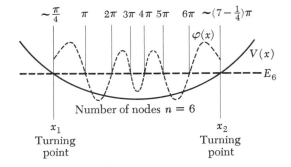

$\sim \frac{\pi}{4}$ π 2π 3π 4π 5π 6π $\sim(7-\frac{1}{4})\pi$

$\varphi(x)$

$V(x)$

E_6

Number of nodes $n = 6$

x_1 x_2
Turning   Turning
point   point

Fig. 23A To illustrate our discussion of the so-called WKB-approximation method for finding stationary states. To find the $(n + 1)$th state, i.e., the nth excited state, we try to select the energy E such that we can fit $(n + \frac{1}{2})$ "half-waves" between the classical turning points. The "local wavelength" at a point depends on the total energy and on the potential at the point.

The solid curve represents the potential, and the dashed curve represents the wave function (between the turning points) for the sixth excited state. Values of the phase $f(x)$ are shown above the turning points, and above the nodes. In this particular case the total change in phase (between the turning points) obeys the relation $\Delta f \cong (n + \frac{1}{2})\pi = (6 + \frac{1}{2})\pi$.

we are closer to the lower limit for the third excited state in Fig. 19A. The formula (24c) therefore represents a compromise.

25   Let us next try to derive an approximate expression for the change in phase of the wave function, as a function of the energy E. Consider first a region in which the potential is constant, and has the value V. In such a region, and for $E > V$, the wave function is of the form

$$\varphi(x) = A \sin\left[(x - x_0) \frac{p}{\hbar} \right] \tag{25a}$$

where A and x_0 are constants, and where

$$p = \sqrt{2m(E - V)} \tag{25b}$$

Comparing (25a) with (24a) we find

$$f(x) = (x - x_0) \left(\frac{p}{\hbar} \right) \tag{25c}$$

As we go a distance dx to the right the change in phase, df, is then

$$df = \left(\frac{p}{\hbar} \right) dx = \frac{1}{\hbar} \sqrt{2m(E - V)}\, dx \tag{25d}$$

Let us now use (25d) as an *approximate* expression for the change of phase with x in the case that $V(x)$ is *not* a constant. This approximation is better justified the more slowly the potential $V(x)$ varies with position. Within this approximation the *total* change of phase between the two turning points x_1 and x_2 is given by

$$\Delta f = \int_{x_1}^{x_2} \frac{df}{dx} dx \cong \frac{1}{\hbar} \int_{x_1}^{x_2} dx \sqrt{2m\,[E - V(x)]} \tag{25e}$$

Let us apply this relation to the case of the $(n + 1)$st stationary state, of energy $E = E_n$. The total change in phase is *also* given approximately by (24c), and if we equate the two expressions for the change in phase we obtain

$$\int_{x_1}^{x_2} dx \sqrt{2m[E_n - V(x)]} \cong (n + \tfrac{1}{2})\pi\hbar \tag{25f}$$

26   The equation (25f) is an equation by which we can determine the energy E_n of the $(n + 1)$st stationary state. To do this we first find the turning points x_1 and x_2, as functions of the energy parameter E, by solving the equations

$$V(x_1) = V(x_2) = E, \qquad x_2 > x_1 \tag{26a}$$

We denote the solutions by $x_1(E)$ and $x_2(E)$. We next evaluate the integral

$$g(E) = \int_{x_1(E)}^{x_2(E)} dx \ \sqrt{2m[E - V(x)]} \tag{26b}$$

which gives us a function $g(E)$ of E. The energies E_n are finally obtained as the solutions to the equation

$$g(E) = (n + \tfrac{1}{2})\pi\hbar \tag{26c}$$

where $n = 0, 1, 2, \ldots$.

The *approximate* method which we have just found for determining the energy levels of a particle in a "potential valley," like the one shown in Fig. 23A, is known as the *WKB-method.*† In many cases it gives fairly accurate results, and it is always useful when we try to form a rough idea of the location of the levels. The nature of the approximation is quite similar to the nature of the approximation which we made in deriving the formula (36b) in Chap. 7 for the transmission coefficient of a potential barrier: the same kinds of integrals in fact appear in both cases.

It is interesting to note that our equation (25f) which we derived within the framework of wave mechanics, is *identical* to the so-called Bohr-Sommerfeld quantum condition of the old Bohr theory. We thus achieve a certain understanding of why the Bohr theory sometimes works quite well, and we can also see why the Bohr theory might sometimes fail badly: the equation (25f) is not rigorously true but is only an approximate relation.

The Harmonic Oscillator. Vibrational and Rotational Excitations of Molecules

27 Let us apply our approximation method to one of the most important eigenvalue problems: that of finding the energy levels of a one-dimensional harmonic oscillator. For this problem the potential $V(x)$ is given by

$$V(x) = \frac{K}{2} x^2 \tag{27a}$$

where K is the "spring constant." If the mass of the particle is m,

---

† Named after G. Wentzel, H. A. Kramers and L. Brillouin. See H. A. Kramers, "Wellenmechanik und halbzahlige Quantisierung," *Zeitschrift für Physik* **39**, 828 (1926).

then the (angular) frequency of oscillation, ω_0, is given within classical mechanics as

$$\omega_0 = \sqrt{\frac{K}{m}} \qquad (27b)$$

To carry out the quantization procedure described in Sec. 26 we must first find the turning points. They are symmetrically placed with respect to the origin, and we can write $x_1 = -x_0, x_2 = x_0$, where, in accordance with (26a), we have

$$x_0(E) = \sqrt{\frac{2E}{K}}, \qquad E = \frac{K}{2} x_0{}^2 \qquad (27c)$$

We next find the function $g(E)$, defined as in (26b) by

$$g(E) = \int_{x_1}^{x_2} dx \ \sqrt{2m[E - V(x)]} = \int_{-x_0}^{x_0} dx \ \sqrt{Km(x_0{}^2 - x^2)} \quad (27d)$$

Introducing a new variable of integration θ by $x = x_0 \sin \theta$ we obtain

$$g(E) = 2 \ \sqrt{Km} \, x_0{}^2 \int_0^{\pi/2} d\theta \cos^2 \theta = \pi E \sqrt{\frac{m}{K}} \qquad (27e)$$

where we have eliminated x_0 by using (27c). Inserting this expression for $g(E)$ into (26c) we obtain the very simple result

$$E_n = (n + \tfrac{1}{2}) \hbar \, \omega_0 \qquad (27f)$$

for the energy E_n of the $(n + 1)$st stationary state of the harmonic oscillator, where $n = 0, 1, 2, \ldots$, is any non-negative integer.

28   It so happens that the rigorous solution of the Schrödinger equation (4b) for the case of a harmonic oscillator, i.e., for a potential function as given in (27a), gives *precisely* the result (27f).

This book is not devoted to the solution of special cases of the Schrödinger equation, and we will not try to solve the harmonic oscillator problem rigorously. Due to a remarkable accident our approximation method actually gives us the correct result, which is more than we had a right to expect.

Figure 28A shows the term scheme (to the left) and the potential function (to the right) for a harmonic oscillator. Note the characteristic equal spacing between the levels. In the figure we have selected the bottom of the potential well as the zero-point of energy: this is, of course, an arbitrary convention.

If the oscillating particle carries a charge we can expect radiative transitions between the energy levels, which, when the radiative processes are taken into account, would then no longer be absolutely

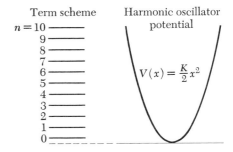

Fig. 28A The potential (at right) and the term scheme (at left) for a harmonic oscillator. Measured from the bottom of the potential "well" the energy of the $(n + 1)$st level is given by $E_n = (n + \tfrac{1}{2})\hbar\omega_0$, where ω_0 is the classical frequency. The WKB-method gives the same result as the rigorous theory.

stable for $n > 0$. One can show that the selection rule for *electric dipole transitions* is that n changes by one unit. The quantum emitted must accordingly be of the classical frequency ω_0 for *any* transition of this kind. This is also what we would predict on the basis of classical theory.

29 The theory of the harmonic oscillator is important in physics because the equations of motion for very many seemingly totally unrelated physical systems are formally equivalent to the equations of motion for a system of harmonic oscillators which interact with each other only very weakly. To a first approximation, in which the interaction between the oscillators is neglected, the quantum theory of such systems is mathematically equivalent to the analytically very simple theory of a system of completely independent harmonic oscillators. The latter kind of system is simple to discuss because each oscillator oscillates by itself as if the other oscillators were not there, and it is clear that if we can describe *one* such oscillator we can describe any number of them.

As examples of systems of this kind we can mention the electromagnetic field, an elastically vibrating solid, and many quantum fields. Furthermore all molecules have vibrational modes which can be described to a good approximation in terms of the theory of harmonic oscillators. Quite generally we can say that the harmonic oscillator theory applies to systems which satisfy *linear*, or *approximately linear*, equations of motion.

30 Figure 30A illustrates the *approximately* harmonic nature of an actual molecular linear oscillator, namely the hydrogen molecule. This molecule has modes of excitation in which the two protons oscillate against each other. These modes may be understood in terms of an effective internuclear potential, shown in the right portion of the figure. The graph shows the potential energy (in eV) of the system as a function of the internuclear separation. The existence of this effective potential, and its functional dependence on the internuclear separation, can be well understood theoretically. We shall discuss this potential in the next section. To study the vibrational states of this, or any other diatomic molecule, we may therefore first find the effective potential, after which we solve the one-dimensional Schrödinger equation with this potential to find the energy levels of the vibrational states.

We have selected the bottom of the potential well as the zero-point of energy, just as in Fig. 28A. As the internuclear separation r goes to zero we can assume that the potential tends to infinity. However, as r tends to infinity the potential tends to a constant,

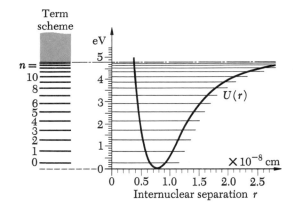

Fig. 30A The right portion of the figure shows the effective internuclear potential $U(r)$ in the hydrogen molecule, and the left portion shows the corresponding term scheme. For the lowest excited states the molecule behaves like a harmonic oscillator. The potential curve near the minimum is of approximately parabolic shape, and the low-lying levels have roughly the same locations as for a harmonic oscillator. (See Fig. 28A.) As the internuclear separation increases, the potential tends to a constant value. The continuum in the term scheme begins at this level, corresponding to the dissociation of the molecule.

The potential $U(r)$ does not describe a "new" kind of force: it is the electromagnetic force in disguise.

+4.8 eV in the figure.   At this energy the molecule dissociates, and
the continuum begins, as indicated in the term scheme to the left
in the figure.   The potential energy is therefore not identical with
the potential energy of a harmonic oscillator, but if we do not go
too high above the bottom of the potential well the curve has an
approximately parabolic shape.   In fact any smooth curve with a
minimum, and a non-vanishing second derivative at the minimum,
has an "approximately parabolic shape" in the neighborhood of the
minimum.   We can therefore expect the system to behave, for
not-too-high excitations, *approximately* like a harmonic oscillator.
We can see the difference between a true harmonic oscillator and
an approximately harmonic oscillator by comparing Figs. 28A and
30A.   The levels are *not* equally spaced in the term scheme in Fig.
30A, but for small excitations they are approximately equally
spaced.   Furthermore there is only a *finite* number of vibrational
states of a molecule.

The dissociation energy of the molecule is the energy which has
to be supplied to the molecule in its ground state to bring about
dissociation.   From Fig. 30A we find that the dissociation energy
of the hydrogen molecule is about 4.5 eV: this is the energy difference between the lower limit of the continuum and the ground
state.

When the molecule is in its ground state, the mean separation
between the nuclei (protons) is about 0.75 Å: the ground state wave
function is clearly concentrated about the value of r corresponding
to the minimum of the potential.

31   Let us now discuss the meaning of the effective internuclear
potential shown in the right part of Fig. 30A.   We are led to this
potential in an approximation scheme for the study of molecular
structure, known as the Born-Oppenheimer approximation.   The
idea is the following.   Since the nuclei (protons) are so much more
massive that the electrons they will move, in the molecule, with a
velocity which is small compared to the velocity of the electrons.
In a *first* approximation we can assume that the nuclei do not move
at all, but stay at a fixed separation r_0 from each other.   To be
specific we shall discuss the hydrogen molecule, but similar considerations apply to other molecules as well.   The problem to solve
in this first approximation is then to find the ground state of the two
electrons in the electrostatic field of the two protons.   Suppose that
we solve this problem for an arbitrary internuclear separation r,
in which case we find the ground state energy $U(r)$ of the *system*,
i.e., including the electrostatic energy of repulsion between the two
protons, as a function of r.   For very small r the energy $U(r)$ is

large and positive, because the electrostatic energy of repulsion between the two protons tends to $+\infty$ as the separation r tends to zero. For very large r the energy $U(r)$ tends to a constant value, U_∞, which is the ground state energy of two hydrogen atoms infinitely far away from each other. It so happens that there is a range of values r for which $U(r) < U_\infty$, as shown in Fig. 30A. The function $U(r)$ has a minimum at the point $r_0 \cong 0.75$ Å. The lowest possible energy of the molecule, under the assumption that the protons do not move, is thus $U(r_0)$, and in this first step in the Born-Oppenheimer approximation this is the ground state energy of the molecule.

32   However, the protons do move, and in the next step in the Born-Oppenheimer approximation this motion is taken into account. This is done by assuming that the protons oscillate against each other, about the "equilibrium separation" r_0. In this (slow) oscillatory motion (which must, of course, be described quantum mechanically) the effective potential energy is given by the function $U(r)$, found in the first step of the approximation scheme.

The function $U(r)$ is thus the effective potential energy in the second step in the Born-Oppenheimer approximation, in which the oscillation of the two protons *against* each other is taken into account. The *fundamental* interaction in terms of which we are to understand molecular structure is thus the electrostatic interaction between the four charged particles in the hydrogen molecule. The effective potential $U(r)$ arises as a consequence of this fundamental interaction, and it therefore does not describe some *new* kind of force. We can say that it is the electrostatic force in disguise. *This is an important point to understand.*

33   The problem of how to find $U(r)$ explicitly is beyond the scope of this book. Let us, however, try to understand, in a very qualitative fashion, how it is that $U(r)$ can have a minimum. To do this we must convince ourselves that there are configurations of the particles in the *molecule* for which the electrostatic energy is smaller (i.e., more negative) than for two infinitely separated hydrogen atoms, although the electron-proton distances are not smaller in the molecule than in the atoms. This is certainly a necessary although not a sufficient condition for molecular binding.

Consider the configuration shown in Fig. 33A in which the two electrons and the two protons are placed at the vertices of a square of side length a. The lines symbolize the electrostatic interactions between the six pairs of particles. For this particular configuration the total electrostatic potential energy E'_{pot} is given by

$$E'_{\text{pot}} = +2\,\frac{e^2}{a\sqrt{2}} - 4\,\frac{e^2}{a} = \frac{e^2}{a}\,(\sqrt{2} - 4) \qquad (33a)$$

This potential energy should be compared with the total *potential* energy E''_{pot} of two hydrogen atoms separated by a very large distance from each other. This potential energy is given by

$$E''_{\text{pot}} = -2\,\frac{e^2}{a_0} \qquad (33b)$$

where a_0 is the Bohr radius. In the particular case that we select $a = a_0$, the *difference* of the two quantities E'_{pot} and E''_{pot} is *negative*, i.e.,

$$\Delta E'_{\text{pot}} = E'_{\text{pot}} - E''_{\text{pot}} = \frac{e^2}{a_0}\,(\sqrt{2} - 2) \cong -1.2 R_\infty \qquad (33c)$$

where R_∞ is the Rydberg constant, $R_\infty = e^2/(2a_0) \cong 13.6$ eV.

We have thus found a special configuration for which ΔE_{pot} is negative. It is clear, however, that there are "neighboring" configurations for which ΔE_{pot} is likewise negative: it is not necessary that the particles be at the vertices of a square.

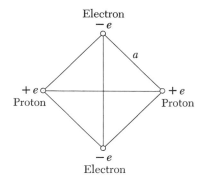

Fig. 33A If $a = a_0$ in the above configuration the *potential* energy will be smaller than the total potential energy of two hydrogen atoms separated by a large distance. The electron-proton distances are then the same for the above "molecule" as in the hydrogen atom, and we can think about the configuration as arising when two hydrogen atoms come together. This example shows that the force between two hydrogen atoms *might* be attractive, but it does not in any way *prove* that a stable molecule can in fact exist.

34 The *total* energy of the hydrogen molecule is the sum of the potential energy and the kinetic energy. If we now recall our discussion in Sec. 14, Chap. 6, of the implications of the uncertainty relation for the structure of the hydrogen atom, we realize that the electrons in the hydrogen molecule must be given "enough space" within the molecule, because otherwise the uncertainty relation requires their momenta, and thereby their kinetic energies, to be large. In our discussion of the hydrogen atom we concluded that if the uncertainty in the position of the electron is of order a_0, which means that it "occupies" a region of linear size a_0, then the kinetic energy will be of order R_∞. The same considerations apply to the hydrogen molecule: if the kinetic energy is to be of this order of magnitude we must allow the electrons to occupy a region of size a_0.

To proceed further we would have to experiment with various regions of confinement for the electrons, and for each choice we would have to compute the potential and kinetic energies, taking into account the requirements of the uncertainty principle. This is a bit involved, and we shall not attempt to do it here. The best way of attacking this problem is to try to invent suitable wave functions which describe both electrons, and then to compute the total energy for these wave functions in accordance with the Schrödinger theory. Since we have not discussed two-particle wave functions we are not prepared to try this approach.† In view of what has been said the reader is perhaps prepared to believe that a minimum exists for the total energy $U(r)$ as a function of the internuclear separation r. Just as in our discussion of the hydrogen atom the minimum in the energy results from a compromise: the electrons must be allowed enough space so that the kinetic energy can be kept small, but yet be confined to a small enough region for the potential energy to be appreciable. Roughly speaking the total potential energy is *negative* and inversely proportional to the "size" of the molecule, whereas the total kinetic energy is *positive* and inversely proportional to the *square* of the size. For some optimal size of the molecule the sum of these two contributions to the energy will have a minimum.

35 Let us now try to estimate the "typical" vibrational frequencies in a (diatomic) molecule. The potential curve is approximately parabolic in the neighborhood of the minimum (at $r = r_0$), and we can try to represent the potential function $U(r)$ by the expression

$$U(r) \sim \left(\frac{r - r_0}{a_0} \right)^2 R_\infty + U(r_0) \tag{35a}$$

This is a reasonable guess. For $r = r_0$ the right side assumes the correct value $U(r_0)$. For $|r - r_0| = a_0$ the potential is larger than $U(r_0)$ by the amount R_∞. Since the size of a molecule is of order a_0, and since the binding energy is of order R_∞, we would expect the potential to behave roughly in this manner.

The right side in (35a) is the potential for a harmonic oscillator. The "spring constant" K for this oscillator is given by

$$K \sim \frac{2R_\infty}{a_0{}^2} = \frac{\alpha^2 mc^2}{a_0{}^2} \tag{35b}$$

Suppose that the effective mass of the oscillator is M. The vibrational frequency ω_v of the molecule is then given by

$$\omega_v = \sqrt{\frac{K}{M}} \sim \alpha^2 \left(\frac{mc^2}{\hbar} \right) \sqrt{\frac{m}{M}} \tag{35c}$$

where we have inserted the expression $a_0 = \alpha^{-1}(\hbar/mc)$ for the Bohr radius. We emphasize that the estimate (35c) is only a crude order of magnitude estimate.

In our discussion in Chap. 2 of characteristic magnitudes in

---

† The first satisfactory theory of molecular binding was presented in W. Heitler and F. London, "Wechselwirkung neutraler Atome und homöopolare Bindung nach der Quantenmechanik," *Zeitschrift für Physik* **44**, 455 (1927).

atomic physics we concluded that we can regard the quantity

$$\omega_e = \alpha^2 \left(\frac{mc^2}{\hbar} \right) \tag{35d}$$

as a "typical" frequency associated with *optical* transitions in an atom or molecule; i.e., transitions in which the electronic configuration changes. We can then write (35c) in the form

$$\omega_v \sim \omega_e \sqrt{\frac{m}{M}} \tag{35e}$$

The quantity M is for all molecules of the order of a nuclear mass, whereas m is the electronic mass. The "typical" electronic frequencies, ω_e, lie in the visible region of the electromagnetic spectrum. As we see the "typical" vibrational frequencies of molecules, ω_v, are smaller by the factor $\sqrt{m/M}$. They will thus be found in the near infrared region, and this prediction is in accordance with observations.

36 Let us find the effective mass M explicitly for a *diatomic* molecule, in which the masses of the two nuclei are M_1 and M_2. The two nuclei oscillate against each other, in such a way that the center of mass lies on the line joining the two nuclei. Let r be the internuclear separation, and let r_1 and r_2 be the distances of the two nuclei from the center of mass, as indicated in Fig. 36A. The kinetic energy of this system is then given by

$$T = \tfrac{1}{2} M_1 \dot{r}_1^2 + \tfrac{1}{2} M_2 \dot{r}_2^2 = \tfrac{1}{2} \left(\frac{M_1 M_2}{M_1 + M_2} \right) \dot{r}^2 \tag{36a}$$

where the dots indicate time derivatives. The potential energy of the oscillator is given by (35a) as a function of r, and the kinetic energy is given by (36a) as a function of $\dot{r}$. The effective mass M of this oscillator is then the coefficient of $\dot{r}^2/2$, i.e.,

$$M = \frac{M_1 M_2}{M_1 + M_2} \tag{36b}$$

and this is the expression which should be inserted into (35c). The quantity M is known as the *reduced mass* of the two-body system.

37 Since we do not have available a precise expression for the "spring constant" K, which we *estimated* in Sec. 35, we cannot find the *precise* vibrational frequency of the diatomic molecule. We can, however, make precise predictions concerning the *isotope effect*. Consider first a molecule in which the nuclear masses are M_1' and M_2', and the vibrational frequency ω_v'. Consider next an

TABLE 35A   *Vibrational Frequencies of Selected Diatomic Molecules*

Molecule	Frequency cycles/sec	Wave number cm^{-1}
C_2	4.921×10^{13}	1641.35
N_2	7.074×10^{13}	2359.61
O_2	4.374×10^{13}	1580.36
NO	5.708×10^{13}	1904.03
CO	6.506×10^{13}	2170.21
IBr	0.805×10^{13}	268.4
S_2	2.176×10^{13}	725.68

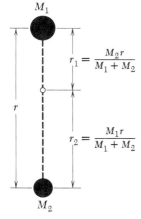

Fig. 36A   Schematic representation of a diatomic molecule. The masses of the nuclei are M_1 and M_2. The small white circle on the line joining the nuclei indicates the center of mass of the system. In the text we consider vibrational excitations in which the nuclei oscillate against each other.

otherwise identical molecule, i.e., a chemically identical molecule, except that we substitute other isotopes, of masses M_1'' and M_2'', for the original nuclei. Let the vibrational frequency for this molecule be ω_v''. The spring constant K is the same for both molecules (within the Born-Oppenheimer approximation), because we found the effective potential $U(r)$ by ignoring the nuclear motion. It follows that the frequencies ω_v' and ω_v'' must be related through

$$\frac{\omega_v'}{\omega_v''} = \sqrt{\frac{M_1''M_2''(M_1 + M_2)}{M_1M_2(M_1'' + M_2'')}} \tag{37a}$$

This prediction has been found to correspond quite accurately to observations. That this is so increases our faith in the essential correctness of the simple ideas which we have presented.

38 Let us now consider the *rotational excitations* of a molecule. With every molecule there is associated a system of discrete rotational states of the molecule, in which the molecule rotates as a whole about some axis. Let us try to estimate the order of magnitude of the energy differences associated with rotational excitations.

For simplicity let us consider a diatomic molecule, as shown schematically in Fig. 36A. Let us assume that in a particular rotational state the molecule rotates with an angular velocity ω_a about an axis passing through the center of mass of the molecule and *perpendicular* to the symmetry axis of the molecule, i.e. to the line joining the two nuclei. We neglect for the moment the vibrational motion, and we thus regard the molecule as a rigid "dumbbell." With the notation of Fig. 36A the speed of the nucleus 1 is then $\omega_a r_1$ and the speed of the nucleus 2 is $\omega_a r_2$. The kinetic energy T_r of the rotational motion is thus given by

$$T_r = \tfrac{1}{2}M_1(\omega_a r_1)^2 + \tfrac{1}{2}M_2(\omega_a r_2)^2 \tag{38a}$$

Expressing r_1 and r_2 in terms of the masses M_1 and M_2, and in terms of the internuclear separation r, as in Fig. 36A, we obtain

$$T_r = \tfrac{1}{2}\left(\frac{M_1M_2}{M_1 + M_2}\right)(\omega_a r)^2 = \tfrac{1}{2}M(\omega_a r)^2 \tag{38b}$$

where M is the reduced mass of the molecule, as defined in Eq. (36b).

The moment of inertia, I, of the molecule with respect to the axis of rotation is given by

$$I = M_1r_1^2 + M_2r_2^2 = Mr^2 \tag{38c}$$

Let us also find the angular momentum J of the molecule with respect to the axis of rotation. It is given by

$$J = M_1 r_1^2 \omega_a + M_2 r_2^2 \omega_a = M r^2 \omega_a = I \omega_a \qquad (38d)$$

We can then write the kinetic energy of the molecule in the form

$$T_r = \frac{J^2}{2I} \qquad (38e)$$

where we have eliminated the angular velocity ω_a from the expression (38b) by the use of the relations (38d).

39 We can guess that the angular momenta which we encounter in molecules will be typically of the order of $\hbar$. It follows that the typical energies associated with rotational excitations will be of the order of magnitude

$$T_r \sim \frac{\hbar^2}{2I} \qquad (39a)$$

Denoting the corresponding frequency by ω_r we can write

$$\omega_r = \frac{T_r}{\hbar} \sim \frac{\hbar}{2I} \qquad (39b)$$

According to (38d) the angular momentum is given by $J = I\omega_a$, and since we assumed $J \sim \hbar$ it follows that $\omega_a \sim \hbar/I$. The angular velocity ω_a and the characteristic rotation frequency ω_r defined by (39b) are thus of the same order of magnitude, as we would expect on the basis of a classical model.

The complete quantum-mechanical theory of the dumbbell molecule leads to a very simple formula for the energy levels. Each rotational state is characterized by a non-negative integral value of the *angular momentum quantum number j*, and the energy of the state is given by

$$E_j = \frac{j(j + 1)\hbar^2}{2I} \qquad (39c)$$

where $j = 0, 1, 2, 3, \ldots$. Although we shall not derive the formula in this book, the author felt it was worthwhile to quote it anyway.

40 The separation between the nuclei in any molecule is of the order of the Bohr radius a_0. We thus estimate the moment of inertia as $I \sim M a_0^2$, and if we insert this expression for I into (39b) we obtain

$$\omega_r \sim \frac{\hbar}{2 M a_0^2} \qquad (40a)$$

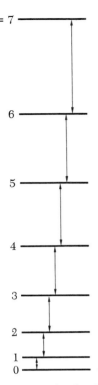

Fig. 39A Term scheme showing the eight first rotational energy levels of a diatomic molecule (regarding the molecule as a rigid dumbbell). According to Eq. (39c) the energy E_j of the state of angular momentum j is given by $E_j = Bj(j + 1)$, where $B = \hbar^2/(2I)$ is the *rotational constant* of the molecule. The vertical arrows show electric dipole transitions in which j changes by one unit.

TABLE 39B   *The Rotational Constant B_e for Selected Diatomic Molecules*

Molecule	B_e (Mc/sec)	r (Å)
BrF	10700	1.76
KCl	3800	2.79
KBr	2400	2.94
$C^{12}O^{16}$	57900	1.13
OH	566000	0.97
NO	51100	1.15

The constant B (see Fig. 39A) is here expressed in terms of the corresponding frequency $B_e = B/h = h/(8\pi^2 I)$, in megacycles per second. The third column gives the internuclear distance r.

It is instructive to rewrite this estimate in terms of the characteristic electronic frequency $\omega_e = \alpha^2(mc^2/\hbar)$. Since the Bohr radius is given by $a_0 = \alpha^{-1}(\hbar/mc)$ we can write (40a) in the form

$$\omega_r \sim \omega_e \left(\frac{m}{M}\right) \tag{40b}$$

as an order of magnitude estimate. (Factors of two are, of course, immaterial in such an estimate.)

Let us compare the characteristic rotational frequencies with the typical vibrational frequencies which we estimated in Sec. 35. Combining the estimates (35e) and (40b) we can write

$$\omega_e : \omega_v : \omega_r \sim 1 : \sqrt{m/M} : (m/M) \tag{40c}$$

Here ω_e is the "typical" electronic transition frequency, ω_v the "typical" vibrational transition frequency and ω_r the "typical" rotational transition frequency. As we see, the rotational transition frequencies are much smaller than both the electronic and vibrational frequencies. They lie in the far infrared (microwave) region.

41 The key idea in the complete explanation of the very complicated optical *band-spectra* emitted by molecules is the idea that every molecule has three different kinds of excitations: *electronic excitations* characterized by the electronic frequency ω_e, *vibrational excitations* characterized by the frequency ω_v, and *rotational excitations* characterized by the frequency ω_r. If we oversimplify the situation we can imagine that we have three systems of energies corresponding to the three different kinds of excitations. The energy of a stationary state of a molecule is thus a sum of three terms: an electronic term, a vibrational term and a rotational term. In making transitions between the various possible energy levels the molecule emits or absorbs photons. In an optical transition the electronic state (configuration) of the molecule changes, and in general the vibrational and rotational states will also change at the same time. The number of possible transition frequencies is therefore very large, and the spectrum shows bands consisting of extremely closely spaced lines. (See Fig. 6B, Chap. 3, for an example.)

It is possible to study the vibrational and rotational spectra separately, i.e., to study transitions in which the *electronic* state of the molecule does not change. After the Second World War new methods for this study were introduced and *microwave spectroscopy* became established as a branch of spectroscopy complementary to the older branch of optical spectroscopy.

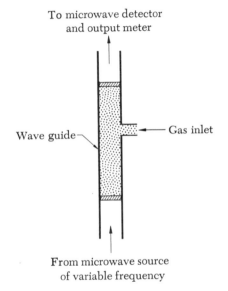

Fig. 41A Very schematic representation of an arrangement which may be used in microwave spectroscopy. The molecules to be studied are present in the form of a gas which fills a portion of a wave guide. Radiation (in the microwave region) passes through the wave guide, and the amount transmitted is measured by the detector and output meter. At the resonant frequencies of the molecule the gas absorbs the microwave radiation, and by measuring the absorption as a function of the frequency the locations of the resonant frequencies can be determined.

By the "microwave region" is understood the wavelength region roughly between 1 mm and 1 meter.

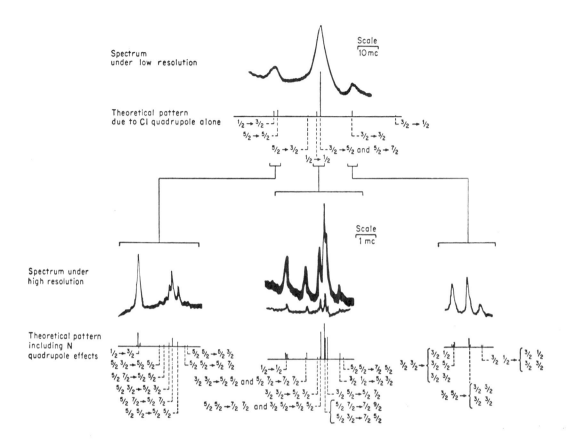

Fig. 41B Microwave spectra under low, and under high resolution, showing the $J = 1$ to $J = 2$ transition in the triatomic molecule $Cl^{35}C^{12}N^{14}$. As we see, this microwave transition "line" exhibits a fine structure: it consists of several closely spaced components. The frequency of the central peak is 23883.30 megacycles/sec. The jagged curves show what is actually measured: the absorption of microwave power as a function of frequency.

The lower spectrum gives a good picture of the high precision which can be achieved in microwave spectroscopy. Note also how well everything can be accounted for theoretically.

This figure appears on page 171 in C. H. Townes and A. L. Schawlow, *Microwave Spectroscopy* (McGraw-Hill Book Co., New York, 1955). See also C. H. Townes, A. N. Holden and F. R. Merrit, "Microwave Spectra of Some Linear XYZ Molecules," *Physical Review* **74**, 1113 (1948). *(Illustration by courtesy of Professor C. H. Townes, Berkeley.)*

Hydrogen-like Systems

42 Let us now consider a three-dimensional problem, namely the problem of finding the energy levels of the hydrogen atom. We shall not really *solve* this problem in this course, but it will be illuminating to consider some aspects of it.

Let us actually consider a somewhat more general problem. A particle of mass m, and charge $-e$, moves in the electrostatic potential due to a nucleus of charge $+eZ$. We shall assume that the nucleus does not move, but stays fixed at the origin. Actually the nucleus would stay fixed only if it were infinitely heavy. However, if the ratio M/m of the nuclear mass M to the "electron" mass m is very large, we can, in a first approximation, regard the nucleus as being infinitely heavy.

The time-independent Schrödinger equation for our problem is then of the form

$$-\frac{\hbar^2}{2m}\boldsymbol{\nabla}^2\,\varphi(\mathbf{x}) - \frac{e^2 Z}{x}\,\varphi(\mathbf{x}) = E\varphi(\mathbf{x}) \qquad (42a)$$

where $x = |\mathbf{x}|$.

43 Let us introduce the new independent variable **y** by

$$\mathbf{x} = \frac{\hbar}{mc\alpha Z}\,\mathbf{y}, \qquad \text{where } \alpha = \frac{e^2}{\hbar c} \qquad (43a)$$

and let us also introduce a new "energy parameter" λ by

$$E = (\alpha Z)^2 mc^2 \lambda \qquad (43b)$$

as well as the wave function $f(\mathbf{y})$ defined by

$$\varphi(\mathbf{x}) = f(\mathbf{y}) \qquad (43c)$$

Rewriting the differential equation (42a) in terms of our new variables and parameters we obtain

$$-\frac{1}{2}\boldsymbol{\nabla}_y{}^2\,f(\mathbf{y}) - \frac{1}{y}f(\mathbf{y}) = \lambda f(\mathbf{y}) \qquad (43d)$$

where $\boldsymbol{\nabla}_y{}^2$ stands for Laplace's differential operator with respect to the variable **y**.

The equation (43d) is the "dimensionless form" of the Schrödinger equation (42a). It is dimensionless in the sense that the physical constants m, e, $\hbar$, c, and Z no longer appear in it. If we can solve (43d) we can re-introduce the old variables by making use of Eqs. (43a) to (43c) and the two equations (43d) and (42a) are clearly completely equivalent.

44 We are thus faced with the purely mathematical problem of solving the equation (43d). We shall not solve this problem but merely give the results, which are as follows:[†]

I. The Schrödinger equation (43d) has square-integrable solutions only when the parameter λ is of the form

$$\lambda_n = -\frac{1}{2n^2} \qquad (44a)$$

where n is any positive integer. This integer is called the *principal quantum number* of a hydrogen-like atom. (It should not be con-

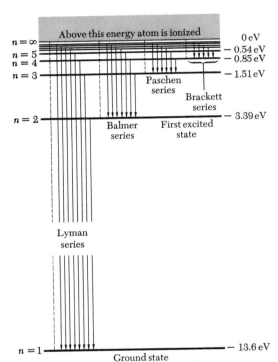

Fig. 45A Term scheme of the hydrogen atom. To a very good approximation the energy E_n of a level of principal quantum number n is given by $E_n = -R_H/n^2$, where $R_H = (1 + m/M_p)^{-1}R_\infty = 13.5976$ eV.

The vertical lines indicate possible electric dipole transitions. These transitions have been arranged into four series, named after early workers in spectroscopy. The Lyman lines all lie in the ultraviolet region. The Balmer series lies in the visible region. See Fig. 1B, Chap. 3, for the appearance of the visible hydrogen spectrum, and for wavelengths of some of the Balmer lines.

[†] The solution for the hydrogen problem is naturally derived in every advanced or intermediate book on quantum mechanics. It was first presented by Schrödinger in his first paper on wave mechanics: "Quantisierung als Eigenwertproblem," *Annalen der Physik* **79**, 361 (1926).

fused with the quantum number n which we introduced in our discussion of the quantum mechanical oscillator.)

II. The continuum begins at $\lambda = 0$. It follows, in view of (43b), that the atom is ionized above the energy $E = 0$.

III. For any given value of n, and with $\lambda = \lambda_n$, the differential equation (43d) has n^2 linearly independent solutions. These solutions can be classified with the help of a quantum number l which describes the spatial symmetry properties of the wave functions. For example, all the solutions for which $l = 0$ are spherically symmetric. The quantum number l ranges for 0 to $(n - 1)$, and for each pair (n,l) the equation has $(2l + 1)$ linearly independent solutions which correspond to different *orientations* of the atom. The quantum number l can also be given the physical interpretation that it measures the angular momentum of the atom, and it is therefore called the *orbital angular momentum quantum number.*[†]

45   In view of these mathematical facts we can now conclude that the possible energy levels of the atom (in its non-ionized state) are given by

$$E_n = -\frac{1}{2}(\alpha Z)^2 \, mc^2 \left(\frac{1}{n^2}\right) \tag{45a}$$

To satisfy the curiosity of the reader we wish to quote *one* explicit solution of the Schrödinger equation (42a), namely the wave function for the ground state. In this case we have $n = 1$, and consequently $l = 0$, which means that the wave function is spherically symmetric. Explicitly the wave function is given by

$$\varphi_{10}(\mathbf{x}) = \sqrt{\frac{Z^3}{\pi a_0{}^3}} \exp\left(-\frac{xZ}{a_0}\right) \tag{45b}$$

where $a_0 = \hbar/(mc\alpha)$.

The reader may wish to verify for himself that the wave function $\varphi_{10}(\mathbf{x})$ does satisfy the wave equation (42a), and that it is normalized to unity, which means that the integral over all of space of the square of the wave function equals unity.

46   Our discussion so far has been based on the assumption that the nucleus remains fixed at the origin. We can very easily generalize our discussion to the case when the nucleus also moves. Let the nuclear mass be M, and let the mass of the electron be m. The reduced mass μ of the nucleus-electron system is then given by

[†] Compare with the discussion in Secs. 30–31 and 54, Chap. 3.

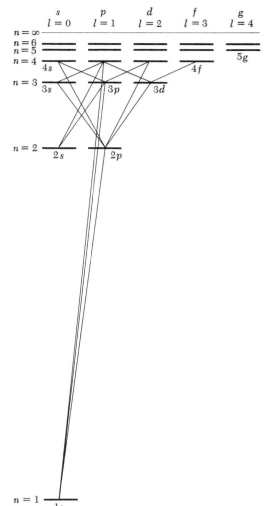

Fig. 45B Term scheme for a hydrogen-like atom. The levels have been arranged into columns corresponding to different values of the orbital angular momentum quantum number l. All electric dipole transitions between levels of principal quantum number four or less are shown. In such transitions l must change by one unit. Note that the state $2s$ cannot decay by an electric dipole transition: the level is metastable.

The above term scheme should be compared to the alkali term schemes in Figs. 28A and 32A, Chap. 3. There are many similarities.

$$\mu = \frac{mM}{m + M} = m \left(1 + \frac{m}{M} \right)^{-1} \qquad (46a)$$

in accordance with our discussion in Sec. 36.

The problem of studying the motion of two particles in their center of mass system, moving under the influence of a force describable by a potential which depends only on the distance between the two particles is completely equivalent to the problem of studying the motion of a single (fictitious) particle, carrying the reduced mass of the system. This particle moves in a *fixed* potential field of force, described by the original potential as a function of the separation of the particles. In order to take the nuclear motion into account we must thus replace the mass m everywhere in our formulas by the reduced mass μ. The energy levels of the system are given by

$$E_n = - \frac{1}{2} (\alpha Z)^2 \mu c^2 \left(\frac{1}{n^2} \right) \qquad (46b)$$

We can also write this as

$$E_n = - \left(\frac{\mu}{m} \right) Z^2 R_\infty \left(\frac{1}{n^2} \right) \qquad (46c)$$

where R_∞ is the Rydberg constant

$$R_\infty = \tfrac{1}{2} \alpha^2 m c^2 \cong 13.6 \text{ eV} \qquad (46d)$$

We should note immediately that in the case of the hydrogen atom (for which $m/M \cong 1/1836$) the reduced mass is very close to the electron mass. As we can see from Eq. (46a) the difference amounts to about 1 part in 2000.

Let us also note that the reduced mass for the deuterium atom is not identical with the reduced mass for the hydrogen atom, and for this reason the deuterium spectrum differs slightly from the hydrogen spectrum. (See Prob. 7, Chap. 2.) The difference is easily observed spectroscopically.

47   Our formulas (46c) describe the energy levels of "hydrogen-like systems" in general, by which we mean bound systems of two oppositely charged particles, provided that the binding is due only to the Coulomb attraction between the particles. Setting $Z = 2$ in (46c) we obtain the energy levels of singly ionized helium, and setting $Z = 3$ we obtain the energy levels of doubly ionized lithium. The correct reduced masses, which are very close to the electron mass, are obtained from Eq. (46a), where M stands for the mass of the helium nucleus, or the lithium nucleus.

"Atoms" in which an electron is replaced by a muon (mu-meson) are known as *muonic atoms*. They are formed when a negative muon is captured by the Coulomb field of a nucleus as the muon is being slowed down in bulk matter. Let us first note that the Bohr-radius of an "atom" is *inversely* proportional to the mass of the "electron." This means that a muonic atom must be about 200 times smaller than an ordinary atom, because the muon mass is about 200 electron masses. Suppose now that a muon is captured by, say, an aluminum atom. Through emission of electro-magnetic radiation the system rapidly goes into a state in which the muon is very close to the aluminum nucleus: i.e., the muon wave packet is much more concentrated around the nucleus than the wave packet of the electrons. The muon and the aluminum nucleus then form a small muonic atom *inside* the electron "cloud," and this muonic atom is clearly a hydrogen-like system.

That muonic atoms indeed form in the manner described has been proved experimentally through the observation of the electro-magnetic radiation emitted by these "atoms."[†] This radiation always lies in the X-ray region, as we can see by considering (46b): the reduced mass is in this case close to the *muon* mass.

One of the subheadings in Chap. 5 was: "There is but one Planck's constant." We note here that the experimental verification of the theoretical prediction for the energy levels of muonic atoms is a very good piece of evidence for the universality of the de Broglie relation.

48   Let us summarize our discussion of hydrogen-like "atoms" as follows. The system consists of two particles, one of charge $-e$, and one of charge $+eZ$. Without explicitly solving the *two-*particle Schrödinger equation which describes the system (and which we have not even written down), we can conclude that its discrete energy levels are given by

$$E_n = (\alpha Z)^2 \, (\mu c^2) \, \lambda_n \tag{48a}$$

where μ is the reduced mass, α the fine structure constant, and where the dimensionless numbers λ_n are the eigenvalues defined by the dimensionless one-particle Schrödinger equation (43d). To find the numbers λ_n is a purely mathematical problem which we have left for a later course, although we have revealed that these numbers are in fact given by $\lambda_n = -1/(2n^2)$.

In other words: if we know the hydrogen spectrum we also know

---

† V. L. Fitch and J. Rainwater, "Studies of X-rays from Mu-Mesonic Atoms," *The Physical Review* **92**, 789 (1953).

the spectra of deuterium, singly ionized helium, doubly ionized lithium, and of all the muonic atoms in which one muon is bound by the electrostatic field of any nucleus. This is so because we were able to find out how the energy levels must depend on the relevant physical parameters such as the charge number Z, and the masses of the two particles. Our discussion illustrates the power of simple dimensional arguments.

Advanced Topic: The Position and Momentum Variables in the Schrödinger Theory†

49 Let us now try to find, within our simple Schrödinger theory, the mathematical objects which in quantum mechanics play the role of the classical position and momentum variables.

Let $\psi(x,t)$ be a Schrödinger wave function, which is normalized to unity. In this and the next sections we are going to consider the wave function at a fixed instant of time t; therefore we may suppress the time variable, and write $\psi(x)$ for brevity.

Since $|\psi(x)|^2$ is a probability density which defines a probability distribution in the physical observable x, the averages of x and x^2 must be given by

$$\text{Av}\,(x) = \bar{x} = \langle\psi|x|\psi\rangle = \int_{-\infty}^{\infty} dx\, x\,|\psi(x)|^2 \tag{49a}$$

$$\text{Av}\,(x^2) = \langle\psi|x^2|\psi\rangle = \int_{-\infty}^{\infty} dx\, x^2\,|\psi(x)|^2 \tag{49b}$$

The notation $\langle\psi|x|\psi\rangle$, which is read "the expectation value of x for the state ψ," is very commonly employed in quantum mechanics.

Now, if $\bar{x}$ denotes the average of x we define the *uncertainty in x*, or the root-mean-square deviation of x, by

$$\Delta x = \sqrt{\text{Av}\left((x-\bar{x})^2\right)} \tag{49c}$$

or

$$(\Delta x)^2 = \int_{-\infty}^{\infty} dx\,(x-\bar{x})^2\,|\psi(x)|^2$$
$$= \text{Av}\,(x^2) - 2\bar{x}\,\text{Av}\,(x) + \bar{x}^2 \tag{49d}$$

from which it follows that

---

† Can be omitted in a first reading.

$$(\Delta x)^2 = \text{Av}\left((x - \bar{x})^2\right) = \text{Av}\,(x^2) - [\text{Av}\,(x)\,]^2 \qquad (49e)$$

We note that the more concentrated the wave function $\psi(x)$ is about the mean position $\bar{x}$, the smaller is Δx. A state for which the position is *precisely* known, i.e., for which $\Delta x = 0$, cannot be realized physically.

The average of any function of x is computed in analogy with the formulas (49a) and (49b) which give the averages of x and x^2. In particular the average of the potential energy is

$$\text{Av}\,(E_{\text{pot}}) = \text{Av}\left(V(x)\right) = \langle\psi\,|\,V(x)\,|\,\psi\rangle$$

$$= \int_{-\infty}^{\infty} dx\;V(x)\,|\psi(x)|^{\,2} \qquad (49f)$$

50 Let us think very carefully about what the above really means. The probability interpretation of the Schrödinger wave function *forces* us to define the *average* of the position variable x as in Eq. (49a). The integral in the right-hand side of this equation thus permits us to find the numerical value of the *average of the quantum-mechanical position variable* x, given the wave function which describes any particular state of the particle. But what is the numerical value of the "quantum-mechanical variable x *itself*"? The answer is that a quantum-mechanical variable does not have a *numerical* value: *it is defined only in terms of a procedure whereby its average can be computed for any given wave function.*

The position variable x is a particularly simple variable in the Schrödinger theory, and for this variable the full implication of the basic principle that quantum-mechanical variables are defined through their averages (for all states) is not immediately obvious. The symbol x also occurs as an independent variable in the wave function, and therefore the definition (49a) might not strike us as being particularly profound. Consider, however, the quantum-mechanical momentum variable (which we shall denote by p). The symbol p does not "occur" in the wave function, and in view of this we may at first wonder whether a momentum variable "exists" at all. To settle this question we shall *define* the quantum-mechanical momentum variable p through a definite prescription whereby the *average* of p can be computed for any given state. The real problem is therefore whether we can define the average momentum in a physically reasonable way.

51 To orient ourselves we first consider a wave function, normalized to unity, which over a very large interval is of the form $\psi(x) = C \exp\,(ix\bar{p}'/\hbar)$. Outside this interval the wave function tends to zero. For such a wave the average momentum must be very close to $\overline{p}'$, and we may write Av $(p) \cong \overline{p}'$. In the interval mentioned above we have

$$-i\hbar \frac{\partial}{\partial x}\, \psi(x) = \overline{p}' \psi(x) \tag{51a}$$

and, since the wave function is normalized to unity, we have

$$\overline{p}' \cong \int_{-\infty}^{\infty} dx\, \psi^*(x) \left(-i\hbar \frac{\partial}{\partial x}\right) \psi(x) \tag{51b}$$

We have here assumed that most of the contribution to the integral comes from the region in which (51a) holds. For a wave function of the special form considered we may thus find the average momentum by evaluating the integral (51b). We shall now assume that this integral gives the average momentum exactly, for *all* (normalized) wave functions. We thus *postulate:*

$$\text{Av}\,(p) = \langle \psi | p | \psi \rangle = \int_{-\infty}^{\infty} dx\, \psi^*(x) \left(-i\hbar \frac{\partial}{\partial x}\right) \psi(x) \tag{51c}$$

for *every* normalized Schrödinger wave function $\psi(x)$. This means that in the Schrödinger theory the momentum variable p is represented by a *differential operator* which acts on the wave function standing to the right of it in the integral in Eq. (51c). In other words:

$$p = -i\hbar \frac{\partial}{\partial x} \tag{51d}$$

52 The square of the momentum variable is then represented by the differential operator

$$p^2 = -\hbar^2 \frac{\partial^2}{\partial x^2} \tag{52a}$$

and the average of the square of the momentum is given by

$$\text{Av}\,(p^2) = \langle \psi | p^2 | \psi \rangle = \int_{-\infty}^{\infty} dx\, \psi^*(x) \left(-\hbar^2 \frac{\partial^2}{\partial x^2}\right) \psi(x) \tag{52b}$$

In complete analogy with the formulas (49c) to (49e) we define the uncertainty Δp in p, by the equations

$$\Delta p = \sqrt{\text{Av}\left((p - \overline{p})^2\right)} \tag{52c}$$

$$(\Delta p)^2 = \text{Av}\left((p - \overline{p})^2\right) = \text{Av}\,(p^2) - [\text{Av}\,(p)]^2 \tag{52d}$$

where $\overline{p} = \text{Av}\,(p)$.

Notice that the same argument which led us to define the average momentum as in Eq. (51c) applies to the definition of the average of p^2, as in Eq. (52b).

53   If we now look at the expressions (49a), (49b), (49f), (51c) and (52b) we discover a common element: the average value of a quantum-mechanical variable Q is given by an expression of the form

$$\text{Av}\,(Q) = \langle\psi\,|\,Q\,|\,\psi\rangle = \int_{-\infty}^{\infty} dx\,\psi^*(x)Q\,\psi(x) \qquad (53a)$$

where Q is either a differential operator acting on the wave function standing to the right of it, or else just x, or x^2 or some other function of x.   This is in fact the general scheme according to which quantum-mechanical variables are defined (in the Schrödinger theory): the *average* of the variable Q is given by an expression such as the right-hand side of Eq. (53a) where Q is a suitable linear operator acting on the wave function standing to the right of it.   (For the position variable the linear operator is simply "multiplication by x.")   Furthermore: the average of Q^2 is obtained by replacing Q in the integral by Q^2, where $Q^2\psi(x)$ is what we obtain if we let Q act twice on $\psi(x)$.

54   We illustrate these ideas through further examples.   Let the mass of the particle be m.   The kinetic energy, E_{kin}, of the particle is then described by the differential operator

$$E_{\text{kin}} = \frac{p^2}{2m} = -\frac{\hbar^2}{2m}\frac{\partial^2}{\partial x^2} \qquad (54a)$$

The *total* energy of the particle is described by the operator H, which is the sum of the operators describing the kinetic and potential energies.   In the Schrödinger theory the energy operator H is thus a differential operator:

$$H = \frac{p^2}{2m} + V(x) = -\frac{\hbar^2}{2m}\frac{\partial^2}{\partial x^2} + V(x) \qquad (54b)$$

in accordance with our discussion in Sec. 10 of this chapter.

55   The reader should note that up to Sec. 51 we have been vague about what we mean by the momentum in the Schrödinger theory. As long as we deal with a wave of the form $\exp\,(ixp/\hbar)$ it is clear that the p occurring in the exponent is the momentum.   We must, however, define the momentum generally, for *all* (normalized) Schrödinger wave functions, and this is precisely what we have done through the relations (51c) and (51d).

We may wonder whether the momentum could have been defined differently. A careful examination of this point shows that our definition is essentially unique in the sense that it is demanded by the requirement that our chosen momentum variable shall have a sensible physical interpretation in accordance with our notion of momentum in classical physics.

56 The plausibility of the definition (51c) of the average momentum can be tremendously strengthened by the following theorem, due to P. Ehrenfest, which we shall state, but not prove here:†

The averages of the quantum-mechanical variables satisfy the same equations of motion as the corresponding classical variables in the corresponding classical description. Specifically this theorem says that

$$\frac{d}{dt}\,\text{Av}\,(x) = \frac{1}{m}\,\text{Av}\,(p) \tag{56a}$$

$$\frac{d}{dt}\,\text{Av}\,(p) = -\text{Av}\left(\frac{dV(x)}{dx}\right) \tag{56b}$$

provided that the Schrödinger wave function, $\psi(x,t)$, with respect to which the above averages are computed, satisfies the Schrödinger equation

$$H\psi(x,t) = i\hbar\,\frac{\partial\psi(x,t)}{\partial t} \tag{56c}$$

where H is the differential operator given in Eq. (54b).

The Schrödinger wave function $\psi(x,t)$ depends on the time t, and this time-dependence is described by the Schrödinger equation (56c). It follows that the averages of x and p will also depend on the time, and one may show that the equations (56a) and (56b) must hold. The proof is not particularly difficult. We carry out the time differentiations inside the integral which defines the average in question. We then eliminate the time derivatives of ψ and ψ^* by making use of the Schrödinger equation (56c) and its complex conjugate form. Rearranging the terms through integration by parts we obtain the results stated in equations (56a) and (56b). The interested reader may wish to carry out this process in detail at this point: we shall not give the detailed proofs as they are a bit tedious.‡

---

† P. Ehrenfest, "Bemerkung über die angenäherte Gültigkeit der klassischen Mechanik innerhalb der Quantenmechanik," *Zeitschrift für Physik* **45**, 455 (1927).

‡ The reader will find proofs in E. Merzbacher, *Quantum Mechanics* (John Wiley and Sons, New York, 1961), p. 41, and L. I. Schiff, *Quantum Mechanics*, 3rd ed. (McGraw-Hill Book Co., New York, 1968), p. 28.

57 The theorem which we just stated, and which can be generalized without difficulty to the three-dimensional case, is of great
importance for our conceptual understanding of quantum mechanics. It explains why classical mechanics can be regarded as a
limiting case of quantum mechanics whenever the uncertainty in
the variables, i.e., the statistical spread in the variables typical for
quantum mechanics, can be ignored. We wish, of course, to have
this kind of correspondence between classical and quantum mechanics, and the fact that Ehrenfest's theorem can be proved for
our momentum variable strongly indicates that our choice is the
correct one.

That classical mechanics should arise as a limiting case of quantum mechanics is the essence of the so-called *correspondence
principle* of Bohr. It is an important principle because if quantum
mechanics is to be a comprehensive description it must be able to
account for *all* physical phenomena, including the phenomena
which can also be described classically. Historically the correspondence principle served as a guide in the early development of
quantum mechanics. It places a constraint, we may say, on possible
new theories, but it should not be thought that it determines these
uniquely. There can be no rules for "quantization," i.e., there
can be no prescription for how one passes from a classical description to a quantum-mechanical description. It obviously makes no
sense to say: "In order to find the *correct* (quantum-mechanical)
equations it is necessary first to state the *wrong* (classical) equations,
after which one passes from the wrong equations to the correct
equations by a certain magical operation." The correct equations
of physics are rather found by clever guessing, guided by known
experimental facts, and these guesses are then subjected to further
experimental tests.

58 For any quantum-mechanical variable Q the quantity

$$\Delta Q = \sqrt{\text{Av}\,(Q^2) - [\text{Av}\,(Q)]^2} \qquad (58a)$$

computed for a given wave function, can be used as a measure of
the precision with which the variable Q is known for the state
described by this wave function. The variable Q has a *precise*
value in a particular state if and only if $\Delta Q = 0$ for this state. As
an example of this idea we can mention that the energy variable H
is precisely known for each stationary state: it has the value E,
where E is the energy of the state. For a non-stationary state we
have $\Delta H > 0$.

An uncertainty relation in general is a restriction on the precision
with which two different variables can be known simultaneously:
it takes the form of an inequality involving $\Delta Q'$ and $\Delta Q''$ for the

two variables Q' and Q''. We now have available a precise definition of Δx, as given by Eq. (49e), and a precise definition of Δp, as given by Eq. (52d). We could, without too much difficulty, prove the *precise* uncertainty relation

$$\Delta x \, \Delta p \geqq \frac{\hbar}{2} \tag{58b}$$

i.e., prove that the inequality (58b) holds for *all* wave functions, and that furthermore there are some wave functions for which (58b) holds as an equality. We shall not undertake this task here since we already have a good qualitative understanding of *why* a relation such as (58b) must hold, and that is sufficient for this course.

References for Further Study

1) For further discussions of simple problems in the Schrödinger theory we refer to the books mentioned at the end of Chap. 7 (items 2).

2) G. M. Barrow: *The Structure of Molecules* (W. A. Benjamin, Inc., New York, 1963), is a readable introduction to molecular structure and molecular spectra. The discussion is elementary, and our preparation suffices for a complete reading of this book.

3) F. O. Rice and E. Teller: *The Structure of Matter* (Science Editions, Inc., 1961), is, as the name of the book indicates, devoted to a general discussion of the structure of matter (from the standpoint of quantum-mechanical ideas). The discussion is elementary, and the book can easily be read with our preparation. The reader may want to supplement our discussion by reading selected portions from this book.

Problems

1 (a) Consider the problem of a particle confined to a potential well with infinitely high walls, as shown in Fig. 4A. Let us study the wave function given by the expression (6b), for $n' = 17$ and $n'' = 18$. Plot the probability density, given by Eq. (6c), for the following times: $t = 0$; $t = t_0/4$; $t = t_0/2$; $t = 3t_0/4$ and $t = t_0$, where $t_0 = (4ma^2)/(35\pi\hbar)$. These plots will suggest to you the periodic motion of a particle back and forth between the walls. The period of this motion is t_0.

(b) Consider the motion of a classical particle of mass m and energy $E_c = \frac{1}{2}(E_{17} + E_{18})$ in the same potential well, and compare the period of this motion with t_0 found above.

(c) The wave packet in part (a) of this problem is not particularly well-confined. In fact it is spread out over about $\frac{1}{2}$ the size of the well. To produce a sharply defined wave packet which resembles the classical point particle more closely, we need a superposition of a large number of eigenfunctions. If the position is to be well-defined, the momentum, and thereby the energy, will be poorly defined. Note now that the energy of the nth level is proportional to n^2, whereas the spacing between two neighboring levels is approximately proportional to n. For a wave packet of high average energy it is therefore possible to have a superposition of a large number of eigenfunctions such that the instantaneous position of the particle is reasonably well-defined, and such that the *fractional* spread in the energy is also small. We here encounter another example of the passage to the classical limit. A wave packet in a potential well can behave like a classical particle provided that its average energy is high compared with the ground state energy.

We cannot here study the passage to the classical limit in complete detail, but let us study just one aspect of the problem. Let $n' = n$, and $n'' = n + 1$. Find the period of the motion of the packet represented by the superposition (6b) and compare this period with the period of a classical particle moving with an energy E such that $E_{n+1} \geqq E \geqq E_n$. Consider, in particular, the limit as $n \to \infty$.

2 For the sake of argument the author makes the following assertion (inspired by some attempts to "explain" quantum mechanics in certain popular writings). The probability density $P(x) = |\psi(x,t)|^2$ for a *stationary* state represented by the wave function $\psi(x,t)$ can be understood as representing the *time-average* of the probability density for a particle moving *classically* in the potential, with the energy of the stationary state. In other words: the particle moves classically, but if we average this motion over a time which is large compared with the natural period of the motion, then we are led to the probability density $P(x)$. For a particle in three-dimensional motion, for instance an electron in the hydrogen atom, we can give a similar interpretation of the absolute square of the wave function representing a *stationary* state. The particle moves classically, but our measuring instruments are too crude to follow the details of the motion, and we therefore observe instead a probability distribution of the electron in the atom, and this can be understood as arising from averaging the classical motion over a large time.

The reader will note that this assertion, interpreted literally, can be immediately refuted. The author therefore backs away a bit. He asserts instead that whereas this interpretation of the square of the wave function is not strictly correct, it is nevertheless a very *useful* way of thinking about the quantum-mechanical motion of a particle: it gives us real insight into what goes on, provided it is understood in an approximate sense.

It is the task of the reader to utterly refute these ideas; both the naive first assertion and the modified second assertion. In so doing the reader should take into consideration the discussion in the beginning of this chapter, as well as our discussion of "double-slit experiments" in Chaps. 4 and 5.

3 The integration in Eq. (22a) extends from $-a$ to $+a$. Suppose that we instead integrate from $-\infty$ to $+\infty$. How does *this* integral depend on the time t, and what is its value at $t = 0$?

4 We should convince ourselves that an attractive potential does not necessarily lead to bound states. To do this we consider the specific example shown in the figure next to Prob. 5. Let B be the depth of the well, let its width be a, and let m be the mass of the particle. Show that if the quantity $G = a^2 B m / \hbar^2$ is smaller than a certain number G_0, then there will be no bound states, whereas there will be at least one bound state whenever $G > G_0$. Find the constant G_0. Note that these considerations only apply to a well for which one wall is infinitely high. For a well such as shown in Fig. 19A, at least one bound state will exist no matter how shallow the well is.

Guided by this example, present arguments why each one of the following conditions is favorable for the occurrence of bound states: (*a*) A large mass m. (*b*) A deep well. (*c*) A wide well. Illustrate your arguments, which should be given for a more general potential well than the one shown in Fig. 19A, with suitably drawn graphs.

On the basis of this example we can understand why two atoms do not always form a stable molecule, despite the fact that the forces between the atoms might be attractive for certain separations between the atoms. (If the force is everywhere *repulsive*, as is sometimes the case, there can naturally not be any bound states.) We can regard the potential shown in the figure of this problem as an idealization of the more realistic molecular potential shown in Fig. 30A

5 As a simple *one-dimensional* model of the deuteron (which is a bound state of a neutron and a proton) let us assume that the neutron-proton potential is as shown in the adjoining figure, with $a = 1.85 \times 10^{-13}$ cm, and with $B = 41.6$ MeV. Find the binding energy of the deuteron in this model, and compare it with the experimental value 2.21 MeV. The good agreement is, of course, *not* a triumph for the theory because the observed binding energy has been used, together with other observational data, to find reasonable values for a and B. The potential which we are considering is not realistic, although it reproduces *some* features of the neutron-proton interaction correctly. The problem of finding the effective potential from "first principles" is unsolved. *Note:* The mass m is the *reduced* mass of the proton-neutron system, $m = M_p / 2$.

6 In the *vibrational* spectrum of hydrogen chloride, HCl, it is found that the spectral lines are actually closely spaced doublets. The short wavelength member of these doublets has an intensity of roughly three times the intensity of the long wavelength member. For the lines which occur in the neighborhood of about 5600 cm⁻¹ (wave numbers) the separation of the two

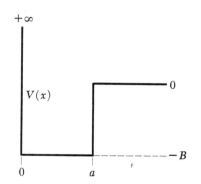

This figure refers to Probs. 4 and 5. In Prob. 5 the solid curve represents the potential energy of the neutron-proton system according to an oversimplified model which, however, is useful in understanding some properties of the deuteron and some features of low energy neutron-proton scattering. The abscissa is the separation between the neutron and the proton.

components is measured to be about 4 cm^{-1}. Give an explanation of this phenomenon, and derive the separation between the components theoretically. Also explain the relative intensities of the two members of the doublets.

7 In the study of the frequencies associated with *rotational* transitions in the iodine-chlorine molecule, the following frequencies (in megacycles per second) have been measured

$$\begin{array}{lll}
\mathrm{ICl}^{35} & 6980\ \mathrm{mc} & 27{,}336\ \mathrm{mc} \\
\mathrm{ICl}^{37} & 6684\ \mathrm{mc} & 26{,}181\ \mathrm{mc}
\end{array}$$

The upper row refers to the molecule containing the isotope Cl^{35}, whereas the lower row refers to the molecule containing the isotope Cl^{37}. The iodine nucleus is the isotope $_{53}$I^{127} in both molecules.

(*a*) Can you explain the frequencies in the lower row, given the frequencies in the upper row?

(*b*) If the sample used in the measurement is prepared from naturally occurring chlorine, all four frequencies will naturally be observed. Can you predict the ratio of the intensities of the lines in the upper row to the intensities of the corresponding lines in the lower row?

(*c*) Consider the isotope effect in general for the rotational levels of a diatomic molecule. Let ω_r' be a rotational transition frequency for a molecule in which the nuclear masses are M_1' and M_2', and let ω_r'' be the corresponding frequency for a chemically identical molecule, which, however, is built of different isotopes, the nuclear masses of which are M_1'' and M_2''. We can again relate ω_r' to ω_r'' without a detailed theory of the molecule. Show that the ratio of the two frequencies is of the form

$$\frac{\omega_r'}{\omega_r''} = \left(\frac{M_1'' M_2'' (M_1' + M_2')}{M_1' M_2' (M_1'' + M_2'')} \right)^k$$

and determine the correct exponent k. Your expression should be carefully compared with the expression (37a), which describes the isotope effect on the vibrational spectrum. The dependence on the isotopic masses is different in the two cases.

8 Consider a "typical" crystal, built of atoms of atomic weight A. Let the crystal be in the form of a cube, of side length L. Derive an order of magnitude estimate for (*a*) the lowest (vibrational) resonant frequency of the crystal; (*b*) the highest resonant frequency of the crystal. Write your results such that the dependence of the frequencies on the fundamental constants α, $\beta = m/M_p$ and $\hbar/mc^2$ as well as on the constants A and $N \sim L/a_0$ (where a_0 is the Bohr radius, and M_p is the mass of the proton) is clearly shown. (*c*) Give some specific numerical examples in which you give the frequencies in megacycles/second.

Isotopic species	$J = 1 \leftarrow 0, v = 0$ Rotational frequencies (Mc/sec)	B_e (Mc
C^{12}O^{16}	115 271.204$\pm$0.005	57 89
C^{13}O^{16}	110 201.370$\pm$0.008	55 3
C^{12}O^{18}	109 782.182$\pm$0.008	55
C^{14}O^{16}	105 871.110$\pm$0.004	53
C^{13}O^{18}	104 711.416$\pm$0.008	5
C^{12}O^{17}	112 359.276$\pm$0.060[b]	

Experimentally measured rotational frequencies for carbon monoxide molecules of different isotopic compositions. The fragment of the table is taken from a paper by B. Rosenblum, A. H. Nethercot, Jr., and C. H. Townes, "Isotopic mass ratios, magnetic moments and sign of electric dipole moment in CO," *The Physical Review* **109**, 2228 (1958). The numbers quoted give a good picture of the accuracy which can be achieved in microwave spectroscopy.

The reader who has solved Prob. 7 might wish to test his results against the data given in the above table. The agreement will be good but not perfect. Our theory, in which the diatomic molecule is regarded as rigid, represents an oversimplification. To account for the experimental numbers to the precision with which they are known requires a more sophisticated theoretical approach.

9 In Sec. 50, Chap. 2, we said that it is possible, in principle, to derive an expression for the ratio of the velocity of sound, c_s, in a crystal to the velocity of light c, such that c_s/c is expressed in terms of four constants only: the fine-structure constant $\alpha \cong 1/137$, the electron-proton mass ratio $\beta = m/M_p$ and the atomic weight A and the atomic number Z of the atoms in the crystal. To derive a *precise* expression for c_s/c is a formidable problem, but we can easily make an order of magnitude estimate which shows the main dependence of c_s/c on α, β and A. Derive such an order of magnitude relation, and test your formula for the case of copper. ($A = 63.6$, $c_s = 4700$ m/sec.)

10 (a) For the potential $U(r)$ shown in Fig. 30A we note that the spacing between neighboring levels *decreases* as the quantum number n increases. Explain in a qualitative way why this is so.

(b) Draw a parabola representing the potential for a strictly harmonic oscillator. On the same graph draw two other potential curves, symmetric with respect to the origin, representing two "almost harmonic" potentials, and such that the radius of curvature of all three curves is the same at the origin ($=$ the minimum of the potential). These two curves are to be drawn such that for the first one of these the spacing between neighboring levels *increases* with the quantum number n, whereas for the second one the spacing between neighboring levels *decreases* with n. It is not necessary to find the levels explicitly, but you should explain why the two curves have the stated properties.

11 As we explained in Sec. 47 the energy levels of doubly ionized lithium are obtained by a simple scaling from the energy levels of singly ionized helium, the scaling factor being close to 9/4. Both ions are hydrogen-like one-electron systems. For the sake of argument the author now wants to maintain that the energy levels of singly ionized lithium should be obtainable by a similar scaling from the energy levels of neutral helium, because both systems are two-electron systems, differing only in the magnitude of the nuclear charge. In other words, the ratios of the wavelengths of corresponding spectral lines ought to be a constant, just as is the case with doubly ionized lithium and singly ionized helium. This is, however, not the case experimentally. The term schemes of neutral helium and singly ionized lithium are quite similar, but not obtainable from each other by a simple scaling. Explain clearly why it is that the simple scaling argument works for the one-electron systems but not for the two-electron systems.

12 The mean life of the $2p$-state in hydrogen is 0.16×10^{-8} sec. What is the mean life of the $2p$-state in singly ionized helium?

13 With reference to the preceding problem: What is the mean life of the 2*p*-state in the muonic atom formed when negative muons are captured in aluminum?

14 Compute the wavelength of the photon emitted when a muonic aluminum atom makes a transition from the 3*s*-state to the 2*p*-state.

15 Find the "Bohr radius" of (*a*) a muonic aluminum atom; (*b*) a muonic lead atom, and compare these radii with the nuclear radii. It is interesting to make this comparison, because if the "Bohr radius" turns out to be comparable with the nuclear radius then we can clearly not regard the nucleus as being a charged point without extension, which means that the energy levels of the muonic atom cannot be given precisely by a formula like (46b). It has been found, experimentally, that the level systems of *heavy* muonic atoms deviate considerably from the prediction (46b). Through a systematic observation of these deviations it has been possible to draw definite conclusions about the charge distribution in nuclei, and about the size of nuclei.

16† Make an attempt to prove Ehrenfest's theorem, mentioned in Sec. 56, along the lines indicated in that section. For further inspiration, see Sec. 50, Chap. 7.

17† (*a*) Apply Ehrenfest's theorem to the case of a harmonic oscillator for which the potential function is given by $V(x) = (K/2) x^2$, and obtain two differential equations satisfied by Av $[x(t)]$ and Av $[p(t)]$. Solve these equation and express Av $[x(t)]$ in terms of Av $[x(0)]$ and Av $[p(0)]$. Compare the solution with the solution of the corresponding classical problem.
  (*b*) For a stationary state Av $[x(t)] = 0$, but for a non-stationary state Av $[x(t)]$ is, in general, an oscillatory non-zero function of the time. Keeping in mind the discussion in Sec. 27, present arguments based on your results in part (*a*) of this problem, to the effect that the energy levels of the harmonic oscillator must be equally spaced, with spacing $\hbar \sqrt{K/m}$. Note that our discussion in Sec. 27 only tells us that the level spacing must be approximately constant, although it so happens that the level spacing is exactly constant and equal to $\hbar \sqrt{K/m}$.

18 Let us consider a "dumbbell" diatomic molecule. We discussed the rotational excitations of such a molecule in Secs. 38–40. Let us suppose that the center of charge of the molecule does not coincide with its center of mass. The molecule will then carry an electric dipole moment, and as it rotates we

---

† These problems refer to an advanced topic.

expect, classically, that it will emit electromagnetic radiation at a frequency equal to the classical angular velocity ω_a.

According to quantum mechanics the energy levels of the molecule are given by Eq. (39c). It is reasonable to assume that the quantum number j changes by one unit if the molecule emits, or absorbs, electric dipole radiation. Express the frequency of the emitted radiation in terms of the angular momentum quantum number j of the initial state of the molecule, and compare the result with the classically derived formula. For large values of j we ought to approach the "classical limit." Is this indeed the case?

Chapter 9

The Elementary Particles

and their Interactions

Chapter 9 The Elementary Particles and their Interactions

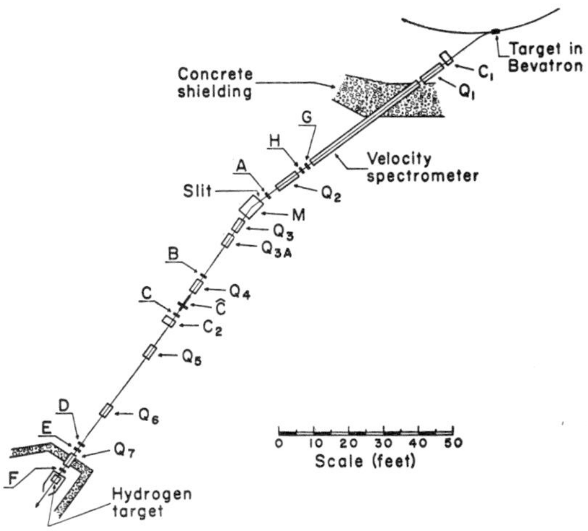

Fig. 2A General plan of an experiment to measure various cross sections for antiproton-proton elastic and inelastic scattering. The antiprotons emerge from the target in the accelerator (top right) and they are deflected and focused on a liquid hydrogen target (bottom left). C_1, C_2 and M are deflecting magnets. Q_1–Q_7 are focusing magnets. A–H are scintillation counters. $\hat{C}$ is a Cerenkov counter. The events taking place in the liquid hydrogen target are observed through counters surrounding the target. (These counters are not shown in the figure.) The purpose of the somewhat elaborate arrangement of counters and magnets is to define the antiproton beam and to discriminate against events in the target due to particles other than antiprotons. The measurements were carried out at antiproton energies of 1.0, 1.25 and 2.0 BeV.

The illustration is taken from R. Armenteros et al., "Antiproton-Proton Cross Sections at 1.0, 1.25, and 2.0 BeV," *The Physical Review* 119, 2068 (1960). This article should be consulted for further details. For results, see Fig. 5A in this chapter. (*Courtesy of The Physical Review.*)

Collision Processes and the Wave Picture

1 In our final chapter we want to discuss some aspects of the most basic and central issues of physics today, which concern the elementary particles and their interactions. In this realm of physics we encounter a host of problems for which we have no solutions at present. We would like to have a theory in terms of which we could "understand" why the various elementary particles exist and why they have the properties they have. We hope, in other words, that it is possible to state a few very basic principles in terms of which we can account for the great multitude of phenomena which have been observed. Is this hope justified? Certainly not on any logical grounds. It could well happen that we are condemned to live with phenomenological theories which summarize the experimental facts in a slightly more economical way than a set of tables and graphs could do, but which lack the comprehensiveness, conceptual simplicity and beauty which we would like to see in a fundamental theory. The author finds it most disagreeable to contemplate this possibility. He prefers to believe that in some sense things are ultimately simple, and he finds some encouragement through meditation over the historical development of physics. As a body of knowledge physics has expanded at a very rapid rate, and the amount of information which we now possess about detailed phenomena is astonishing. What is even more astonishing, however, is the fact that we can account for the details as well as we do in terms of fairly simple theories. In saying this the author does not want to imply that theoretical physics is a trivial subject, but he does feel that the underlying basic principles of our theories (such as they are perceived today) are characterized by a *conceptual* simplicity which is quite striking. As we have indicated before there does not exist today a "simple" comprehensive theory of the elementary particles. In this chapter we shall try to give the reader some idea of the approaches which have been tried, and of the issues and problems which we encounter in this realm of physics.

2 The bulk of our knowledge about the elementary particles derives from collision experiments. It is therefore appropriate that we here say something about the interpretation of such experiments. In a scattering experiment a beam of A-particles from an accelerator

Fig. 2B Photograph of the liquid hydrogen target used in the experiment described in Fig. 2A. The hydrogen is in the container in the center of the device. The antiprotons are incident perpendicularly to the plane of the picture. *(Photograph by courtesy of the Lawrence Radiation Laboratory, Berkeley.)*

impinges upon a target of B-particles (in the form of a solid, liquid or gas.) We observe the particles which emerge from each collision of an A-particle with a B-particle. We say that the collision is *elastic* if no new particles appear in the collision: the A-particle is then merely scattered by the B-particle. If other particles appear we speak of an *inelastic* process.

The observational results are commonly expressed in terms of various *cross sections*. Let us first consider the simplest of these, namely the *total cross section*. We denote this quantity by σ_T. To define σ_T operationally we imagine that the target is a very thin plane layer of randomly distributed B-particles. Let the (on the

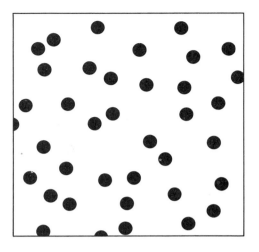

Fig. 3A We can express the effectiveness by which the B-particles (in the target) remove A-particles from the incident beam by a total cross section σ_T. With each B-particle is associated a circular disc of area σ_T, such that an A-particle (imagined to be a point) interacts with the B-particle if and only if it hits the disc. The figure above shows these imaginary discs for a very thin layer of B-particles. If there are n B-particles per unit area the total "blocked" area within a unit area will be $n\sigma_T$. The probability that an A-particle passes through such a layer is accordingly $(1 - n\sigma_T)$. The above figure should not, of course, be taken literally. The B-particles are not in reality small discs or spheres.

average) uniform density of particles in the layer be n particles per unit area. The total cross section is then defined by

$$\sigma_T = \frac{P}{n} \qquad (2a)$$

where P is the probability that an A-particle incident perpendicularly on the layer undergoes *some* interaction with one of the B-particles such that it is removed from the incident beam. In this definition it is essential that the layer be sufficiently thin so that the observed probability P will be small compared to unity. (We shall elaborate on this point in Sec. 4.)

3 We can think about the total cross section in terms of the following model. A circular disc of area σ_T is assigned to each B-particle. The discs are oriented perpendicularly to the incident beam of A-particles, and we imagine that they have the property that an A-particle which hits a disc is removed from the beam, whereas the A-particle is unaffected if it misses the discs. Consider again our thin target layer of n B-particles per unit area. The total area covered by the discs contained in a region of area F is equal to $nF\sigma_T$. This means that a fraction $n\sigma_T$ of the layer is "opaque" and a fraction $(1 - n\sigma_T)$ is "transparent." The probability that an A-particle in the incident beam is removed from the beam is accordingly $P = n\sigma_T$. The relation (2a) can thus be interpreted in this manner, but the reader should understand that the opaque discs exist only in our imagination. The cross section is a very convenient measure of the tendency of particles A and B to interact with each other, but it should not be thought that it refers to geometric properties of either one of the particles.

4 Let us consider the generalization of the relation (2a) to the case when the target layer is not necessarily thin. Let $P(n)$ be the probability that an A-particle is removed from the beam if it hits a layer of B-particles uniformly distributed with a projected surface density n. The quantity $T(n) = 1 - P(n)$ is then the probability for transmission through the layer. Suppose that we place a layer of projected surface density n_1 behind another layer of projected surface density n_2. The surface density of the combined layer is then $(n_1 + n_2)$. The probability that a particle passes through *both* layers is obviously given by

$$T(n_1 + n_2) = T(n_1)T(n_2) \qquad (4a)$$

This equation must hold for all positive real numbers n_1 and n_2. Its general solution is

$$T(n) = \exp\left(-Cn\right) \qquad (4b)$$

where C is a real constant.   We thus have

$$P(n) = 1 - \exp(-Cn) \tag{4c}$$

Notice now that

$$\lim_{n \to 0} \frac{P(n)}{n} = C \tag{4d}$$

and if we compare this relation with the relation (2a) (which is supposed to hold for very small n) we conclude that $C = \sigma_T$.   We thus have

$$P(n) = 1 - \exp(-n\sigma_T), \qquad T(n) = \exp(-n\sigma_T) \tag{4e}$$

As we see the intensity of the transmitted beam decreases exponentially with the thickness of the target.   In practice we might have targets in the form of thin foils of various thicknesses.   To measure the total cross section we perform a simple attenuation measurement: we determine (with counters) the fractional decrease in the intensity of the transmitted beam as a function of the thickness of the foil.   The cross section is then computed with the help of the relations (4e).

5   In a similar manner we define other kinds of cross sections.   We can imagine, for instance, that the A-particle can react with the B-particle to produce a C- and a D-particle:

$$A + B \to C + D \tag{5a}$$

The *reaction cross section* $\sigma_{AB \to CD}$ for this process is then defined by

$$\sigma_{AB \to CD} = \sigma_T P_{AB \to CD} \tag{5b}$$

where $P_{AB \to CD}$ is the probability that the reaction (5a) takes place when an A-particle is removed from the beam through an interaction with a B-particle in the target.   Let us assume that (5a) is the only *reaction*, i.e., the only *inelastic* process, which can take place.   A particle can, however, also be removed from the beam by an *elastic* scattering, in which the A- and the B- particles remain after the collision.   We define the *elastic cross section* σ_e by

$$\sigma_e = \sigma_T P_e \tag{5c}$$

where P_e is the probability that a collision event which removes a particle from the beam is elastic.   The three cross sections are thus related by

$$\sigma_T = \sigma_e + \sigma_{AB \to CD} \tag{5d}$$

since we obviously have $P_e + P_{AB \to CD} = 1$.

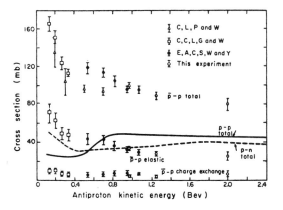

Fig. 5A Graph showing antiproton-proton cross sections as functions of the kinetic energy of the antiproton.   The three experimental points denoted by open circles were obtained in the experiment described in Fig. 2A.   The cross section for proton-proton scattering are shown on the same graph for comparison.   Note that the total antiproton-proton cross section is about twice as large as the total proton-proton cross section.

This graph is taken from R. Armenteros et al., "Antiproton-Proton Cross Sections at 1.0, 1.25, and 2.0 BeV," *The Physical Review* **119**, 2068 (1960).   *(Courtesy of The Physical Review.)*

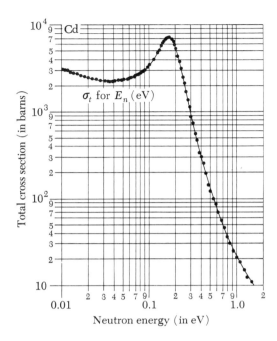

Fig. 6A Curve showing total cross section of neutrons against cadmium as a function of the neutron energy. Note that this curve refers to naturally occurring cadmium. The cross section is therefore an average of the cross sections of the different isotopes. Such a cross section is of limited interest from the standpoint of basic theory which is concerned with cross sections for individual isotopes. The average cross section is however a useful concept in engineering applications. Because of its large cross section for low-energy neutrons cadmium is commonly used to control the reaction rate in nuclear reactors.

The above curve is a portion of a graph in a compilation by H. H. Goldsmith, H. W. Ibser, and B. T. Feld, "Neutron Cross Sections of the Elements," *Reviews of Modern Physics* **19**, 259 (1947).

6 To express cross sections in nuclear and elementary particle physics the units *barn* (b) and *millibarn* (mb) are employed, where

$$1 \text{ barn} = 10^{-24} \text{ cm}^2, \qquad 1 \text{ mb} = 10^{-3} \text{ barn} \qquad (6a)$$

Fig. 6A shows the total neutron cross section for cadmium as a function of the kinetic energy of the neutron, and Fig. 6B shows an analogous graph for neutron collisions with silver. Note that these curves refer to the *chemical elements*, and they are therefore averages for the different naturally occurring isotopes.

A look at these graphs makes it immediately clear that the total cross section has nothing to do with the "geometric" properties of the nuclei. Note, for instance, the dramatic dependence of the cross section on the energy. In cadmium the cross section drops from the peak value of 7200 barns at the neutron energy 0.176 eV to the value 20 b at 1 eV. The cross section curve for silver similarly shows a strong energy dependence, with a very pronounced resonance peak at about 0.52 eV.

Let us also consider the magnitudes of the cross sections. The silver and cadmium nuclei are of roughly the same size. On the basis of the formula

$$r \cong A^{1/3} \times (1.2 \times 10^{-13} \text{ cm}) \qquad (6b)$$

which gives the radius r of a nucleus in terms of its mass number A, we estimate that the radii are $r \sim 5.8$ fermi (since $A \sim 110$.) The corresponding *geometric* cross section πr^2 is accordingly about 1.0 barn. This is 7000 times smaller than the peak cross section in Fig. 6A.

The reader should also look at Figs. 24A and 24B in this chapter. Fig. 24B shows the elastic cross sections for the scattering of positive and negative pions against protons. Fig. 24A shows what is essentially the reaction cross section for the reaction $Al^{27} + p \rightarrow Si^{28} + \gamma$. Note the many very sharp resonance peaks.

7 The cross sections which we have discussed (measured as functions of the energy) provide us with some information about the interactions between the particles in a collision process. We obtain much more information if we also measure the *angular distribution* of the particles emerging from the collision region. We consider, for simplicity, the elastic scattering of the A-particles in the beam by the B-particles in the target. We measure the intensity of the scattered A-particles in different directions with the help of a counter placed at various positions at a fixed distance from the target. The intensity of the incident beam is kept constant during this sequence of measurements. We express the results in terms of

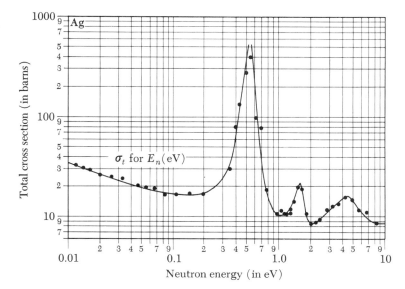

Fig. 6B Curve showing total cross section of neutrons against naturally occurring silver as a function of the neutron energy. Note the pronounced resonance peak. A glance at this figure, and the analogous Fig. 6A, shows immediately that the cross section has no particular relationship to the *size of* a nucleus. The curves for silver and cadmium are very different, and both show a rapid variation with energy. The wave theory of collisions can account very well for the general nature of the observed cross-section curves.

The above curve is a portion of a graph appearing in the compilation by H. H. Goldsmith, H. W. Ibser, and B. T. Feld, "Neutron Cross Sections of the Elements," *Reviews of Modern Physics* **19**, 259 (1947). This paper should be consulted for references to the early work in this field.

a *differential cross section* $\sigma_e(E;\theta,\varphi)$. This quantity is a function of suitable polar angles θ and φ used to specify the direction of observation. It is also a function of the energy E, as we have indicated explicitly. The differential cross section is so defined that $\sigma_e(E;\theta,\varphi)d\Omega$ equals the probability that an incident A-particle is scattered into a solid angle of magnitude $d\Omega$, centered about the direction defined by the angles θ and φ, for the case when the target is a layer of B-particles of unit surface density. For the same counter held at a fixed distance from the target, but at different directions, the counting rate is directly proportional to the differential cross section.

In most scattering situations in practice the differential cross section depends only on the energy and on the angle between the incident beam and the direction of the scattered A-particle. If we denote this angle by θ we can write the differential cross section as $\sigma_e(E;\theta)$ since it does not depend on the other polar angle.†

The total elastic cross section is obtained by integrating the differential cross section over all directions. If it is independent of the angle φ, as above, we thus have

$$\sigma_e(E) = \int d\Omega \, \sigma_e(E;\theta) = 2\pi \int_0^\pi d\theta \, \sin\theta \, \sigma_e(E;\theta) \qquad (7a)$$

---

† Our notation, which is fairly standard, is not entirely satisfactory since the same symbol is used for the differential cross section and for the "total" cross sections discussed earlier. The two kinds of cross sections are distinguished from each other by the explicit presence or absence of the angular variables in the symbol for the quantity.

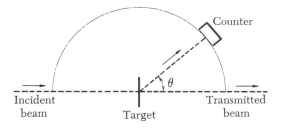

Fig. 7A Very schematic representation of a scattering experiment. A beam of particles from an accelerator impinges upon a target. The relative number of particles scattered in various directions is determined by a counter. The figure shows the detection of particles at the scattering angle θ. In an experiment of this kind we can determine the differential cross section of the scattering process.

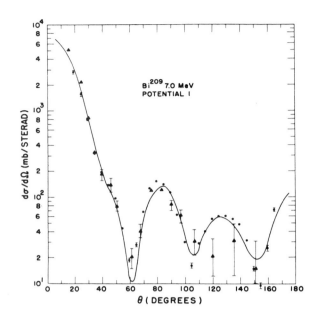

The differential cross section for the elastic scattering of neutrons from the bismuth isotope Bi²⁰⁹. The graph shows the experimental points together with a theoretical curve based on a particular model. The abscissa is the scattering angle, and the ordinate is the differential cross section in units of millibarns per unit solid angle. The kinetic energy of the neutrons was 7 MeV.

This graph is taken from C. D. Zafiratos, T. A. Oliphant, J. S. Levin, and L. Cranberg, "Large-Angle Neutron Scattering from Lead at 7 MeV," *Physical Review Letters* **14,** 913 (1965). *(Courtesy of Physical Review Letters.)*

We could define a differential cross section for an inelastic process in an analogous manner.

8 The various cross sections, as functions of the energy, constitute the primary data which we obtain from scattering experiments. We are then faced with the problem of inferring something about the nature of the perhaps unknown interactions from these data. Or, we might already have a theory, in which case we compute the expected cross sections within the theory and then compare our predictions with the experimental results.

As we said before, most of our information about the elementary particles comes from the analysis of scattering experiments. Special mathematical methods have been developed for this analysis. It would take us too far to discuss these here. Needless to say, the problem of finding the "forces" from the cross sections is far from trivial in practice although it is straightforward (in a sense) in principle.

9 If we were to interpret a scattering event classically we would say that the incident particle is deflected in the force field of the target particle. Quantum mechanically we regard scattering as a manifestation of the diffraction of waves. This is in fact how we discussed electron diffraction in Chap. 5. Our explanation of the observed phenomena was that the incident electron wave is diffracted by all the atoms in the crystal. In certain directions the diffracted waves can interfere constructively, and these are the directions in which we observe intensity maxima. Scattering is thus a manifestation of the diffraction of de Broglie waves by *obstacles*, i.e., by the atoms in the crystal.

Now the reader will note that our description of electron diffraction has an objectionable "unsymmetric" feature. We say that the incident electron waves are diffracted by "obstacles." But the "obstacles" are also physical particles, and we know that all physical particles are waves. It is clearly not consistent to arbitrarily regard some particles as waves, and other particles as *classical* "obstacles." What we see in the electron diffraction experiments is the interaction of the incident electron waves with the wave packets which represent the atoms in the crystal. If we want to be consistent we must say that *scattering arises as a consequence of the interaction of waves with waves.*

Later in this chapter we shall explore this idea further. Let us note at this point that this new insight in no way invalidates our discussion of electron diffraction. The important thing is that the incident wave encounters *something*, and that its interaction with this something leads to a diffraction of the wave. As long as we focus our attention only on the incident particle it is quite immaterial what it is that the particle encounters, be it a "classical obstacle," or a concentrated wave packet.

10 Let us now try to present the barest outline of a wave theory of scattering. We consider the simplest possible case in which the wave representing an A-particle is scattered (diffracted) elastically in a fixed centrally symmetric field of force. We can imagine that the force field is derivable from a potential which tends rapidly to zero with increasing distance from the center of the force field. Our problem is somewhat analogous to the barrier problems which we discussed in Chap. 7. The A-particle finds itself in a region in which the potential function varies with position and as a result an incident plane wave will be diffracted by the potential.

According to the model which we are considering we represent the B-particle in the target by a spherically symmetric potential, although we know that the B-particle should also be described as a wave. It is the case, however, that the correct quantum mechanical description of two-particle scattering is *mathematically equivalent* to our model. Our model is therefore not bad at all. If we think carefully about what we are doing we will realize that we have done similar things before. In discussing alpha-radioactivity in Chap. 7 we described the situation in terms of the motion of a "quantum-mechanical" alpha-particle in a potential field of force. In discussing molecular vibrations we considered the motion of a single particle under the influence of an approximately harmonic molecular potential. In each of these cases we replaced the real problem, which always involves the motion of at least two particles, by a model problem in which a single particle moves in a potential which describes its interactions with all the other particles.

11 Suppose a plane wave of the form

$$\psi_i(\mathbf{x},t) = C \exp\left(i\mathbf{x} \cdot \mathbf{p}_i - i\omega t\right) \tag{11a}$$

represents an A-particle incident on a single B-particle (located at the origin $\mathbf{x} = 0$). Here $\mathbf{p}_i$ is the momentum of the wave, ω is the energy,† and C is a normalization constant. The wave will be diffracted by the B-particle. We shall try to guess the form of the wave function which describes the diffracted wave at a *very large* distance from the origin. We shall argue that the function

---

† We employ units such that $\hbar = 1$.

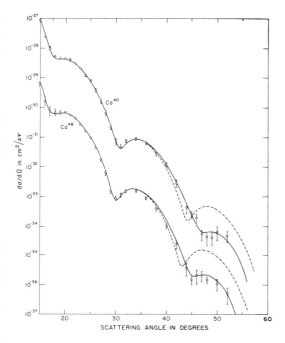

Graphs showing the differential cross sections for the elastic scattering of electrons from two different isotopes of calcium. The ordinate is the differential cross section in units of cm² per unit solid angle, provided that the data for Ca⁴⁸ are multiplied by 10 and the data for Ca⁴⁰ are divided by 10. (The curves are very similar, which is why they have been separated by the use of these scale factors.) The electrons were of energy 750 MeV.

The electromagnetic interaction between the electron and the nucleus is responsible for the scattering, and the purpose of the measurement was to explore the charge distribution in the nuclei. Note the tremendous variation (by a factor of 10^9) of the differential cross section with scattering angle.

Graph taken from J. B. Bellicard et al., "Scattering of 750-MeV Electrons by Calcium Isotopes," *Physical Review Letters* **19**, 527 (1967). (*Courtesy of Physical Review Letters.*)

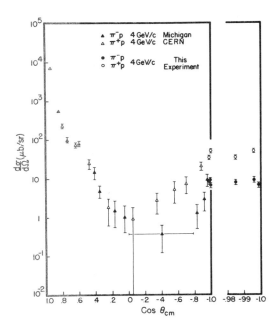

A plot of the elastic pion-proton differential cross sections for a pion momentum of 4 BeV/c. The abscissa is the cosine of the scattering angle in the center-of-mass system. The ordinate is the differential cross section in units of microbarns per unit solid angle. The cross section in the neighborhood of the backward direction (i.e. for cos θ_{cm} close to -1) is shown at right on an expanded horizontal scale. Note that data are presented both for positive and negative pions: the identification of the different experimental points is given in the top portion of the graph.

This figure is taken from W. R. Frisken et al., "Backward Elastic Scattering of High-Energy Pions by Protons," *Physical Review Letters* **15**, 313 (1965). (*Courtesy of Physical Review Letters.*)

$$\psi_s(\mathbf{x},t) \cong C f(\theta) \frac{1}{x} \exp (ixp - i\omega t) \qquad (11b)$$

is a reasonable choice. We denote the distance from the origin by x and the magnitude of the incident momentum by p, i.e., $x = |\mathbf{x}|$, $p = |\mathbf{p}_i|$. The function $f(\theta)$ is some function of the angle θ between the direction of the incident momentum $\mathbf{p}_i$ and the direction of the position vector $\mathbf{x}$ (from the origin to the "point of observation").

Let us now study various features of the wave function ψ_s to see whether it can represent the scattered wave. The amplitude of the scattered wave is proportional to the amplitude C of the incident wave, and our guess thus reflects the reasonable assumption that the response is *linear*. The frequency ω of the scattered wave is the same as the frequency of the incident wave. This means that the energy of the A-particle is conserved, as it has to be since we wanted to consider *elastic* scattering in the fixed field of force of the B-particle.

The factor $\exp (ixp - i\omega t)$ obviously describes a spherical wave which propagates *outward*. At any one point the phase velocity is directed away from the origin along the radius vector. A wave representing a scattered particle must clearly have this feature. The factor $1/x$ in the expression (11b) describes the decrease in the amplitude of the scattered wave with distance. The *intensity* of the wave is proportional to the absolute square of the wave function. The intensity of the scattered wave measures the outgoing probability flux (or if we like, the particle flux in a sequence of repeated measurements) and this quantity *must* decrease as $1/x^2$ with distance. The amplitude must therefore decrease as $1/x$, as we have assumed.

12 As we see, simple physical considerations demand that the wave function describing the scattered wave be of the form stated in (11b). The function $f(\theta)$ is called the *scattering amplitude*. It obviously describes the angular distribution of the scattered particles. To relate the scattering amplitude to the differential cross section we argue as follows. Consider a small surface-region containing the point $\mathbf{x}$ on the surface of the sphere centered at the origin and passing through $\mathbf{x}$. Let its area be dF. The probability dP that the scattered particle passes through this area must then be proportional to the product of dF and the absolute square of the wave function $\psi_s(\mathbf{x},t)$. We can accordingly write

$$dP = k \, |\, \psi_s(\mathbf{x},t)\,|^2 dF = k \, |\, C\,|^2 \, |f(\theta)|^2 \left(\frac{dF}{x^2}\right) \qquad (12a)$$

where k is some fixed constant of proportionality. Since $dF/x^2 = d\Omega$ is the magnitude of the solid angle subtended by the small area with respect to the origin we can also write

$$dP = k \, |C|^2 \, |f(\theta)|^2 \, d\Omega \qquad (12b)$$

and we note that dP is the probability that the scattered particle emerges within the small cone of solid angle $d\Omega$.

Consider next the incident wave, given by (11a). We imagine a (circular) disc of *unit* area, perpendicular to the incident momentum $\mathbf{p}_i$ and centered about the origin. The probability that the incident particle passes through this disc must be equal to

$$P_i = k|\psi_i|^2 = k|C|^2 \qquad (12c)$$

where k is the *same* constant which occurs in (12a) and (12b).

Considering the expressions (12b) and (12c) we can say the following. In a sequence of repeated scattering experiments (in which the A-particles always have the same initial momentum $\mathbf{p}_i$) the ratio of the number of scattered particles emerging within the cone of solid angle $d\Omega$ to the number of particles incident on the unit disc is equal to

$$\frac{dP}{P_i} = |f(\theta)|^2 \, d\Omega \qquad (12d)$$

If we now recall our discussion of the differential cross section $\sigma_e(\theta)$ in Sec. 7 we realize that the ratio dP/P_i is just the product of the differential cross section with $d\Omega$. We thus obtain the important relation

$$\sigma_e(\theta) = |f(\theta)|^2 \qquad (12e)$$

which says that the differential cross section is simply the absolute square of the scattering amplitude.

13 If we want to find a theoretical expression for the scattering amplitude $f(\theta)$ we must, of course, solve our diffraction problem explicitly. This means that we have to find a solution of the Schrödinger equation, or perhaps of some other equation, which applies to the problem at hand. In our model we must find a solution to the Schrödinger equation with the potential which the A-particle sees because of the presence of the B-particle. The wave equations of quantum mechanics have an infinity of solutions, and it is also necessary that we find the *right one*, which can be interpreted as describing the scattering situation. The condition which we must impose is that for *large* distances from the origin the wave function must be of the form

$$\psi(\mathbf{x},t) \cong C \exp(i\mathbf{x} \cdot \mathbf{p}_i - i\omega t) + Cf(\theta)\frac{1}{x}\exp(ixp - i\omega t) \quad (13a)$$

This means that far away from the scattering center we have the plane "incident wave" together with the outgoing scattered wave. We shall not attempt to solve such problems here. It is possible to prove, under very general conditions, that for *every* choice of the incident momentum $\mathbf{p}_i$ there exists a *unique* solution of the wave equation which has the asymptotic form (13a). The scattering amplitude is thus uniquely determined for a given incident momentum and for a given interaction (potential). It will in general depend on the magnitude p of the incident momentum; and if we want to emphasize this we can write the scattering amplitude as $f(p;\theta)$. Once we have determined the scattering amplitude we find the differential cross section by (12e).

14   Let us consider the important and simple special case when the scattering amplitude is *independent* of the scattering angle θ, i.e., $f(\theta) = f = $ a constant. The differential cross section is then given by $\sigma_e(\theta) = |f|^2 = $ a constant, and the angular distribution is spherically symmetric. This situation tends to occur for low-energy scattering. It is easy to understand qualitatively the reason for this. The angular distribution tends to be complex, i.e., rapidly varying as a function of θ, when the wavelength of the incident wave is small compared to the size of the "object" by which the wave is diffracted. We can think about the diffraction as taking place all over the object, so that each "part" of the object sends out a diffracted wave. In a given direction these waves can interfere constructively or destructively, depending on their relative phases. If the wavelength is small compared to the object a small change in the direction of observation can have an appreciable effect on the relative phases, and the differential cross section can thus vary rapidly with the angle θ. However, if the wavelength is large compared to the object these "geometrical" interference effects do not occur, and the scattering amplitude is a slowly varying function of direction. In the extreme low-energy limit, in which the wavelength is *very* long compared to the size of the scattering object, the scattering amplitude is independent of angle, and the scattering is spherically symmetric.

15   In the case $f(\theta) = f = $ a constant, the scattered wave

$$\psi_s(\mathbf{x},t) = \frac{Cf}{x}\exp(ixp - i\omega t) \quad (15a)$$

depends on the incident wave *only* through the parameter C which expresses the amplitude of the incident wave at the origin. In particular the scattered wave is independent of the direction of the incident momentum $\mathbf{p}_i$. This is what we would expect if the scattering object is very small compared to the wavelength.

Suppose that we replace the plane wave given in (11a) by its (uniform) average over all directions of $\mathbf{p}_i$. That is, we consider a new scattering problem in which the incident wave is of the form

$$\psi_{i0}(\mathbf{x},t) = \frac{1}{4\pi} \int_{\bigcirc} d\Omega_p \, C \exp\left(i\mathbf{x} \cdot \mathbf{p}_i - i\omega t\right) \tag{15b}$$

This integral over all directions is easily evaluated if we select the angle θ between $\mathbf{x}$ and $\mathbf{p}_i$ as one of the polar angles of $\mathbf{p}_i$. We obtain

$$\psi_{i0}(\mathbf{x},t) = \frac{1}{4\pi} \int_0^{2\pi} d\varphi \int_0^{\pi} d\theta \, \sin\theta \, C \exp\left(ixp \cos\theta - i\omega t\right)$$

$$= \frac{C}{2ixp} \left(\exp(ixp) - \exp(-ixp)\right) \exp(-i\omega t) \tag{15c}$$

If the scattered wave does not depend on the direction of the incident momentum, an incident wave of the form ψ_{i0} will produce the *same* scattered wave as the plane wave in (11a). We can think about the wave ψ_{i0} as the *spherically symmetric part* of the incident plane wave. Only this part of the incident wave leads to the spherically symmetric wave ψ_s given in (15a).

16   The spherically symmetric part ψ_{i0} of the incident plane wave has an interesting form. If we look at the expression (15c) we notice that the wave is the sum of an *outgoing* spherical wave and an *ingoing* spherical wave. A plane wave "contains" two waves of this kind because it describes motion toward the origin as well as motion away from the origin. The amplitudes of the two waves are identical in magnitude. This *must* be the case, because otherwise the outward flux would be different from the inward flux. Since we have assumed elastic scattering (in which the A-particles are preserved) the inward and outward fluxes (of A-particles) must be the same.

Let us now consider the spherical average of the expression (13a) in the case $f(\theta) = f = $ a constant. This average is given by

$$\psi_0(\mathbf{x},t) = \psi_{i0}(\mathbf{x},t) + \psi_s(\mathbf{x},t)$$

$$= \frac{C}{2ixp} \left([1 + 2ipf] \exp(ixp) - \exp(-ixp)\right) \exp(-i\omega t) \tag{16a}$$

We can interpret this expression as the asymptotic form of the wave function which describes the scattering situation in which the spherical wave ψ_{i0} plays the role of the incident wave. Inspection of (16a) shows that the wave $\psi_0(\mathbf{x},t)$ also has both an ingoing and an outgoing part. If the scattering process is elastic the absolute values of the amplitudes of these two waves *must* be equal. This leads to the important condition

$$|1 + 2ipf| = 1 \qquad (16b)$$

on the scattering amplitude f.

It is convenient to write the general solution of the equation (16b) in the form

$$f = \frac{1}{2ip}\left(e^{2i\delta} - 1\right) \qquad (16c)$$

where δ is any *real* number. The quantity δ is called the (*s*-wave) *phase shift*. It is in general a function of the magnitude p of the momentum.

17 Let us investigate how large the elastic cross section can be in the case of spherically symmetric scattering. The differential cross section equals $|f|^2$, and the total elastic cross section σ_e is obtained by integrating the differential cross section over all directions. We thus have [taking (16c) into account]

$$\sigma_e = \frac{\pi}{p^2}\,|e^{2i\delta} - 1|^2 \qquad (17a)$$

For a fixed p this quantity assumes its maximum value if δ is of the form $\delta = (n + \tfrac{1}{2})\pi$, where n is any integer. The maximum value equals

$$(\sigma_e)_{\max} = \frac{4\pi}{p^2} \qquad (17b)$$

The above formula refers to a system of units in which $\hbar = 1$. It is very easy to "restore" the constant $\hbar$. It must occur squared in the numerator because dimensionally the cross section is an area. With the cgs or MKS system of units we thus have

$$(\sigma_e)_{\max} = 4\pi\left(\frac{\hbar}{p}\right)^2 \qquad (17c)$$

The maximum cross section for spherically symmetric scattering is thus $(1/\pi)$ times the square of the de Broglie wavelength h/p of the incident particle. For small momenta this cross section can be very large. On the basis of the wave picture of scattering we can thus easily understand the large cross sections mentioned in Sec. 6, which may have caused the reader some worry.

18   As we have said the phase shift δ is a function of the magnitude p of the incident momentum.   Since the incident energy ω is a monotonic function of p we can just as well regard δ as a function of the energy.   Let us therefore write $\delta(\omega)$ for the phase shift to emphasize that it depends on the energy.

Whenever the phase shift, as a function of the energy, passes through one of the values $(n + \frac{1}{2})\pi$ the cross section assumes the maximum value given in (17b).   We say that the scattering is *resonant* at such a point.   Let us study the behavior of the scattering amplitude and the cross section in the immediate neighborhood of a resonance.   We denote the energy at the resonance by ω_0, and we thus have $\delta(\omega_0) = (n_0 + \frac{1}{2})\pi$ for some integer n_0.

We shall rewrite (16c) using the defining relations

$$\cot(\delta) = \frac{\cos(\delta)}{\sin(\delta)} = \frac{i(e^{i\delta} + e^{-i\delta})}{(e^{i\delta} - e^{-i\delta})} \tag{18a}$$

for the cotangent.   As the reader should immediately convince himself we can write

$$f(\omega) = \frac{1}{2ip}\left(e^{2i\delta(\omega)} - 1\right) = \frac{(1/p)}{\cot[\delta(\omega)] - i} \tag{18b}$$

At the point $\omega = \omega_0$ we have $\cot[\delta(\omega_0)] = 0$.   We can then try to expand $\cot[\delta(\omega)]$ in powers of $(\omega - \omega_0)$ in the neighborhood of the point $\omega = \omega_0$.   Retaining only the linear term we have

$$\cot[\delta(\omega)] \cong -\frac{2}{\Gamma}(\omega - \omega_0) \tag{18c}$$

where, following established custom, we have written the derivative of $\cot[\delta(\omega)]$ at ω_0 as $-2/\Gamma$.

We shall assume that the phase shift increases with energy in the neighborhood of the resonance.   This means that $\cot[\delta(\omega)]$ decreases with energy, and the parameter Γ which we have introduced in (18c) is positive.   Inserting the approximate expression (18c) (which is only valid near the resonance) into (18b) we obtain

$$f(\omega) \cong -\frac{1}{p}\left(\frac{\Gamma/2}{(\omega - \omega_0) + i\Gamma/2}\right) \tag{18d}$$

and

$$\sigma_e(\omega) \cong \frac{4\pi}{p^2}\left(\frac{(\Gamma/2)^2}{(\omega - \omega_0)^2 + (\Gamma/2)^2}\right) \tag{18e}$$

The reader will recognize the formula (18e) as the Breit-Wigner resonance formula (21d) in Chap. 3, which we derived there through a different line of reasoning. The quantity Γ is the width of the resonance. In Chap. 3 we associated excited levels with resonances, and we shall adhere to this idea here. The quantity $1/\Gamma = \tau$ is then the mean life of the excited level which manifests itself as a resonance.

What Is Meant by a Particle?

19 Before we consider the problem of interactions further it is in order for us to examine our notions of what a particle is. Let us imagine that we try to formulate reasonable qualifications for membership in the Set of Particles.

A particle is, in some sense, a "single" coherent object which has a definite identity, and which can be localized in a limited region of space at a given time. It is characterized by definite physical attributes, and we may tentatively require that it should have a definite mass, a definite charge, a definite intrinsic angular momentum, etc., and that it should be absolutely stable when alone in space.

20 Under these rules we would admit the proton, the electron, the positron, the neutrinos, the photon and the stable nuclei. This admission procedure raises, however, some immediate problems. The neutral atoms and all the ions in their ground states also satisfy the criteria and should thus be admitted. The same applies to all molecules and molecular ions, in their ground states, and our set becomes distressingly large if we include these objects as well, which in all fairness we have to do. On the other hand we would have to reject an object such as the alpha-active radium nucleus $_{88}\text{Ra}^{226}$ on the grounds that it is not stable. This is not satisfactory because we must admit that it is *almost* stable (half-life 1622 years), and from the standpoint of the chemist the radium atom is just as respectable as the barium atom. Worse still, we would have to reject the neutron. The neutron is the twin brother of the proton, and we regard it as one of the building blocks of nuclei. Inside a stable nucleus the neutron is as stable as the proton, although it does decay when alone in space. Its mean life is, however, 17 minutes, which is a very long time on a nuclear or atomic time scale (i.e. long compared to 10^{-24} or 10^{-8} seconds). In an experiment in which the phenomenon under study takes place in a time very small compared to 17 minutes the neutron behaves just like a stable particle. For instance, we can perform diffraction experiments with neutrons against a crystal.

Finally it can be stated against our admission policy that it could well happen that some of the "stable" nuclei which we have admitted are in fact unstable, although their lifetimes are so long that we have not yet become aware of their instability. This might lead to a later expulsion of objects which we have already happily admitted.

21   In view of the above we realize, as sensible men, that we must modify our admission criteria. We will now admit objects which are "just a little bit unstable," and under the new rules we welcome the neutron and the radium nucleus. This means that we also give up our requirement that a particle shall have a definite mass, because as we have learned in Chap. 3, if a system has a finite lifetime τ, then the energy (in this case the rest energy of the particle) is defined only within an uncertainty of the order of $\hbar/\tau$. In other words, if the mean life of a particle is τ, then the uncertainty in its rest mass must be of the order

$$\Delta m \sim \frac{\hbar}{\tau c^2} \tag{21a}$$

For the neutron this uncertainty is extremely small; it amounts to less than 10^{-27} amu.

22   Once we have given way on the requirement of absolute stability we find it very difficult to decide just how unstable we will allow a particle to be. The muon has a half-life of about 10^{-6} seconds which is short on a macroscopic scale but very long on the nuclear time scale. The same is true of the charged pions, which have half-lives of about 10^{-8} seconds. These particles must therefore be admitted. The neutral pion has a mean-life of the order of 10^{-16} seconds. This is still long compared to 10^{-24} seconds, and furthermore the neutral pion is obviously related to the charged pions. We accordingly admit the neutral pion as well, together with such particles as the K-mesons and the hyperons. The mean-lives of the K-mesons and hyperons are generally of the order of 10^{-10} seconds. Note that the corresponding uncertainties in their rest masses, as given by Eq. (21a), are still very small compared to their rest masses.

23   We must now decide whether we should admit all the excited states of atoms, molecules and nuclei. In favor of their admission it can be argued that many of the excited states have lifetimes which

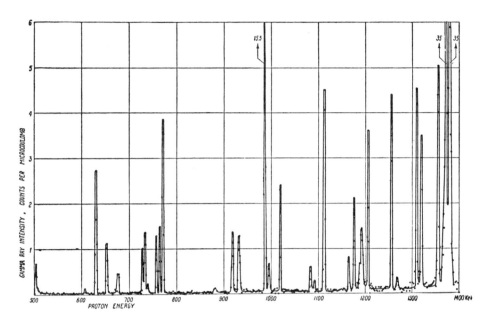

Fig. 24A Yield curve for the reaction $Al^{27} + p \rightarrow Si^{28} + \gamma$, from a paper by K. J. Broström, T. Huus, and R. Tangen, "Gamma-Ray Yield Curve of Aluminum Bombarded with Protons," *Physical Review* **71**, 661 (1947). The ordinate is a measure of the cross section for the reaction. The abscissa is the kinetic energy of the incident protons, in keV, in the lab frame of reference. The sharp peaks are resonances. They reveal the existence of excited states in the silicon nucleus produced in the reaction. (*Courtesy of The Physical Review.*)

are very long compared to the lifetime of the neutral pion, or in fact long compared to the lifetime of the neutron. Some of the excited states decay through the emission of material particles, and some decay through the emission of photons. Is it fair to exclude the excited states if we admit the "ground state" of $_{88}Ra^{226}$, which also decays through the emission of a particle? Furthermore: perhaps some of the hyperons should be regarded as "excited states" of the nucleon? (The hyperons all decay into other particles, one, and only one, of which is a nucleon.) We find it very difficult to resist these pressures, and we accordingly admit the "excited states."

24 At this point we realize that our membership has well exceeded the million mark, which we find disagreeable. If our original intention was to form a fairly small and manageable society of respectable particles, this aim has now been thwarted. Furthermore, our latest concession, to admit the "excited states," raises some serious doubts about our entire admission policy. To see this, let us consider the experimental determination of an excited state, i.e., an energy level above the ground state of a system. In Chap. 3 we explained how the excited states manifest themselves as resonances in scattering processes. An example of this is the resonant scattering of light by an atom. If we measure the effectiveness of an atom as a scatterer of light, as a function of the frequency of the light, we find sharp maxima at the frequencies corresponding to the energy differences between the excited states and the ground state. This phenomenon is, however, not restricted to the scattering of light: we also encounter it in the scattering of material particles. Figure 24A shows an example. The ordinate is a measure of the cross section, and the curve thus shows the experimentally measured cross section, as a function of the energy, for the absorption of protons be aluminum. The sharp peaks in the cross section reveal the locations of the excited states in the silicon nucleus produced in the reaction.

The width T of a resonance peak measures the uncertainty in the energy of the corresponding excited state. As long as the resonances are very sharp the interpretation of the resonances as manifestations of excited states is clear-cut. We have agreed that such excited states are "particles." Let us now look at Fig. 24B which shows the cross section for the scattering of pions on protons, as a function of the energy. The cross section for positive pions shows one pronounced peak, as well as a slight "bump" at a higher energy. The cross section for negative pions shows three moderately well defined peaks. Do all these peaks correspond to particles? The inclination of many physicists today is to say that they do. The masses of these "particles" (?) are simply the abscissas of the maxima.

25 The dilemma which we are facing is where to draw the line. We certainly do not want to say that every small "bump" in a curve showing a cross section as a function of energy corresponds to a particle, but on the other hand, any rule according to which a resonance must be "sufficiently" narrow if we are to accept it as defining a particle is somewhat arbitrary. In other words: if an object is to be admitted into the set of particles then its lifetime cannot be *too* small, but where do we draw the line?

Let us re-examine our aims. Perhaps nothing is really gained by trying to define *precisely* what we mean by a particle in general. Our attempts have led us to a class of objects with millions of members, containing, among other particles, such qualitatively distinct objects as pions and protein molecules. According to common English usage these objects can reasonably be called particles, but we can hardly expect to learn anything profound about fundamental interactions if we try to treat pions and protein molecules as equals in our basic theory. Some of the particles are obviously composite systems, and we should describe them as such in our

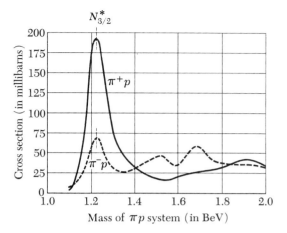

Fig. 24B The two curves show the observed cross section in the scattering of positive and negative pions against protons. The ordinate is the total cross section in millibarns, and the abscissa is the *total* energy of the pion and the proton in the center-of-mass system of reference. It is convenient to express the energy in this manner, because the location of a prominent peak directly corresponds to the mass of the "particle" or resonant state.

Note the large peaks at an energy of about 1.238 BeV. This energy corresponds to a *pion kinetic energy* of about 195 MeV in the laboratory system of reference in which the incident pion collides with a proton at rest.

We have denoted these resonances by the symbol $N^*_{3/2}$. The notation $\Delta(1238)$ also occurs frequently in the literature.

TABLE 26A *The Leptons*

	Particle	Charge	Mass MeV
e⁻	electron	$-e$	0.511
e⁺	positron	$+e$	0.511
μ^-		$-e$	105.7
μ^+	muons	$+e$	105.7
ν_e	e-neutrino	0	0
$\bar{\nu}_e$	e-antineutrino	0	0
ν_μ	μ-neutrino	0	0
$\bar{\nu}_\mu$	μ-antineutrino	0	0

The muons are unstable and decay according to: $\mu^\pm \to e^\pm + \bar{\nu} + \nu$. (Of the neutrinos one is presumably a μ-neutrino, and the other an e-neutrino.) The mean life of the muons is 2.20×10^{-6} sec. The other particles are stable. The leptons all have spin angular momentum $\frac{1}{2}$.

TABLE 26C *The Principal Baryon Octet*

	Particle	Mass MeV	Mean life sec	Principal decay modes
p	proton	938.256	stable	—
n	neutron	939.550	1.01×10^3	$pe^-\bar{\nu}$
Λ	lambda-hyperon	1115.58	2.51×10^{-10}	$p\pi^-$ $n\pi^0$
Σ^+		1189.47	0.81×10^{-10}	$p\pi^0$ $n\pi^+$
Σ^0	sigma-hyperons	1192.56	$< 10^{-14}$	$\Lambda\gamma$
Σ^-		1197.44	1.65×10^{-10}	$n\pi^-$
Ξ^0	cascade	1314.7	3.0×10^{-10}	$\Lambda\pi^0$
Ξ^-	particles	1321.2	1.7×10^{-10}	$\Lambda\pi^-$

These particles all have spin angular momentum $\frac{1}{2}$ and baryon number $+1$. There exists an anti-baryon octet consisting of the anti-particles of the above particles. The anti-particles have the same masses, spins, and mean lives, but opposite charges and baryon number.

TABLE 26B *The Principal Meson Octet*

	Particle	Mass MeV	Mean life sec	Principal decay modes
π^+	charged pions	139.60	2.61×10^{-8}	$\mu^+ \nu_\mu$
π^-				$\mu^- \bar{\nu}_\mu$
π^0	neutral pion	134.98	0.89×10^{-16}	$\gamma\gamma$ $\gamma e^+ e^-$
K^+	charged	493.8	1.23×10^{-8}	$\mu^\pm \nu$ $\pi^\pm \pi^0$ $\pi^\pm \pi^+ \pi^-$
K^-	K-mesons			
$K^0 \}$	neutral $\{K_1$	497.9	0.87×10^{-10}	$\pi^+ \pi^-$ $\pi^0 \pi^0$
$\bar{K}^0 \}$	K-Mesons $\{K_2$		5.68×10^{-8}	$\pi^0 \pi^0 \pi^0$ $\pi^+ \pi^- \pi^0$ $\pi \ \mu \ \nu$ $\pi \ e \ \nu$
η	eta-meson	548.6	? $< 7 \times 10^{-20}$ $> 7 \times 10^{-21}$	$\gamma \ \gamma$ $\pi^0 \ \pi^0 \ \pi^0$ $\pi^0 \ \gamma \ \gamma$ $\pi^+ \ \pi^- \ \pi^0$ $\pi^+ \ \pi^- \ \gamma$

The above mesons have spin angular momentum 0 and baryon number 0. The two neutral K-mesons K^0 and $\bar{K}^0$ behave in their decays as if they were "mixtures" of two particles K_1 and K_2 of different lifetimes and very slightly different masses.

theory: we should "explain" them in terms of the interactions of their more elementary constituents.

From a practical point of view we can think about a hierarchy of increasingly more elementary particles. Depending on the kind of physical phenomenon we wish to consider our notion of the "elementary constituents" of a composite system changes. It is common English usage to say that a molecule is a bound state of atoms, that an atom is a bound state of a nucleus and a number of electrons, and that a nucleus is a bound state of protons and neutrons. The proton, neutron and electron are, however, not in any obvious way bound states of anything else: they might well be among the ultimate elementary particles. As such they are objects of particular interest in basic theory.

26 Let us consider the subset of the (truly?) Elementary Particles among all our particles. Our first principle will be that we do not admit any object which is "obviously composite" into this smaller and more exclusive set. In Chap. 1 we discussed some experimental criteria for the composite or elementary nature of a particle. All atoms, all molecules and all nuclei heavier than the proton are clearly composite and hence excluded from our new set. This leaves us with about a hundred particles which "are not obviously composite." We admit the proton, the neutron, the anti-proton, the anti-neutron, the long-lived hyperons and their anti-particles, the pions, the K-mesons, the muons, the neutrinos, the electron and the positron, and the photon. Except for the proton, the anti-proton, the electron, the positron, the photon and the neutrinos, all these particles are unstable, but in view of our earlier discussion we shall not use absolute stability as a condition for admissibility.

The elementary particles are divided into four classes. The photon is the sole member of the first of these. The other classes are the *leptons,* the *mesons* and the *baryons* (including the anti-baryons). Tables 26A-C list some of the properties of the leptons and the most respectable mesons and baryons.† (See also Table B in the Appendix.)

27 In Figs. 27A–B the mesons and baryons listed in Tables 26B–C are shown in diagrams which strongly resemble the term schemes discussed in Chap. 3. Each particle is represented by a short horizontal bar on a graph in which the ordinate is the rest mass (in MeV) and the abscissa is the electric charge. (The center of the bar indicates the charge of the particle.)

According to current ideas our particle diagrams should be regarded as entirely analogous to term schemes for atoms. Each diagram corresponds to a "multiplet" of closely related particles which might in some sense be regarded as different states of the "general" particle of the multiplet.

Fig. 27C shows the anti-baryon multiplet of eight particles which are the anti-particles of the eight baryons in Fig. 27B. The anti-particles of the mesons shown in Fig. 27A are contained in the same diagrams: we say that the meson octet is self-conjugate. The negative pion is thus the anti-particle of the positive pion and the

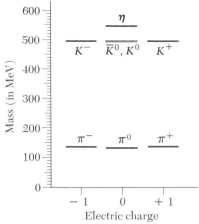

Fig. 27A Mass spectrum of the meson octet to which the pions and K-mesons belong. These particles all have baryon number zero and spin angular momentum zero. The two neutral K-mesons K^0 and $\overline{K}^0$, indicated by the double line in the term scheme, have the same mass to the accuracy of this drawing. Particle-antiparticle pairs are symmetrically situated with respect to the vertical line corresponding to zero charge. The particles π^0 and η are their own anti-particles. $\overline{K}^0$ is the anti-particle of K^0.

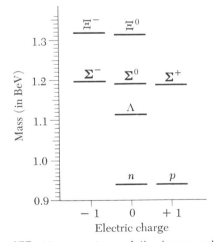

Fig. 27B Mass spectrum of the baryon octet to which the proton (p) and neutron (n) belong. These particles all have baryon number $+1$ and spin angular momentum $\frac{1}{2}$. The diagram can be interpreted as a term scheme, showing the eight different states of the "general particle" associated with this multiplet.

---

† The nomenclature for elementary particles appears to have been invented to give a certain classical Greek flavor to the subject. Although the author's knowledge of the classical languages is extremely limited, he nevertheless feels that he has very good reason to suspect that the linguistic principles which underlie the construction of the "Greek"-sounding terms are not entirely correct.

Fig. 27C Mass spectrum of the anti-baryon octet which consists of the anti-particles of the particles shown in Fig. 27B. The particles in the multiplet at the right all have baryon number -1 and spin angular momentum $\frac{1}{2}$.

If we reflect the baryon diagram with respect to the vertical line corresponding to zero charge we obtain the anti-baryon diagram, and vice versa.

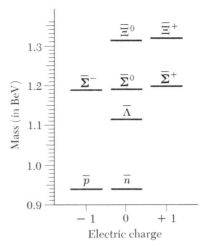

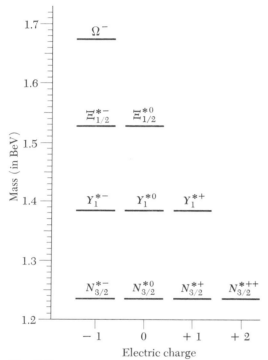

Fig. 27D Term scheme showing a multiplet of ten baryons which includes the most prominent resonances in pion-nucleon scattering. (See Fig. 24B for the pion-nucleon cross sections.) The resonances are denoted $N^*_{3/2}$ in the diagram. (The symbol Δ is also employed in the literature for these particles.) All the particles shown have baryon number $+1$ and spin angular momentum $\frac{3}{2}$.

The regularities in the above pattern of particles are very striking. At this time the details are not well understood. That the pion-nucleon resonances should be included in a multiplet of ten particles can be understood on the basis of the symmetry principle known as the Eightfold Way. The omega-minus particle was in fact predicted theoretically before it was found experimentally. The mean life of this particle is 1.5×10^{-10} sec. All the other particles in the diagram have *extremely* short lifetimes.

negative *K*-meson is the anti-particle of the positive *K*-meson. The particles denoted K_0 and $\overline{K}_0$ form a particle-antiparticle pair. The neutral pion and the eta-meson are their own anti-particles.

Fig. 27D shows a multiplet of ten baryons which contains the resonance marked $N^*_{3/2}$ in Fig. 24B. The status of these particles (resonances) is perhaps still open to some doubt, but most physicists would today be willing to accept them among the elementary particles.

28 We classify the interactions occurring in nature into *strong interactions* (the "nuclear forces" belong to this class), *electromagnetic interactions*, *weak interactions* and *gravitational interactions*. The mesons, baryons and anti-baryons all interact strongly with each other. The photon and the leptons are not affected by the strong interactions: their behavior is governed by the electromagnetic and weak interactions. The strongly interacting particles (which are nowadays often referred to as *hadrons*) also participate in the electromagnetic and weak interactions. Many of the unstable ones among them decay "via" the weak interactions, and consequently have lifetimes which are very long on the nuclear time scale.

The interactions of the elementary particles are governed by a number of very striking conservation laws and symmetry principles. One of these conservation laws says that the total electric charge is conserved in *all* interactions.† There is an analogous conserva-

---

† Charge conservation is a *basic* principle of electromagnetic theory. See Berkeley Physics Course, Vol. II, *Electricity and Magnetism*, p. 4, for a discussion.

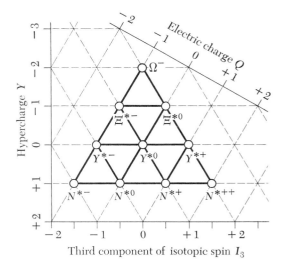

Fig. 29D Eightfold Way symmetry diagram for the baryon decuplet to which the main pion-nucleon resonances belong. The diagram shows the assignments of hypercharge to the particles in the decuplet. The masses of these particles are shown in Fig. 27D.

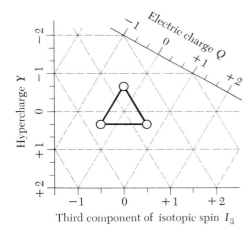

Fig. 31A If quarks indeed exist they would have an Eightfold Way symmetry diagram as shown above. The particles in this triplet would all have baryon number $+\frac{1}{3}$, and presumably spin angular momentum $\frac{1}{2}$. Note that two of the particles carry electric charge $-\frac{1}{3}$, whereas the third carries charge $+\frac{2}{3}$. The corresponding triplet of anti-quarks has a symmetry diagram which is obtained from the above diagram by a reflection with respect to the line corresponding to zero electric charge. The anti-quarks would have baryon number $-\frac{1}{3}$. The mesons in the octet shown in Fig. 29A could be regarded as bound states of a quark and an anti-quark. The baryons in the octet shown in Fig. 29B could be regarded as bound states of three quarks.

If the quark theory has any relation to reality, there must exist at least one stable fractionally charged particle. No such particle has ever been detected experimentally, and it could well be that fractionally charged particles do not exist. At this time the idea of quarks is pure speculation.

Mann that the particle denoted Ω^- in Figs. 27D and 29D exists.†

31 The elementary particles which we have mentioned (plus some additional ones) all appear to be "equally elementary": none of them appears to be "composite." Some people have nevertheless speculated that there might exist still more elementary entities. It has thus been suggested by Gell-Mann that the mesons and baryons might be composite systems made of hitherto unknown particles which he proposed to call *quarks*. This is not a wild and irresponsible suggestion: Gell-Mann had noticed that certain properties of the mesons and baryons, and in particular the symmetry principles which govern their interactions, might be explained in an esthetically pleasing manner if quarks (and anti-quarks) indeed existed. According to Gell-Mann's ideas these particles would carry the charges $\pm e/3$ and $\pm 2e/3$, where e is the charge of the proton, and in this respect they would differ strikingly from all known particles. The symmetry diagram for quarks is shown in Fig. 31A.

The search for quarks goes on, but to date none has been found. Compared with nucleons quarks must be very heavy: otherwise they would have been "seen" in accelerator experiments. We can conclude from this that if a nucleon is indeed a bound state of

---

† V. E. Barnes *et al.*, "Observation of a Hyperon with Strangeness Minus Three," *Physical Review Letters* **12**, 204 (1964). (It may be mentioned that this paper lists 33 authors!)

(three) quarks, then the binding energy must be very large compared to the mass of the nucleon. The nucleon would thus be a very tightly bound system, and in this respect it would differ radically from the bound systems which are familiar to us, namely atoms, molecules, and nuclei. (The binding energy of an atom, a molecule, or a nucleus is *small* compared to the mass of the system.) It is therefore safe to say that whereas nucleons might one day be found to be composite, they are certainly not composite in the same sense as, for instance, the deuteron is composite.

Basic Ideas of Quantum Field Theory

32   Let us next consider some theoretical attempts to understand the interactions of particles. We shall pursue the idea, which we arrived at in Sec. 9, that scattering phenomenon should be regarded as a manifestation of the interaction of waves with waves. The classical idea of two particles exerting a force on each other corresponds in quantum mechanics to the idea that the de Broglie waves of the particles interact. What does this mean? It means that the presence of a de Broglie wave of *one* particle influences the propagation of a de Broglie wave of *another* particle. This can happen only if the medium in which the de Broglie waves propagate is non-linear, i.e. the medium "responds" non-linearly. In a linear medium, for which the propagation of waves is described by a *linear* differential equation, any linear superposition of two waves is another possible wave, and the presence of one wave does not affect the behavior of another wave.

33   Let us discuss the nature of the *vacuum*, or empty space. When electromagnetic theory was developed in the nineteenth century the vacuum was known under another name, namely as the "ether." When we consider waves it is natural to ask what it is that "oscillates," and a physicist of the last century would have said that it is the ether which oscillates. The behavior of electromagnetic waves in the ether is described by Maxwell's equations. It appeared natural to the physicists of the time to try to understand electromagnetism in terms of *mechanical* models, and to regard electromagnetic waves as somewhat analogous to elastic waves in a solid. Much effort went into the construction of such interpretations. The mechanical properties of the ether certainly turned out to be quite unlike the properties of any bona fide solid or fluid, but that circumstance in itself should not be held against the theory.

One may, however, raise grave objections against the mechanical theory of the ether on epistemological grounds: consideration of the mechanical properties of the ether is an unnecessary activity which adds nothing to our understanding of electromagnetism. Maxwell's equations *by themselves*, without any mechanical interpretation, tell us everything about classical electromagnetic theory that has any experimental significance. For instance, if we wish to describe the propagation of radio waves from one antenna to another, it suffices to solve Maxwell's equations with the appropriate boundary conditions, and it is of no consequence whether we have a mechanical model for the wave propagation or not. Physicists gradually came to realize that in the study of electromagnetism all that matters are Maxwell's equations. Consequently the attempts to construct mechanical models were abandoned, and the question of "what it really is that oscillates" was recognized to be an operationally meaningless question.

34   The development of the theory of special relativity did much to hasten the demise of the mechanical ether theory. Let us recall the reasons for this. If the ether has any properties at all akin to those of an ordinary solid, or fluid, then we certainly expect that there exists an inertial frame with respect to which the ether is at rest, at least locally. On the other hand, every relevant experiment seems to indicate that there is *no* way of determining an absolute state of motion relative to the ether: all inertial frames are completely equivalent to each other. This latter statement is, of course, one of the cornerstones of the theory of special relativity. If it is really true, which we believe firmly, it means that the moving ether has the same physical characteristics as the ether at rest, and this is certainly a property which is foreign to any ordinary solid or fluid. In view of this fundamental "unmechanical" property of the ether, it seems senseless to try to assign other mechanical properties to it.

35   Today the *mechanical* ether has been banished from the world of physics, and the word "ether" itself, because of its "bad" connotations, no longer occurs in textbooks on physics. We talk ostentatiously about the "vacuum" instead, thereby indicating our lack of interest in the *medium* in which the waves propagate. We no longer ask what it is that "really oscillates" when we study electromagnetic waves or de Broglie waves. All we wish to do is to formulate *wave equations* for these waves, through which we can predict experimentally observable phenomena. As we have already said, these wave equations must be non-linear if they are to describe

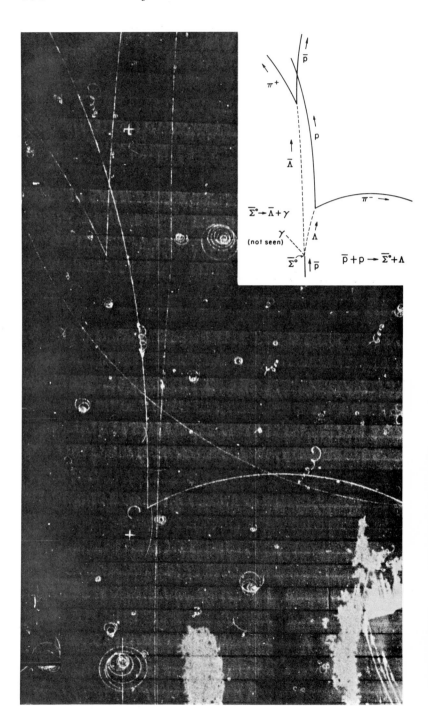

Bubble chamber picture showing the production and decay of an antisigma-zero particle.   The insert shows the reactions, and identifies the various tracks.   The neutral particles (represented by dotted lines in the schematic figure) of course leave no visible tracks. The tracks of the charged particles are curved because the chamber is in a magnetic field perpendicular to the plane of the figure.

The production reaction, in which the antisigma-zero particle and a lambda particle result in a collision of an anti-proton with a proton, is a *strong* interaction. The antisigma-zero particle decays by an *electromagnetic* interaction into an anti-lambda particle and a gamma ray.   The other decay processes seen in the picture are all manifestations of the *weak* interactions. *(Photograph by courtesy of Lawrence Radiation Laboratory, Berkeley.)*

interacting particles. The formulation of such wave equations and the extraction of experimental predictions from them are the objectives of *quantum field theory*, which has a claim to being the fundamental theory of the elementary particles. In this theory the waves are described by *quantum fields*, and in a sense the theory is a quantum-mechanical generalization of the classical theory of waves.

The idea of describing interacting particles in terms of quantum fields is attractive in many respects, and we may properly try to understand the broad features of this theory. A complete discussion requires somewhat complex mathematical tools which we do not have at our disposal at this time, and we must therefore omit all details.

36 Let us consider, quite generally, the problem of describing an interaction between two (or several) particles. To orient ourselves we first consider this problem within the framework of classical physics. In a *non-relativistic* theory we might introduce position-dependent forces acting between the particles. The force acting on one particle depends on the position of that particle, as well as on the positions of the other particles at the same time. The action of the force is in this case instantaneous: if the position of one particle is suddenly changed the corresponding change in the force is perceived instantaneously by the other particle.

We believe that every fundamental theory of nature must be in accordance with the principles of special relativity. We note that an interaction of the kind described above stands in glaring contradiction with these principles. No signal can be propagated by a velocity larger than *c*, and it follows that the action of the force cannot be instantaneous. If the position, or state of motion, of one particle suddenly changes it must take some time before this change is perceived by the other particle, and the minimum time it must take is the time it takes for a light signal to pass between the two particles.

It is not at all a trivial matter to formulate a relativistically invariant theory of interacting classical particles. A profound change in the non-relativistic idea of an instantaneous action at a distance is required.

37 One possible way out of this dilemma is through the introduction of a (classical) *field*. Each particle is the source of a field which can propagate in space, but never with a velocity greater than *c*, and this field may then influence the motion of other particles. In a relativistic *classical* theory of this kind we are thus led

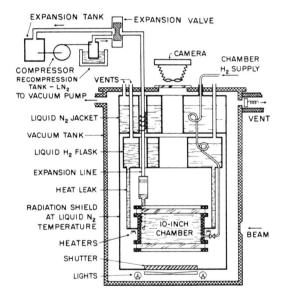

Schematic picture of a liquid hydrogen bubble chamber. The chamber is activated by a sudden decrease in the pressure of the hydrogen liquid. The temperature of the liquid is above the boiling point at the reduced pressure, but boiling does not start immediately: the liquid stays for a short interval of time in a superheated stage. The passage of a charged particle through the liquid leads to local vaporization along the path. A visible track consisting of very tiny gas bubbles is formed, and it is photographed with the camera above the chamber. The pressure is then raised again and all tracks disappear. The chamber is ready for the next exposure. *(Illustration by courtesy of Lawrence Radiation Laboratory, Berkeley.)*

to consider *both* particles and fields. The interaction of charged particles through the mediation of the electromagnetic field is a good example of such a theory: the charges are the sources of the electromagnetic field, and the electromagnetic field in turn influences the motion of the charged particles.

38 Let us look upon the problem of particle interactions from another angle. In the classical non-relativistic theory in which we describe the interaction through a force acting instantaneously the future behavior of an isolated system of several particles is uniquely determined if we are given the positions and velocities of all the particles at some instant of time. In other words, if there are N particles present the state of motion of the system is determined by $6N$ parameters: the system has a finite number of degrees of freedom. In a relativistic theory, on the other hand, in which the interaction is described by a field, it is not enough to merely specify the positions and velocities of all the particles at a given instant of time. We must *also* specify the state of the field. Classical electromagnetic theory illustrates this very clearly: the electromagnetic field is by no means uniquely determined by only the positions and velocities of all the charged particles at an instant of time. Among the initial conditions we must also include the specification of the electric and magnetic fields everywhere in space. The state of the electromagnetic field requires, however, an infinite number of parameters for its description, and our system is no longer a system of a finite number of degrees of freedom. This is clearly a profound distinction between the relativistic and non-relativistic theories.

39 There is another feature of the (classical) relativistic theory which we should note: part of the total energy of the system will, at any instant of time, reside in the field. This is necessarily the case in a theory in which the interaction is mediated through a field. Consider, for instance, two particles A and B interacting with each other. Suppose particle A undergoes a sudden collision with a third particle C which does not interact directly with particle B. The state of motion of A will then change, and in due time this change will manifest itself in a change in the field due to particle A

Bubble chamber picture showing the production and decays of a (neutral) lambda particle and a neutral K-meson. The various tracks are identified in the insert. Only the charged particles leave visible tracks, and these are curved because the chamber is in a magnetic field. The *strong* production reaction is: $p + \pi^- = \Lambda^0 + K^0$. The decay interactions are all *weak*. The negative muon emitted in the decay of the K^0 decays into an electron, a neutrino, and an antineutrino. The last two particles are neutral and cannot be seen. *(Photograph by courtesy of Lawrence Radiation Laboratory, Berkeley.)*

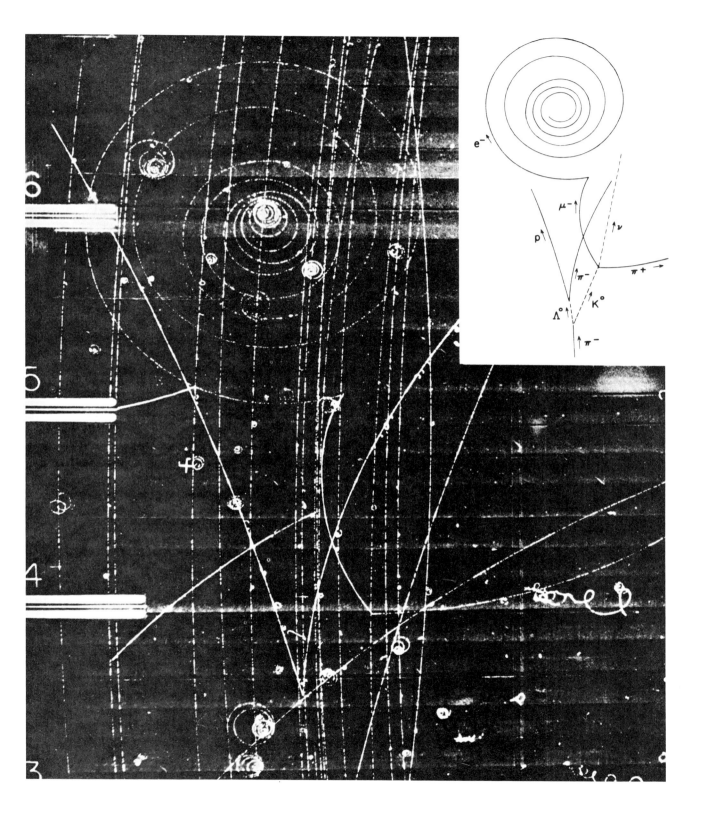

at the location of particle B. The state of motion of particle B will therefore also eventually change, and in particular the kinetic energy of the particle may change. There is thus an exchange of energy between the two particles A and B, mediated through the field. If we wish to have a theory in which it makes sense to speak of the total energy at an instant of time, and if we wish to preserve the principle that the total energy of an isolated system is a constant of motion, then we may ask where the energy which is eventually going to be given to B is to be found during the time which elapses between the instant A collides with C, and the instant when the resulting change in the state of motion of A is first felt at B. We are forced to conclude that this energy must reside in the field.

40 This line of reasoning leads to an interesting further conclusion. Suppose that the situation is otherwise the same, but that particle B is absent. At the instant A collides with C the field due to A undergoes a change: a certain quantity of energy is transferred to the field. This quantity of energy must be the same as in the case when B is present, because particle A cannot very well "know" that particle B is not there at all, ready to receive the energy. Now, if B is not there, where does the energy transferred to the field go? It has to go somewhere, and one possibility is that it is radiated away. This is in fact the case in electromagnetic theory: If a charged particle A collides with another particle C (which we may assume to be uncharged) the particle A will emit an electromagnetic wave, and this wave will carry away energy "to infinity" if there is no other particle present to absorb part of this energy.

We are therefore led to the very general expectation that if the interaction between the particles is mediated through a field, then this field can also manifest itself in the form of freely propagating, energy-carrying waves.

41 Let us now look at the problem of particle interactions from the standpoint of quantum mechanics. Our discussion in the earlier chapters has conditioned us to believe that there is a wave associated with every particle, and that conversely every wave has some particle aspects. We may say that the quantum-mechanical wave is really the same thing as the quantum-mechanical particle: it is a single object which is neither quite a classical particle, nor quite a classical wave packet. This now leads us to a most remarkable unification in our concepts. In classical physics we introduce two distinct kinds of objects, namely on the one hand the particles, and on the other hand the fields which mediate the interactions between the particles.   In quantum physics we can avoid this unsatisfactory dualism by treating the "particles" on the same footing as the fields.   We formulate a field theory which describes the propagation of wave-fields, which are de Broglie waves of particles.   At the same time the field theory describes the inter-action between the waves, and thereby, in a sense, describes the effective forces between the particles.

This is clearly a very attractive idea, and it is the basic idea of quantum field theory.   In the Schrödinger theory the forces be-tween the particles have to be introduced ad hoc.   Given these forces we can make predictions about the motion of the particles, but the Schrödinger theory does not provide us with any "explan-ation" of why the forces are what they are.   In quantum field theory, on the other hand, the existence and nature of the forces is intimately connected with the existence of the particles: we have a unified description of particles, waves and forces.   Quantum electrodynamics, which is an example of a field theory, provides us with an illustration of these features.   It describes the forces between electrons (and positrons) mediated by the electromagnetic field, and it also describes the electromagnetic quanta (photons) which can be emitted by the interacting electrons.

42   Let us review the main features of a quantum field theory. To describe particles and their interactions we introduce quantum fields.   The fields are functions of position and time and they de-scribe, we may say, the *local* state of the vacuum.†   The wave aspects of matter are built into the theory from the very beginning: the solutions to the equations of quantum field theory are waves. The waves also have particle aspects.   A well-localized particle corresponds to a concentrated wave packet: the particle is most likely to be found in those regions in space-time in which the field amplitude is large.

The field equations are non-linear equations and they can thus describe *interactions* between wave packets (particles).   The non-linearity naturally manifests itself only when the field amplitudes are large: if the amplitudes are small the waves propagate approx-imately as in a linear theory.   If two wave-packets corresponding to two particles overlap in a region in space at an instant of time, the non-linearities become manifest and the two waves influence

---

† The fields are actually not "ordinary" complex-valued functions of position and time.   They are mathematical objects known as "operator-valued distributions." However, for our purposes we can think about them as ordinary functions (representing "sound waves in the non-linear ether").

each other. In the classical picture this corresponds to an interaction between the two particles. On the other hand the waves do not interact much if they do not overlap significantly, and to this corresponds the classical picture that two particles have a very weak interaction if the separation between the two is large.

43 Quantum field theory is in an essential way a *many-particle theory:* we have a single unified formalism through which we can describe states of the world in which there is any number of particles of a given kind present. The phenomenon of creation and destruction of particles is a natural feature of quantum field theory. It comes about through the non-linear nature of the field equations. Two wave packets (corresponding to two particles) may overlap and interact and give rise to new wave packets (corresponding to new particles). Thus, for example, if two electrons collide (i.e. come close to each other) an electromagnetic wave may be emitted. We say that a photon has been created.

44 Many more or less comprehensive quantum field theories have been formulated in accordance with these ideas. The theory of quantum electrodynamics is a theory of this kind, and this theory has scored some spectacular successes in the description of the electromagnetic interactions of charged particles, and in particular in atomic physics. The specific proposals for other field theories, designed to describe the weak and the strong interactions, have been much less successful. These theories have provided us with some understanding of a few very general properties of elementary particles, but beyond that they have not really led to any useful experimental predictions. In the case of the strong interactions it has turned out that the scheme of successive approximations on which the predictions of quantum electrodynamics have been based does not work. The reason for the success of electrodynamics is undoubtedly the smallness of the fine structure constant, i.e. the weakness of the electromagnetic interactions. The strong interactions are stronger in an essential way. It has there-

Bubble chamber photograph showing the production and subsequent decay of a lambda-antilambda pair. The drawing in the upper right corner identifies the tracks of the various particles. An incoming anti-proton collides with a proton and produces the lambda-antilambda pair. The latter particles do not produce visible tracks since they are neutral. The lambda decays into a negative pion and a proton (via the weak interaction) and the antilambda decays into a positive pion and an anti-proton. The anti-proton subse-quently collides with a proton and annihilates into pions, of which four are charged and leave visible tracks.

This photograph is shown in the middle of our discussion of quantum fields to remind the reader that one of the objectives of quantum field theory is to give us a theoretical understanding of events such as those seen in the photograph. *(Photograph by courtesy of Lawrence Radiation Laboratory, Berkeley.)*

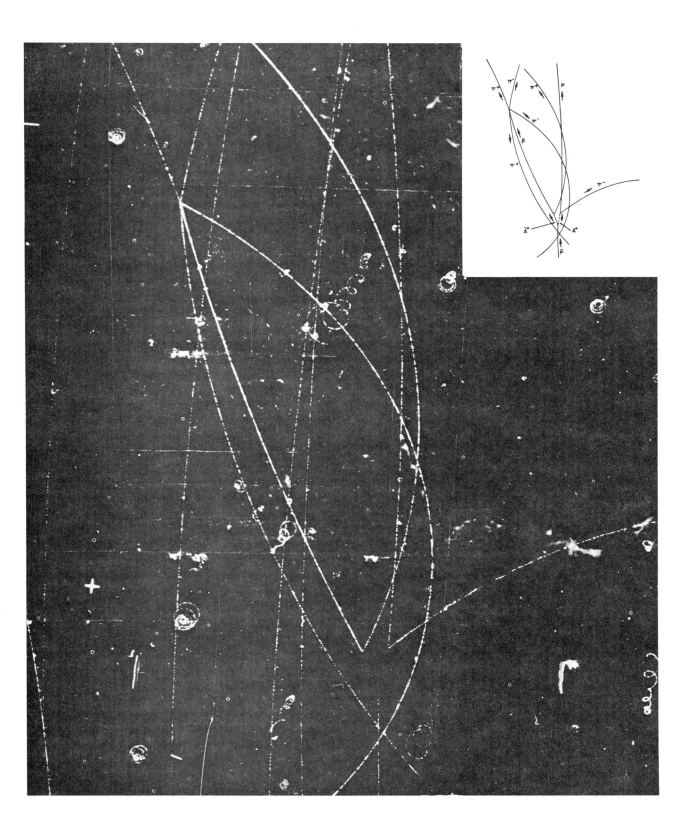

fore not been found possible to solve the field equations which have been proposed and we cannot tell whether these equations are really correct. Most likely they are not. There is actually an infinite latitude in the selection of equations, and our only guiding principle in the past has been the "principle of simplicity." In quantum electrodynamics we have been further guided, in a decisive way, by the classical analog of charged billiard balls interacting with the electromagnetic field.

45   The fact that we have been unable to overcome the considerable mathematical difficulties which arise in the theory and actually *solve* the specific field equations which have been proposed to describe the strong interactions has naturally led to a certain disillusionment with field theory as such, and voices have been heard urging abandonment of all attempts along these lines.

More weighty objections, of an epistemological character, can be raised against quantum field theory. One may say that it is objectionable that many of the *primary concepts* of this theory do not have any immediate operational significance. It is thus unclear how the fields themselves are to be measured: only in certain specific cases has this problem found an even remotely satisfactory solution. What do we mean by the field in a very small region, say of linear dimensions 10^{-100} cm? How, and with what instruments, are we to measure the fields in such a region? Who has really, in any sense of the word, measured distances smaller than 10^{-13} cm?

These are certainly serious objections. Against them one may say that it is not really necessary that every concept in a theory should have an *immediate* operational significance. Even if it is hard to see how distances of 10^{-100} cm could ever be "measured" it may still be possible to retain our space-time coordinates as describing the "arena" in which physical events take place. On the other hand it may also be that a future satisfactory theory of elementary particle interactions must be based on the abandonment of some of our notions of space and time. The quantum field theories describe in *detail* what happens at every point in space at every instant of time, and this may be too much: it may be beyond our knowledge in principle.

46   Considerations such as these led Werner Heisenberg to try to formulate, in 1943, the so-called S-matrix theory of particle interactions. In this theory, which we shall not discuss here, Heisenberg tried to admit only those concepts into the theory which have a clear operational significance, thereby following the same principle which in 1925 led him to his formulation of matrix mechanics. We might say that the theory is concerned only with the *outcomes* of collision processes and not with the detailed sequence of events taking place *during* the process. So far these attempts have not led to a satisfactory theory.

At this time there does not exist any *fundamental* theory of the strong interactions. Many attempts have made but the results do not appear particularly convincing. Whether the future ultimate theory will be a field theory, or an S-matrix theory, or perhaps an entirely new kind of theory created by one of the readers is futile to speculate about.

Pions and Nuclear Forces

47 It is out of the question that we could discuss quantum field theory in any detail in this book: to do this effectively requires fairly advanced mathematical tools. On the other hand we have just seen that the basic ideas of this theory are not complicated at all. Before we leave this subject we shall consider a problem which was first successfully attacked by Hideki Yukawa in 1934.

The problem is concerned with the following question. Is there a particle associated with the nuclear force, i.e. a quantum of the nuclear force field? If so, what are the properties of this particle? Can we find the particle experimentally?

We know that there is a particle associated with the electro-magnetic forces acting between charged particles, namely the photon. We also know that the forces which hold the nuclei together cannot be electromagnetic in origin. These forces are much stronger than the electromagnetic forces, and they are furthermore distinguished by their short range. At distances beyond, say, 10^{-12} cm these forces tend rapidly to zero and become non-existent for all practical purposes beyond a distance of 10^{-11} cm. If we now accept the ideas of quantum field theory we must expect that the nuclear force field can also manifest itself as freely propagating waves, and we may look for the corresponding particles. Just as photons are emitted when two charged particles collide we can expect the quanta of the nuclear force field to be emitted in a sufficiently violent collision of two nucleons.

48 The reader has very likely heard that these particles do exist, and that they are none other than the *pions*. In the days of Yukawa's work, however, no mesons were known and his suggestion that they existed was truly a prediction. He knew the two

Hideki Yukawa. Born 1907 in Tokyo. Yukawa studied physics at Kyoto University, from which he graduated in 1929. After holding positions as a lecturer at Kyoto University and Osaka University he was appointed professor of theoretical physics at Kyoto University in 1939. After World War II Yukawa spent some time in the United States as a member of the Institute for Advanced Study in Princeton, and as a professor of physics at Columbia University. He returned to Japan in 1955 to become director of the newly established Research Institute for Fundamental Physics in Kyoto, and to resume his position as a professor of physics at Kyoto University. Yukawa was awarded the Nobel prize in 1949 for his work on mesons and field theory. *(Photograph by courtesy of Physics Today.)*

outstanding properties of nuclear forces, namely their strength and short range, and he asked himself the same questions we have asked. On the basis of his knowledge of the properties of the nuclear forces he was able to predict the existence of the quanta, and to predict that their mass should be approximately 200 electron masses. In this study he was undoubtedly guided by the analogy with the electromagnetic interaction.

There is an amusing twist to the experimental discovery of Yukawa's mesons. Around 1937 particles of a mass of about 200 electron masses were discovered in the cosmic radiation and these particles were naturally thought to be identical with Yukawa's quanta. Further work revealed, however, that these particles, now known as *muons,* or *mu-mesons,* interact only very weakly with matter (i.e. with nuclei). Consequently they could not very well be the particles responsible for the strong nuclear forces. This mystery was eventually solved, largely through the work of C. F. Powell and co-workers in 1947, when another kind of particle was discovered in the cosmic radiation.† This particle was the pion. It has a mass of about 280 electron masses; it interacts strongly with nuclei and it must without doubt be identified with Yukawa's quantum.

By 1948 the development of particle accelerators had reached a stage where it became possible to produce pions copiously in high-energy nucleon-nucleon collisions. The properties of pions have been extensively studied experimentally, and it is now known that they play an essential role in all phenomena involving strong interactions.

49 Let us now try to "repeat" Yukawa's feat.‡ We shall regard the force between two stationary nucleons as analogous to the electrostatic force between two stationary charged particles, and we shall attempt to solve our problem on the basis of this assumed analogy. It must be admitted that the analogy is far from perfect, but this line of reasoning nevertheless leads us to the correct fundamental relationship between the mass of the pion and the nature of the force between two nucleons.

---

† C. M. G. Lattes, H. Muirhead, G. P. S. Occhialini, and C. F. Powell, "Processes involving charged mesons," *Nature* **159**, 694 (1947). Also, C. M. G. Lattes, G. P. S. Occhialini, and C. F. Powell, "Observations on the tracks of slow mesons in photographic emulsions," *Nature* **160**, 453 (1947).

‡ This will not qualify us for the Nobel prize. It is easy to do something when we know that it can be done and has been done before. The trick is to be the first to do it. Yukawa's theory appears in the paper: "On the Interaction of Elementary Particles," *Proceedings of the Physico-Mathematical Society of Japan,* **17**, 48 (1935).

We argue as follows. Maxwell's equations describe freely propagating electromagnetic waves, in the absence of any sources. The *same* equations also describe the electrostatic field of a stationary point charge, and thereby the potential energy of interaction of two stationary point charges. In fact the electrostatic potential due to one of the stationary charges satisfies the wave equation everywhere outside the charge, and this solution of the wave equation has the special properties that it is *spherically symmetric* and *static*, i.e., independent of the time. Suppose, therefore, that we consider the wave equation satisfied by freely propagating mesons, and look for solutions of this equation which are spherically symmetric and static. Hopefully this will give us the potential of the nuclear force field due to a single nucleon situated at the origin. We denote the potential by $V(r)$. The interaction energy of two stationary nucleons separated by a distance r will then be proportional to $V(r)$, where the constant of proportionality describes the strength of the coupling of the nucleon to the pion field.

50 The wave equation satisfied by the pion de Broglie wave function $\psi(\mathbf{x},t)$ is the Klein-Gordon equation, which we have already derived and discussed in Chap. 5. If we denote the mass of the pion by m_π, and employ units such that $\hbar = c = 1$, the wave equation reads

$$\frac{\partial^2}{\partial t^2}\,\psi(\mathbf{x},t) - \boldsymbol{\nabla}^2\psi(\mathbf{x},t) = -m_\pi{}^2\,\psi(\mathbf{x},t) \tag{50a}$$

where $\boldsymbol{\nabla}^2$ stands for the Laplacian operator

$$\boldsymbol{\nabla}^2 \equiv \frac{\partial^2}{\partial x_1{}^2} + \frac{\partial^2}{\partial x_2{}^2} + \frac{\partial^2}{\partial x_3{}^2} \tag{50b}$$

The wave equation (50a) describes the behavior of de Broglie waves of mesons in the absence of sources. In accordance with our program let us now try to find a static, spherically symmetric solution of this equation which could describe the meson field outside a nucleon located at the origin. The source is in this case a point source, namely the nucleon at the origin, and the wave equation (50a) need not be satisfied *at* the origin, but it must be satisfied *outside* the origin. We regard the solution as a potential function, and we shall denote it by $V(r)$. This solution is thus independent of the time, and consequently the term involving the second time derivative in equation (50a) can be omitted. Our equation assumes the form

$$\boldsymbol{\nabla}^2\,V(r) = m_\pi{}^2\,V(r) \tag{50c}$$

51 The function $V(r)$ is a function of $r = \sqrt{x_1{}^2 + x_2{}^2 + x_3{}^2}$ only, and we next have to find the action of Laplace's differential operator on such a function. We note first that

$$\frac{\partial r}{\partial x_1} = \frac{x_1}{r} \tag{51a}$$

By the chain rule of differentiation we then have

$$\frac{\partial V(r)}{\partial x_1} = \frac{dV(r)}{dr}\frac{\partial r}{\partial x_1} = \left(\frac{x_1}{r}\right)\frac{dV(r)}{dr} \tag{51b}$$

Differentiating once more with respect to x_1 we obtain

$$\frac{\partial^2 V(r)}{\partial x_1{}^2} = \frac{\partial}{\partial x_1}\left(\frac{x_1}{r}\frac{dV(r)}{dr}\right) = \frac{1}{r}\frac{dV(r)}{dr} + \frac{x_1{}^2}{r}\frac{d}{dr}\left(\frac{1}{r}\frac{dV(r)}{dr}\right) \tag{51c}$$

which gives us

$$\boldsymbol{\nabla}^2\, V(r) = \frac{3}{r}\frac{dV(r)}{dr} + r\frac{d}{dr}\left(\frac{1}{r}\frac{dV(r)}{dr}\right) \tag{51d}$$

With a trivial rearrangement of the right side we can write (51d) in the form

$$\boldsymbol{\nabla}^2\, V(r) = \frac{1}{r^2}\frac{d}{dr}\left(r^2\frac{dV(r)}{dr}\right) \tag{51e}$$

This important equation thus describes the action of Laplace's differential operator on a function $V(r)$ which is a function of r only.

52 Our differential equation which is now an ordinary, linear, second-order differential equation is therefore of the form

$$\frac{1}{r^2}\frac{d}{dr}\left(r^2\frac{dV(r)}{dr}\right) = m_\pi{}^2 V(r) \tag{52a}$$

It so happens that this equation can be solved in closed form in terms of elementary functions, and as the reader should verify by carrying out the differentiations, two linearly independent solutions are

$$\frac{1}{r}\exp\left(-rm_\pi\right), \qquad \frac{1}{r}\exp\left(+rm_\pi\right) \tag{52b}$$

The general solution is obtained as a linear combination of the two above special solutions. Now we note that the second solution would correspond to a potential which *increases* beyond any limit as r increases, and this solution would describe an inter-nucleon force which increases with the distance. It is clearly not acceptable physically, and we conclude that the potential must be proportional to the first solution (52b), and we thus have

$$V(r) = C' \frac{1}{r} \exp\left(-r m_\pi\right) \tag{52c}$$

where C' is a constant.

Our rejection of the second solution illustrates again an important principle which we have encountered before: not every solution of the wave equations of quantum mechanics has a physical meaning. The wave functions which are physically meaningful not only have to satisfy the wave equation, but they must also satisfy a number of *boundary conditions,* one of which is that the solution must not increase indefinitely at infinity.

53   We have now reached our goal: the potential energy $U(r)$ of two stationary nucleons separated by the distance r is given by

$$U(r) = \frac{C}{r} \exp\left(-\frac{r}{\lambdabar_\pi}\right) \tag{53a}$$

where $\lambdabar_\pi = 1/m_\pi$, and where C is a constant which describes the strength of the coupling.

Due to the exponential factor the potential $U(r)$ decreases very rapidly as r increases. Very roughly one may say that the *range* of the potential is $\lambdabar_\pi$: as we go much beyond this distance the potential eventually becomes completely negligible. We have already examined this point through some numerical examples, in Sec. 38, Chap. 2.

We know today that the mass of the pion is 140 MeV. The quantity $\lambdabar_\pi = 1/m_\pi$ is nothing but the Compton wavelength of the pion. (In cgs units we would have $\lambdabar_\pi = \hbar/m_\pi c$.) Numerically we have $\lambdabar_\pi = 1.4 \times 10^{-13}$ cm, and this is thus the "range" of the nuclear force field. At the time of Yukawa's prediction he knew, from a variety of experiments, that the range of the force field is about 10^{-13} cm, and he was thus able to *predict* that the mass of the hypothetical meson should be about 100 MeV, which is about 200 electron masses.

We should note that the range is inversely proportional to the mass of the particle, in this case the pion. A massless particle, like the photon, mediates a force of "infinite range": the potential $U(r)$ given by Eq. (53a) then becomes the Coulomb potential. This potential, of course, still decreases with distance but it does not decrease exponentially. We can thus claim, with justification, that we have gained a certain understanding of the connection between the pion and the properties of the nuclear force field.

54   We use this opportunity to explain a commonly employed terminology.   Many physicists would say that the interaction between two nucleons arises through the *exchange of a pion*.   Similarly they would say that the Coulomb interaction of two charged particles arises through the *exchange of a photon*.   These statements actually merely mean that the interaction between two nucleons can be found just as we have found it: i.e., that the same wave equation which describes the propagation of free pions (or photons) also describes forces mediated by the pions (or photons).   When hearing this terminology the reader need not have any visions of billiard balls being exchanged between two nucleons: it is merely a figure of speech.   Once this is understood there is of course no harm in talking about "exchanges of particles"; it is a common practice.   If we follow custom we can describe our discoveries as follows.   The forces between two particles which can interact with each other through their interactions with a third particle can be said to arise through an exchange of the third particle.   The range of the resulting force is inversely proportional to the mass of the particle exchanged.

55   There is one point which we should clarify as the reader may otherwise be bothered by it.   Earlier in this chapter we talked much about the *non*-linear nature of the equations of quantum field theory.   Nevertheless we found the Yukawa potential, given by the expression (53a) by solving a *linear* wave equation.   The reader may therefore have doubts as to whether we have done the right thing.   In fact these doubts are to a certain extent justified.   The linearized theory which we have studied is to be regarded as an approximation, valid whenever the meson field, or the potential $V(r)$, is not too large.   The Yukawa potential should therefore be correct at large distances, i.e., outside the pion Compton wavelength, but could possibly be wrong at *very* small distances.   The truth is that we do not know at present what the interaction is at very small distances, but we have no reason to doubt that at distances larger than say 10^{-13} cm the effective force is of the same

Bubble chamber photograph showing the annihilation of a proton and an anti-proton into pions.   The main event takes place in the middle of the field of view.   The anti-proton is incident from below, and its path is shown by the almost straight "dotted" track.   Eight charged pions are produced in the annihilation.   One of these, the one whose track is initially directed against the direction of the incoming anti-proton, decays into a muon and a neutrino.   The muon subsequently decays into a positron and two neutrinos.   The muon track is hard to distinguish from the pion track, but the beginning of the positron track can be clearly seen.

The chamber is in a magnetic field perpendicular to the plane of the picture.   The tracks of negative particles turn in the clockwise direction and the tracks of positive particles in the opposite direction.   Slowly moving particles leave dense tracks, whereas the tracks of very fast particles tend to have a "dotted" appearance.   *(Photograph by courtesy of Lawrence Radiation Laboratory, Berkeley.)*

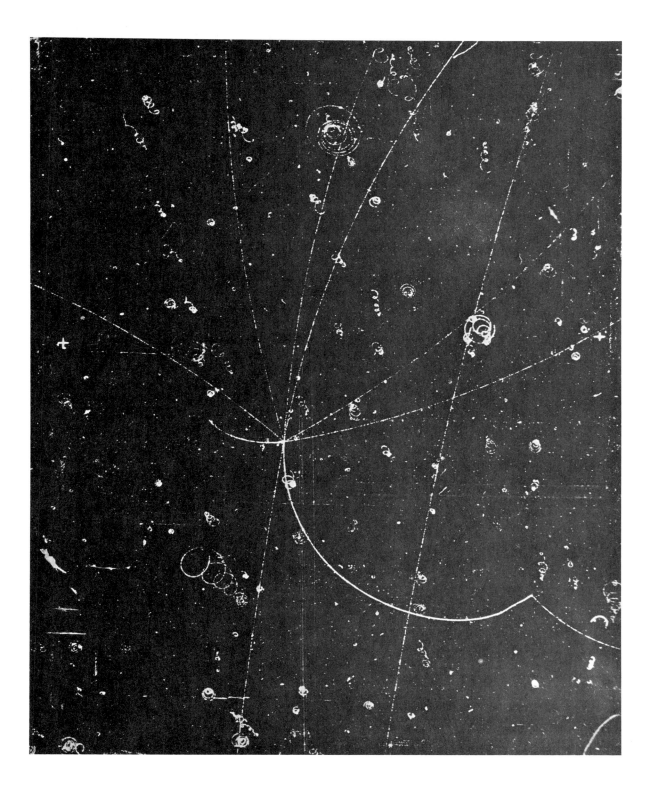

general type as the one given by the Yukawa potential. The fact that we have employed a linear approximation therefore does not invalidate our main conclusion which was that the range of the force is inversely proportional to the mass of the particle exchanged.

Concluding Remarks

56 In the preceding chapters we have learned how to think quantum mechanically about many physical phenomena. Our discussion is certainly very incomplete, which should surprise no one since this is really intended to be an *introductory* book. There are many important general principles which we have not discussed at all, and our discussion of the systematic application of the principles which we have learned is deficient both in scope and depth. We should remember, however, that quantum physics is a subject which has been studied intensely during the last 40 to 50 years, and that an enormous amount of knowledge has been accumulated in this field. No introductory book could possibly survey this vast field, and there is therefore much more to be learned through the simple process of attending lectures and reading books. No reader should feel either discouraged or offended by this statement: it is a simple statement of fact.

We have, however, made a good beginning. We have learned about the wave nature of all physical particles. We have seen how many physical phenomena which cannot be understood on the basis of classical physics can be understood on the basis of the wave picture, and we have also gained a certain understanding of how the classical laws of physics can arise as "limiting cases" of the quantum-mechanical laws. We have learned about the energy levels associated with every physical system, and we have learned how the occurrence of these levels might be understood within quantum mechanics. In the course of our studies we have gained a certain general familiarity with the (at first sight) strange world of microphysics. We have learned something about the order of magnitude of physical quantities in microphysics, and we have learned how to make simple estimates, based on simple models.

The reader who has worked his way through the book to the end of the final chapter has had a glimpse of some of the most central issues of modern physics. He has learned that physics is not a closed subject: many basic problems exist for which no solutions are in sight at this time.

The author concludes by wishing the reader Happy Future Studies in Quantum Physics.

References for Further Study

1) D. H. Frish and A. M. Thorndike: *Elementary Particles* (D. van Nostrand Co., Inc., 1964). This little book (148 pages) presents a simple, up-to-date account, and is hereby recommended Experimental techniques in this field are nicely described.

2) Note the following articles in the *Scientific American*:

a) F. J. Dyson: "Field Theory," April 1953, p. 57.

b) G. F. Chew, M. Gell-Mann and A. H. Rosenfeld: "Strongly Interacting Particles," February 1964, p. 74.

c) M. Gell-Mann and E. P. Rosenbaum: "Elementary Particles," July 1957, p. 72.

d) G. Feinberg and M. Goldhaber: "The Conservation Laws of Physics," Oct. 1963, p. 36.

e) F. J. Dyson: "Mathematics in the Physical Sciences," September 1964, p. 128.

f) W. B. Fowler and N. P. Samios: "The Omega-Minus Experiment," October 1964, p. 36.

g) K. W. Ford: "Magnetic Monopoles," December 1963, p. 122.

h) G. W. Gray: "The Ultimate Particles," June 1948, p. 26.

i) H. A. Bethe: "What holds the Nucleus Together?" Sept. 1953, p. 58.

j) R. E. Marshak: "The Nuclear Force," March 1960, p. 98.

k) S. Penman: "The Muon," July 1961, p. 46.

Problems

1 (*a*) Compute the transmission probability for neutrons of energy 0.1 eV incident normally on a cadmium foil of thickness 0.1 mm. The density of cadmium is 8.7 gm/cm^3. Obtain neutron cross section from Fig. 6A.

(*b*) Similarly compute the transmission probability for neutrons of energy 1 eV incident normally on a 1 cm thick sheet of cadmium.

2 The total cross section for the interaction of a K^+-meson with a proton is about 15 mb when the kinetic energy of the K-meson (incident on a proton at rest) is 400 MeV. What is the average number of interactions per cm of path for a K-meson of this energy in liquid hydrogen (in a bubble chamber)? The density of liquid hydrogen is 0.071 gm/cm^3.

3 The cross section for the production of an electron-positron pair when a gamma quantum of energy 10 MeV collides with a lead atom is about 14 barns. What is the probability that a pair is produced when a gamma ray of this energy is incident normally on a lead plate of thickness 2.5 mm? The density of lead is 11.3 gm/cm^3.

4 At a gamma ray energy of 100 keV the cross section for Compton scattering was measured in one experiment as 0.49 barn. At this energy, which is considerably smaller than the rest energy of an electron, a simple non-relativistic classical computation comes close to the right value. Try to carry out such a computation to see how close you can come. In Compton scattering a gamma ray is scattered by a "free" electron originally at rest. (We discussed the Compton effect in Chap. 4, but we did not discuss the cross section for the scattering.) Suppose a plane wave of amplitude A and frequency ω is incident on an electron originally at rest. The electron will be set in oscillatory motion in the direction of the electric vector of the wave. Let the amplitude of this oscillation be x. The quantity x is obviously proportional to the amplitude A of the wave, and in addition x depends on the frequency ω and on the mass and charge of the electron. The oscillating electron acts like an electric dipole of dipole moment ex. This dipole emits electromagnetic radiation at a total rate W. (We quoted the formula for this rate in Sec. 48, Chap. 3.) You should thus be able to compute what fraction of the energy incident through a unit area (containing the electron) will be scattered by the electron. Re-express your result in terms of a cross section for the scattering: this is the Compton cross section. The Compton cross section for an atom is the product of the cross section for an electron with the number of electrons in the atom.

5 (*a*) In Sec. 17 we presented a simple theory for the maximum cross section for the case that the scattering is spherically symmetric. It is of interest to compare this theory to the experimentally measured π^+-p cross section shown in Fig. 24B. We simplify the problem by regarding the proton as infinitely heavy. The relevant energy in the scattering is then the kinetic energy of the positive pion, which is about 195 MeV (in the laboratory frame of reference) at the location of the prominent resonance denoted $N^*_{3/2}$. Carry out such a comparison. You will find that the order of magnitude is right, but that the experimental cross section differs from the theoretical cross section by "a factor of order unity." The explanation for the discrepancy is simply that the scattering is *not* spherically symmetric. Our simple theory would have to be modified to include other possible angular distributions. With such a modification it is found that the experimental cross section at maximum agrees well with the theoretical prediction.

(*b*) On the basis of the curve shown in Fig. 24B, estimate the mean life of the N^*-"particle."

6 Using our simple theory in Secs. 17–18 for resonant scattering, estimate the cross section for the resonant absorption of gamma rays of energy 14.4 keV by the Fe^{57} nucleus. (This estimate is relevant to the experimental results presented in Fig. 16A, Chap. 4.) Assuming that the absorbing iron nuclei are in a foil 1 mil thick, what is the probability that a gamma ray passes through the foil?

Note that our simple theory does not really apply to photons, among other things because photons have spin angular momentum one. You can therefore not expect to obtain a numerically correct value for the cross section. The dependence of the maximum cross section on the wavelength is however given correctly by our theory, and your estimate is therefore useful as an order of magnitude estimate.

7   The maximum cross section for resonant scattering of light by an atom can obviously be very large because of the long wavelength of visible light. Let us consider the case of resonant scattering of yellow light of wavelength 5896 Å by sodium atoms.

(*a*) In the spirit of the preceding problem, estimate the maximum cross section at resonance.

(*b*) In an actual experiment we might have sodium vapor in a glass vessel as our "target" in the scattering experiment. (Consider, for instance, the experimental arrangements described in Prob. 3, Chap. 3.) The sodium atoms will not all have the same velocity, and as a result the Doppler shift will broaden the absorption line. The mean life of sodium in the $3p_{1/2}$-state is about 10^{-8} sec. From this you can compute the width of the line for an isolated sodium atom at rest. Suppose that the incident light has this linewidth. Suppose further that the atoms in the absorbing vessel have an average random velocity corresponding to a temperature of 200°C. Estimate the *effective* scattering cross section which an atom in the vessel presents to the photons in the incident beam.

(*c*) Using the estimate of the effective cross section obtained in part (*b*) above, find the number of sodium atoms per cm³ which have to be present in the vessel in order that the intensity of the incident light be reduced by a factor two in passing through a layer of gas 1 cm thick. Needless to say such a gas is completely transparent to light of any other wavelength than the resonant wavelength.

8   Consider the particles forming the baryon octet for which the mass spectrum is shown in Fig. 27B and the Eightfold Way symmetry diagram in Fig. 29B. Of these particles one is stable. Among the remaining unstable particles one decays through an electromagnetic interaction (it has a notably shorter lifetime than the other particles) and the others decay through the weak interaction. See whether you can account for these features of the octet in terms of the conservation laws for baryon number, charge, and hypercharge which we have mentioned. To do this you should investigate all possible decays you can think of into particles which have been mentioned in the text, taking into account the experimentally determined masses of these particles. For example: you might begin by asking whether the Σ+ particle could decay into a K+-meson and something else. You will soon discover that the possibilities are severely limited, and that there are not too many cases to consider.

The problem is thus to show in detail that the conservation laws which we have discussed imply that none of the particles can decay through the strong interaction, and that only one can decay through the electromagnetic interaction.

9 The symmetry diagrams in Figs. 29A–D show the values for the various particles of a quantity called the third component of isotopic spin (denoted I_3). We have mentioned that this quantity is also conserved in all strong and electromagnetic interactions.

Investigate whether this conservation law implies anything more than the other conservation laws which we have mentioned and which concern conservation of charge, hypercharge, and baryon number.

10 In the literature on the elementary particles an attribute called "strangeness" is often used to characterize the strongly interacting particles. Each such particle can be assigned a strangeness quantum number S, which we can define by $S = Y - B$, where Y is the hypercharge and B the baryon number. According to this rule the pions and the nucleons have strangeness zero: they are not strange but "familiar."

(a) In what kind of interactions is total strangeness conserved?

(b) There is a simple linear relation between strangeness S, electric charge Q, baryon number B, and third component of isotopic spin I_3. Find this relation (as it reveals itself in the symmetry diagrams in Figs. 29A–D).

11 We want to produce lambda-particles in proton-proton collisions. What is the minimum kinetic energy which a proton incident on a proton at rest must have for this to be possible?

12 (a) In Sec. 11 we guessed that a scattered wave must be of the form

$$\psi_s(\mathbf{x},t) = Cf(\theta) \frac{1}{x} \exp\,(ixp - i\omega t) \tag{a}$$

at large distances from the scattering center. Show that for the special case of spherically symmetric scattering, in which $f(\theta) = f$ is independent of the scattering angle θ, the wave function given in (a) is actually a solution of the Klein-Gordon equation (except for the point $\mathbf{x} = 0$) in empty space. It will be helpful to consider our discussion in Secs. 51–52 in this connection.

(b) Show that for an arbitrary $f(\theta)$ the expression in (a) is an *approximate* solution of the Klein-Gordon equation. You should show that if this wave function is substituted into the Klein-Gordon equation, then the equation is satisfied except for an error term which tends to zero as $1/x^2$ when x tends to infinity.

Appendix

For tables of units and conversion factors, see inside back cover of book.

For a table of crude values of important physical constants, see inside front cover of book.

TABLE A   *General Physical Constants**

Planck's constant:	$h = 2\pi\hbar = (6.62559 \pm 0.00015) \times 10^{-27}$ erg sec
	$\hbar = \dfrac{h}{2\pi} = (1.05449 \pm 0.00003) \times 10^{-27}$ erg sec
Velocity of light:	$c = (2.997925 \pm 0.000001) \times 10^{10}$ cm sec^{-1}
Electronic charge:	$e = (4.80298 \pm 0.00006) \times 10^{-10}$ esu
	$= (1.60210 \pm 0.00002) \times 10^{-19}$ coul
Gravitational constant:	$G = (6.670 \pm 0.005) \times 10^{-8}$ dyne cm^2 gm^{-2}
Fine structure constant:	$\alpha = \dfrac{e^2}{\hbar c} = (7.29720 \pm 0.00003) \times 10^{-3}$
	$\dfrac{1}{\alpha} = 137.0388 \pm 0.0006$
Avogadro's number:	$N_0 = (6.02252 \pm 0.00009) \times 10^{23}$ mol^{-1}
Boltzmann's constant:	$k = (1.38054 \pm 0.00006) \times 10^{-16}$ erg ($^\circ$K)$^{-1}$
Faraday constant:	$N_0 e = (96487.0 \pm 0.5)$ coul mol^{-1}
Universal gas constant:	$R = N_0 k = 8.314 \times 10^7$ erg ($^\circ$K)$^{-1}$ mol^{-1}
	$= 1.986$ cal ($^\circ$K)$^{-1}$ mol^{-1}
Mass of electron:	$m = (9.10908 \pm 0.00013) \times 10^{-28}$ gm
	$= (5.48597 \pm 0.00003) \times 10^{-4}$ amu
	$= (0.511006 \pm 0.000002)$ MeV/c^2
Atomic mass unit:	(amu) $= (1.66043 \pm 0.00002) \times 10^{-24}$ gm
	$= (931.478 \pm 0.005)$ MeV/c^2
Proton mass:	$M_p = (1.67252 \pm 0.00003) \times 10^{-24}$ gm
	$= (1.00727663 \pm 0.00000008)$ amu
	$= (938.256 \pm 0.005)$ MeV/c^2

* Most of the data in this table are taken from an article by E. R. Cohen and J. W. M. DuMond, "Our Knowledge of the Fundamental Constants of Physics and Chemistry in 1965," *Reviews of Modern Physics* 37, 537 (1965).

TABLE A *General Physical Constants* (*continued*)

Neutron mass:	$M_n = (1.0086654 \pm 0.0000004)$ amu $= (939.550 \pm 0.005)$ MeV/c^2
Compton wavelength of electron:	$\lambda_e = \dfrac{h}{mc} = (2.42621 \pm 0.00002) \times 10^{-10}$ cm $\mathit{X}_e = \dfrac{\hbar}{mc} = (3.86144 \pm 0.00003) \times 10^{-11}$ cm
First Bohr radius:	$a_0 = \dfrac{\hbar^2}{me^2} = \alpha^{-1}\mathit{X}_e$ $= (5.29167 \pm 0.00002) \times 10^{-9}$ cm
"Classical radius" of electron:	$\dfrac{e^2}{mc^2} = \alpha \mathit{X}_e = (2.81777 \pm 0.00004) \times 10^{-13}$ cm
Nonrelativistic ionization potential of hydrogen with infinite proton mass:	$R_\infty = \tfrac{1}{2}\alpha^2 mc^2 = (13.60535 \pm 0.00013)$ eV
Rydberg constant for infinite proton mass:	$\tilde{R}_\infty = \dfrac{\alpha}{4\pi a_0} = \dfrac{R_\infty}{hc} = (109737.31 \pm 0.01)$ cm^{-1}
Rydberg constant for hydrogen:	$\tilde{R}_\mathrm{H} = (109677.576 \pm 0.012)$ cm^{-1}
Bohr magneton:	$\mu_B = \dfrac{e\hbar}{2mc}$ $= (9.27314 \pm 0.00021) \times 10^{-21}$ erg gauss^{-1}
Frequency associated with 1 eV:	$(2.41804 \pm 0.00002) \times 10^{14}$ cycle/sec
Wave number associated with 1 eV:	(8065.73 ± 0.08) cm^{-1}
Temperature associated with 1 eV:	(11604.9 ± 0.5) °K

TABLE B   *The Most Stable Elementary Particles**

| Particle | Spin | Mass MeV | Mean life sec | Important Decays† | | |
				Partial mode	Branching fraction	Q MeV
γ photon	1	0	stable	stable		
LEPTONS						
ν_e e-neutrino	$\frac{1}{2}$	0 (< 0.2 keV)	stable	stable		
ν_μ μ-neutrino		0 (< 2 MeV)				
$e^\mp$ electron-positron	$\frac{1}{2}$	0.511006	stable	stable		
$\mu^\mp$ muons	$\frac{1}{2}$	105.659	2.20×10^{-6}	$e\nu\nu$	100%	105
BARYONS ‡						
p proton	$\frac{1}{2}$	938.256	stable	stable		
n neutron		939.550	1.01×10^3	$pe^-\nu$	100%	0.78
Λ lambda-hyperon	$\frac{1}{2}$	1115.58	2.51×10^{-10}	$p\pi^-$	66%	30
				$n\pi^0$	34%	41
				$p\mu\nu$	1.4×10^{-4}	72
				$pe\nu$	0.88×10^{-3}	177
Σ^+ sigma-hyperons	$\frac{1}{2}$	1189.47	0.81×10^{-10}	$p\pi^0$	53%	116
				$n\pi^+$	47%	110
				$p\gamma$	1.9×10^{-3}	251
Σ^0		1192.56	$< 1.0 \times 10^{-14}$	$\Lambda\gamma$	100%	77
Σ^-		1197.44	1.65×10^{-10}	$n\pi^-$	100%	118
				$ne^-\nu$	1.3×10^{-3}	257
				$n\mu^-\nu$	0.6×10^{-3}	152
				$\Lambda e^-\nu$	0.6×10^{-4}	81
Ξ^0 cascade particles	$\frac{1}{2}$	1314.7	3.0×10^{-10}	$\Lambda\pi^0$	100%	7
Ξ^-		1321.2	1.74×10^{-10}	$\Lambda\pi^-$	100%	5
				$\Lambda e^-\nu$	3.0×10^{-3}	205
Ω^- omega-minus	$\frac{3}{2}$	1674	1.5×10^{-10}	$\Xi\pi$	~50%	221
				$\Lambda\overline{K}$	~50%	66

TABLE B *The Most Stable Elementary Particles (continued)*

Particle	Spin	Mass MeV	Mean life sec	Important Decays† Partial mode	Important Decays† Branching fraction	Important Decays† Q MeV
MESONS						
$\pi^{\pm}$ charged pions	0	139.58	2.608×10^{-8}	$\mu\nu$	100%	34
				$e\nu$	1.24×10^{-4}	139
				$\mu\nu\gamma$	1.24×10^{-4}	34
				$\pi^0 e\nu$	1.0×10^{-8}	4.08
π^0 neutral pion	0	134.98	0.89×10^{-16}	$\gamma\gamma$	98.8%	135
				γe^+e^-	1.2%	134
$K^{\pm}$ charged kaons (K-meson)	0	493.8	1.235×10^{-8}	$\mu\nu$	63.4%	388
				$\pi^{\pm}\pi^0$	21.0%	219
				$\pi^{\pm}\pi^-\pi^+$	5.6%	75
				$\pi^{\pm}\pi^0\pi^0$	1.7%	84
				$\mu^{\pm}\pi^0\nu$	3.4%	253
				$e^{\pm}\pi^0\nu$	4.8%	358
K^0 neutral kaons	0	497.9				
K_1			0.87×10^{-10}	$\pi^+\pi^-$	69.3%	219
				$\pi^0\pi^0$	30.7%	228
K_2			5.68×10^{-8}	$\pi^0\pi^0\pi^0$	23.5%	93
				$\pi^+\pi^-\pi^0$	11.5%	84
				$\pi\mu\nu$	27.5%	253
				$\pi e\nu$	37.4%	358
				$\pi^+\pi^-$	0.15%	219
				$\pi^0\pi^0$	0.36%	228
η eta-meson	0	548.6	$< 7 \times 10^{-20}$ $> 0.7 \times 10^{-20}$	$\gamma\gamma$	31.4%	549
				$\pi^0\pi^0\pi^0$	21.0%	144
				$\pi^0\gamma\gamma$	20.5%	414
				$\pi^+\pi^-\pi^0$	22.4%	135
				$\pi^+\pi^-\gamma$	4.6%	269

† Q stands for the kinetic energy released in the decay.

‡ To each one of the baryons corresponds an anti-baryon, not listed separately.

* The data in this table are taken from a survey article by A. H. Rosenfeld et al., "Data on Particles and Resonant States," *Reviews of Modern Physics* **39**, 1 (1967). Data for many more particles are presented in this article. Additional information about the particularly stable particles is also given. Some of the rarer known decay modes have been omitted from our table.

TABLE C   *The Chemical Elements*

Element	Symbol	Atomic number	Atomic mass* amu	Element	Symbol	Atomic number	Atomic mass* amu
Actinium	Ac	89	(227)	Erbium	Er	68	167.26
Aluminum	Al	13	26.9815	Europium	Eu	63	151.96
Americium	Am	95	(243)	Fermium	Fm	100	(253)
Antimony	Sb	51	121.75	Fluorine	F	9	18.9984
Argon	Ar	18	39.948	Francium	Fr	87	(223)
Arsenic	As	33	74.9216	Gadolinium	Gd	64	157.25
Astatine	At	85	(210)	Gallium	Ga	31	69.72
Barium	Ba	56	137.34	Germanium	Ge	32	72.59
Berkelium	Bk	97	(247)	Gold	Au	79	196.967
Beryllium	Be	4	9.0122	Hafnium	Hf	72	178.49
Bismuth	Bi	83	208.980	Helium	He	2	4.0026
Boron	B	5	10.811	Holmium	Ho	67	164.930
Bromine	Br	35	79.909	Hydrogen	H	1	1.00797
Cadmium	Cd	48	112.40	Indium	In	49	114.82
Calcium	Ca	20	40.08	Iodine	I	53	126.9044
Californium	Cf	98	(251)	Iridium	Ir	77	192.2
Carbon	C	6	12.01115	Iron	Fe	26	55.847
Cerium	Ce	58	140.12	Krypton	Kr	36	83.80
Cesium	Cs	55	132.905	Lanthanum	La	57	138.91
Chlorine	Cl	17	35.453	Lawrencium	Lw	103	(257)
Chromium	Cr	24	51.996	Lead	Pb	82	207.19
Cobalt	Co	27	58.9332	Lithium	Li	3	6.939
Copper	Cu	29	63.54	Lutetium	Lu	71	174.97
Curium	Cm	96	(247)	Magnesium	Mg	12	24.312
Dysprosium	Dy	66	162.50	Manganese	Mn	25	54.9380
Einsteinium	Es	99	(254)	Mendelevium	Md	101	(256)

* The numbers within parentheses in the atomic mass column are the mass numbers of the most stable isotopes of radioactive elements.

TABLE C *The Chemical Elements* *(continued)*

Element	Symbol	Atomic number	Atomic mass* amu	Element	Symbol	Atomic number	Atomic mass* amu
Mercury	Hg	80	200.59	Samarium	Sm	62	150.35
Molybdenum	Mo	42	95.94	Scandium	Sc	21	44.956
Neodymium	Nd	60	144.24	Selenium	Se	34	78.96
Neon	Ne	10	20.183	Silicon	Si	14	28.086
Neptunium	Np	93	(237)	Silver	Ag	47	107.870
Nickel	Ni	28	58.71	Sodium	Na	11	22.9898
Niobium	Nb	41	92.906	Strontium	Sr	38	87.62
Nitrogen	N	7	14.0067	Sulfur	S	16	32.064
Nobelium	No	102	(255)	Tantalum	Ta	73	180.948
Osmium	Os	76	190.2	Technetium	Tc	43	(98)
Oxygen	O	8	15.9994	Tellurium	Te	52	127.60
Palladium	Pd	46	106.4	Terbium	Tb	65	158.924
Phosphorus	P	15	30.9738	Thallium	Tl	81	204.37
Platinum	Pt	78	195.09	Thorium	Th	90	232.038
Plutonium	Pu	94	(244)	Thulium	Tm	69	168.934
Polonium	Po	84	(209)	Tin	Sn	50	118.69
Potassium	K	19	39.102	Titanium	Ti	22	47.90
Praseodymium	Pr	59	140.907	Tungsten	W	74	183.85
Promethium	Pm	61	(145)	Uranium	U	92	238.03
Protactinium	Pa	91	(231)	Vanadium	V	23	50.942
Radium	Ra	88	226.0254	Xenon	Xe	54	131.30
Radon	Rn	86	(222)	Ytterbium	Yb	70	173.04
Rhenium	Re	75	186.2	Yttrium	Y	39	88.905
Rhodium	Rh	45	102.905	Zinc	Zn	30	65.37
Rubidium	Rb	37	85.47	Zirconium	Zr	40	91.22
Ruthenium	Ru	44	101.07				

* The numbers within parentheses in the atomic mass column are the mass numbers of the most stable isotopes of radioactive elements.

Index

the
ell-